INTEGRATION OF FUNDAMENTAL POLYMER SCIENCE AND TECHNOLOGY—4

The proceedings of the international meeting on polymer science and technology, Rolduc Polymer Meeting—4 held at Rolduc Abbey, Limburg, The Netherlands, 23–27 April 1989

INTEGRATION OF FUNDAMENTAL POLYMER SCIENCE AND TECHNOLOGY—4

Edited by

P. J. LEMSTRA

Eindhoven University of Technology, Eindhoven, The Netherlands

and

L. A. KLEINTJENS

DSM-Research, Geleen, The Netherlands

ELSEVIER APPLIED SCIENCE
LONDON and NEW YORK

ELSEVIER SCIENCE PUBLISHERS LTD
Crown House, Linton Road, Barking, Essex IG11 8JU, England

Sole Distributor in the USA and Canada
ELSEVIER SCIENCE PUBLISHING CO., INC.
655 Avenue of the Americas, New York, NY 10010, USA

WITH 49 TABLES AND 187 ILLUSTRATIONS

© 1990 ELSEVIER SCIENCE PUBLISHERS LTD
Softcover reprint of the hardcover 1st edition 1990

British Library Cataloguing in Publication Data

Rolduc Polymer Meeting (*4th, 1989*)
 Integration of fundamental polymer science
 and technology— 4
 1. Polymer science
 I. Title II. Lemstra, P. J. II. Kleintjens, L. A.
 547.7

 ISBN-13: 978-94-010-6831-4   e-ISBN-13: 978-94-009-0767-6
 DOI: 10.1007/978-94-009-0767-6

Library of Congress CIP data applied for

PREFACE

The aim of the Rolduc Polymer Meetings is to stimulate interdisciplinary discussions between academic and industrial polymer scientists and engineers. Experts are invited to review selected topics and to initiate discussions relating to future trends and developments.

The general theme of these meetings is 'Integration of Fundamental Polymer Science and Technology'. In order to serve this goal, all participants are accommodated in Rolduc Abbey, a well-preserved medieval monument in Limburg (The Netherlands) to provide an optimum atmosphere for the exchange of ideas.

About 350 participants took part in the 4th Rolduc Polymer Meeting, which was held from 23 to 27 April 1989. This volume contains invited and selected contributed papers on topics such as solution properties, chemistry, emulsion polymerization, liquid crystalline polymers, structure/morphology and blends/composites.

We are fully aware of the fact that the reader will not find an integrated presentation of lectures in this volume. Unfortunately, it is impossible to put down in writing the atmosphere of this and previous meetings. However, we hope that the reader will be stimulated to present his own views in forthcoming meetings after reading these proceedings.

We wish to thank all contributors to this volume.

P.J.L.
L.A.K.

CONTENTS

Part 3: Emulsion Polymerization

Part 5: Structure/Morphology

Part 6: Blends/Composites

Part 7: Miscellaneous

Part 1

SOLUTION PROPERTIES

BRIDGING TREATMENT OF POLYMER SOLUTIONS IN GOOD SOLVENTS

WALTER H. STOCKMAYER
Department of Chemistry
Dartmouth College
Hanover, New Hampshire 03755, USA

ABSTRACT

The "bridging" function for connecting dilute and concentrated regimes in polymer solution thermodynamics has heretofore been applied only to poor solvents and two-phase systems. A comparison of the bridging treatment in an athermal solution (good-solvent system) with a modern cross-over formula due to Ohta and Oono shows that the former gives a rather satisfactory representation of the osmotic pressure.

INTRODUCTION AND HISTORICAL REMARKS

Over forty years ago Flory (1,2) called attention to the inherently large fluctuations of local concentration in dilute polymer solutions and the resulting difficulties in applying the then standard mean-field theories (2-5) of polymer-solution thermodynamics to the dilute regime. Accordingly, separate dilute-solution theories based on a virial expansion for the osmotic pressure, as formally justified by McMillan and Mayer (6), were constructed and embellished during the next several decades (7-10).

Early efforts at production of a theory to encompass all concentrations were those of Fixman (11,12) and Edwards (13). In 1974, the writer and several of his colleagues (14) proposed on simple physical grounds a "bridging function," depending exponentially on concentration, to connect previous separately established expressions for dilute and concentrated regimes into a single tractable analytic recipe. A somewhat similar "switching function" was used by Chapela and Rowlinson (15) to help represent the thermodynamic properties of methane and carbon dioxide both near and far from the critical point; and such an exponential device is visible in a much earlier classical equation of state for fluids (16). A further study of this type, for non-polymeric fluid mixtures, is due to dePablo and Prausnitz (17).

The major objective of Koningsveld et al. (14) was improvement of the description of spinodals and binodals, particularly for the well-studied cyclohexane/polystyrene system. Modest success was achieved, but measurements of distribution-ratios for individual polymer species between the two phases (18,19) were only poorly reproduced by the bridging scheme. A more sophisticated and fundamental bridging treatment was later described by Irvine and Gordon (20), with spinodals and phase equilibria again serving as experimental tests. In this as in the earlier versions the crucial variable in the bridging function is the concentration ratio c/c^*, where c^* is the overlap threshold at which (roughly) the polymer coils are beginning to overlap and interpenetrate (2,21).

The developments described above were focused in applications exclusively on poor-solvent systems. The purpose of the present work is to probe the applicability of the bridging approach to polymer solutions in good solvents. This seemed to be a useful exercise, since it could either encourage or discourage efforts to effect further refinements or improvements in the method. Our limited results are in fact encouraging.

In recent years there have been many investigations of polymer solutions in good solvents, spurred initially by the work of deGennes and his colleagues. It was recognized as useful for high molecular weights to distinguish not two but three different concentration regimes, roughly delineated as follows: (a) dilute, $c<c^*$, coils well separated; (b) semi-dilute, $c>c^*$, coils overlapping but polymer volume fraction still low; (c) concentrated, very high polymer concentrations. The boundaries between adjacent regions are of course very diffuse and are further blurred by polydispersity. An important feature of the semi-dilute regime was recognized by desCloizeaux (22), and led to his well-known scaling law for the osmotic pressure: in semi-dilute solutions the osmotic pressure must be independent of chain length and (in sufficiently good solvents) the reduced osmotic pressure,

$$Z \equiv \pi M/cRT$$

must depend only on the ratio c/c^*. These two requirements produce the desCloizeaux scaling law (21,22):

$$\pi \sim c^{9/4}$$

The semi-dilute regime comes to an upper end when Z no longer depends solely on c/c*. An example is seen in the extensive measurements of Noda et al. (23) on solutions of polystyrene in benzene and toluene. For sufficiently low molecular weights, c* may be so high that the semi-dilute regime never appears.

Theoretical expressions for the osmotic pressure spanning both dilute and semi-dilute regions in good solvents have been offered by several groups of investigators (24-28), and semi-empirical prescriptions are also available (29,30). None of these, however, leads to reliable results in highly concentrated solutions. We shall use the Ohta-Oono formulation (27) in the test calculation presented below.

BRIDGING MODEL

We follow in general the notation of Koningsveld et al. (14) or of Irvine and Gordon (20). The free enthalpy of mixing ΔG at constant temperatures and pressure to form a binary solution consisting of n_1 moles of solvent and n_2 moles of polymer is written

$$\Delta G/NRT = (1-\phi)\ln(1-\phi) + m^{-1}\phi\ln\phi + \Gamma \tag{1}$$

where the total number of moles of segments is

$$N = n_1 + mn_2; \tag{2a}$$

$$m \equiv V_2/V_1 , \tag{2b}$$

the V's being molar volumes. The function Γ consists of three parts, $\Gamma = \Gamma_1 + \Gamma_2 + \Gamma_3$. The first of these is the usual interaction function, given to a very good approximation by the expressions

$$\Gamma_1 = g_0\phi(1-\phi) \tag{3}$$

with

$$g_0 = (\beta_0 + T^{-1}\beta_1)/(1-\gamma\phi) \tag{4}$$

The concentration-dependent term in the denominator takes account of the circumstance that surfaces of contact control the nearest-neighbor interactions; for a lattice model, one would put $\gamma = 2/z$, where z is the lattice coordination number.

The first two terms on the right-hand side of eq. 1 come from the standard Flory-Huggins mixing entropy, and at one time it was considered relatively unimportant to augment these by taking account of higher approximations (4,31-33) to the lattice combinatory problem. However, it proves to be helpful to include such corrections, and we follow Irvine and Gordon (20) by using the expression first given by Huggins:

$$\Gamma_2 = \gamma^{-1}(1-\gamma)\phi\ln(1-\gamma) - \gamma^{-1}(1-\gamma\phi)\ln(1-\gamma\phi) \tag{5}$$

New theoretical advances (34) and Monte Carlo calculations (35,36) show that the Huggins formula, eq. 5, accounts for a substantial portion of the correction to the Flory-Huggins approximation.

Finally, we take the bridging term Γ_3 essentially from Irvine and Gordon, with an amendment to be introduced later:

$$\Gamma_3 = m^{-1}\phi(1-\phi)\exp(-\lambda\phi) \; ; \tag{6a}$$
$$\lambda = \lambda_o m^{1/2} \tag{6b}$$

(Irvine and Gordon also considered the case of a polydisperse macromolecular solute, but we deal here only with the strictly binary case.) Basically, the exponential factor in eq. 6a is equal to $\exp(-c/c*)$ where $c*$ is the overlap concentration. When this quantity is estimated under poor-solvent conditions for polystyrene, the factor λ_o in eq. 6b is expected to be of the order of unity.

We now consider the special case of a very good solvent for which $\Gamma_1 = 0$. (This is not an athermal case, because in general β_o is not zero.) Under such conditions, $c*$ is no longer proportional to $m^{-1/2}$ but more nearly proportional to $m^{1-3\nu} \sim m^{-0.8}$. However, we ignore this change in the first calculations presented here.

We define the reduced osmotic pressure by

$$Z \equiv \pi M/cRT = -m\Delta\mu_1/RT\phi \; , \tag{7}$$

and, after differentiating eq. 1 to obtain the solvent chemical potential change $\Delta\mu_1$, we obtain

$$(Z-1)/\phi = m\phi^{-2} [\gamma^{-1}\ln(1-\gamma\phi) - \ln(1-\phi)]$$
$$- [1 + \lambda(1-\phi)]\exp(-\lambda\phi) \; . \tag{8}$$

The corresponding osmotic second virial coefficient is proportional to

$$[(Z-1)/m\phi]_{\phi=0} = \frac{1}{2}(1-\gamma)-m^{-1} - \lambda_o m^{-1/2} \; , \tag{9}$$

which in general does not match the experimental data. We therefore revert
to the earlier procedure of Koningsveld et al. (14), who used the observed
second virial coefficient as a fitting parameter, and replace eq. 8 by

$$(Z-1)/\phi = m\phi^{-2}[\gamma^{-1}\ln(1-\gamma\phi)-\ln(1-\phi)]$$
$$- [1-L(1-\phi)]\exp(-\lambda\phi) , \qquad (8')$$

which has the low-concentration limit

$$[(Z-1)/m\phi]_{\phi=0} = \frac{1}{2}(1-\gamma)-m^{-1}(1+L) \qquad (9')$$

SAMPLE CALCULATION

For our numerical example we choose m = 1000, corresponding to a polystyrene
with M $\cong$ 10^5 g/mole, and we put γ = 1/3, as would be the case for a simple
cubic lattice (z = 6). The parameters λ and L of eqs. 8' and 9' are free
for matching the theoretical formula of Ohta and Oono (27). The latter may
be written

$$Z-1 = (X/2)\exp\{[X^{-1} + (1-X^{-2})\ln(1+X)]/4\} , \qquad (10)$$

where X is proportional to polymer concentration. The low-concentration
limit of eq. 10 is

$$[(Z-1)/\phi]_{\phi=0} = (X/2\phi)\exp(1/8) \qquad (11)$$

On the basis of experimental data (23) we choose X = 300ϕ to give an appropriate second virial coefficient, based on eq. 11. The resulting course of
(Z-1)/ϕ from eq. 10 is shown as the full curve in Fig. 1. Note that a
finite limit equal to 625 is reached at ϕ = 1. This violates Henry's law
for the solvent, as is true also of other expressions designed to describe
only dilute and semi-dilute regimes. The appropriate limiting behavior
(Z $\rightarrow$ ∞ as ϕ $\rightarrow$ 1) is obtained only if an appropriate term in $\ln(1-\phi)$ is
present, as in eq. 8. This term cannot appear if the solvent is treated
only as a continuous background.

Two other curves are shown in Fig. 1. The dashed curve is based on
eq. 8' with λ = 27.1 and L = 150. It is seen to match the full curve very
well over the entire dilute and semi-dilute range; moreover, it has the
proper limiting behavior in the concentrated regime. The inflection points
seen at intermediate concentrations are surely artifacts of the model, and
should not be taken seriously. We are pleased to observe, however, that
the value λ = 27.1 corresponds to λ_o = 0.85 according to eq. 6b, and is

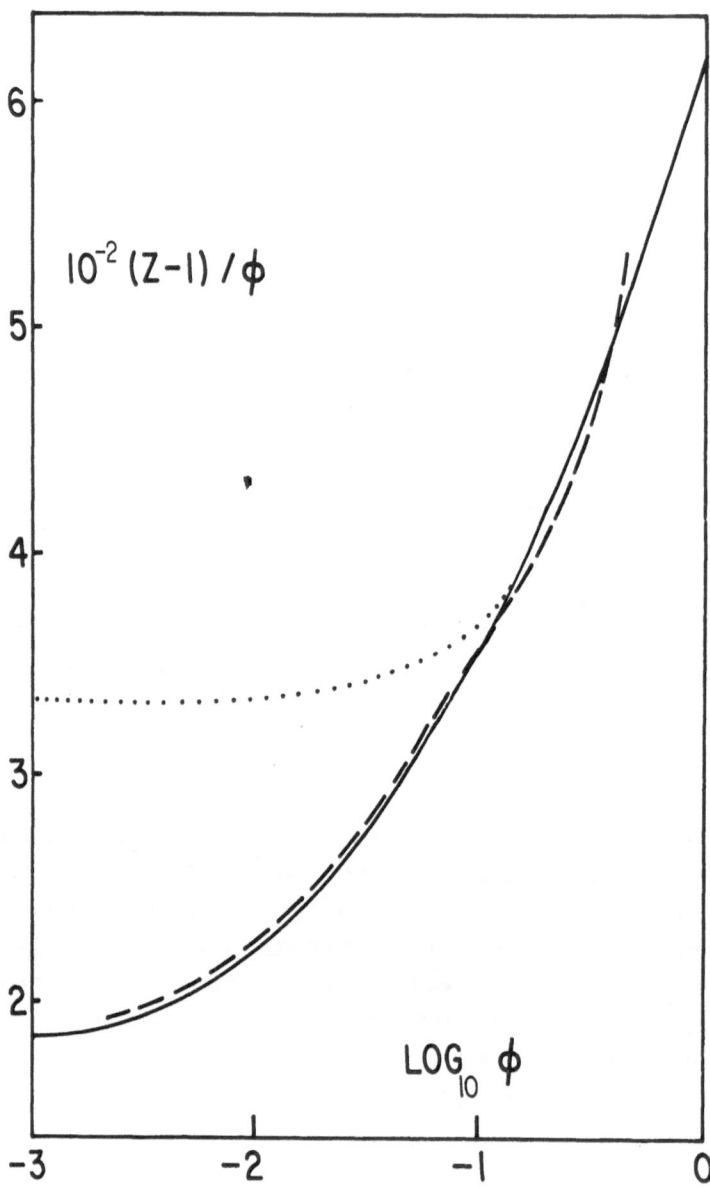

Figure 1. Comparison of reduced osmotic pressures based on eq. 10 (full curve) and eq. 8' (dashed curve). The dotted curve is based on eq. 8 or 8' without the last ("bridging") term.

therefore of the physically predicted magnitude.

The dotted curve in Fig. 1, with a low-concentration limit of about 330, corresponds to complete omission of the bridging term Γ_3. If the Huggins combinatory contribution Γ_2 were also omitted, the limit would be about 500.

Our conclusion , based on the good match between dashed and full curves, is that the exponential bridging device has some theoretical basis and is capable of being useful. This conclusion may optimistically also be extended to the case of polymer blends.

ACKNOWLEDGMENTS

Some of these calculations were made when the writer was a guest at the Institute for Chemical Research of the University of Kyoto, and he thanks Professor Michio Kurata for the hospitality accorded there. This work has been supported by the National Science Foundation, USA, under grant No. DMR 86-08633, Polymers Program, Division of Materials Research.

REFERENCES

1. Flory, P.J., _J. Chem. Phys._, 1945, _13_, 453.

2. Flory, P.J. _Principles of Polymer Chemistry_, Cornell University Press, Ithaca, 1953, Chaps. VII and XII.

3. Flory, P.J., _J. Chem. Phys._, 1942, _10_, 51.

4. Huggins, M.L., _Ann. N.Y. Acad. Sci._, 1942, _41_, 1.

5. Staverman, A.J., _Rec. trav. chim._, 1941, _60_, 640.

6. McMillan, W.G., and Mayer, J.E., _J. Chem. Phys._, 1945, _13_, 276.

7. Zimm, B.H., _J. Chem. Phys._, 1946, _14_, 164.

8. Flory, P.J., and Krigbaum, W.R., _J. Chem. Phys._, 1950, _18_, 1086.

9. Casassa, E.F., and Markovitz, H., _J. Chem. Phys._, 1958, _29_, 493.

10. Yamakawa, H., _Modern Theory of Polymer Solutions_, Harper & Row, New York, 1971, Chap. IV.

11. Fixman, M., _J. Chem. Phys._, 1960, _33_, 370.

12. Fixman, M., and Peterson, J.M., _J. Amer. Chem. Soc._, 1964, _86_, 3524.

13. Edwards, S.F., Proc. Phys. Soc. (London), 1966, 88, 265.

14. Koningsveld, R., Stockmayer, W.H., Kennedy, J.W., and Kleintjens, L., Macromolecules, 1974, 7, 73.

15. Chapela, G., and Rowlinson, J.S., J. Chem. Soc., Faraday I, 1974, 70, 584.

16. Benedict, M., Webb, G.B., and Rubin, L.C., J. Chem. Phys., 1940, 8, 334.

17. DePablo, J.J., and Prausnitz, J.M., A. I. Ch. E. Journal, 1988.

18. Breitenbach, J.W., and Wolf, B.A., Makromol. Chem., 1967, 108, 263.

19. Kleintjens, L.A., Koningsveld, R., and Stockmayer, W.H., Brit. Polym. J., 1976, 8, 144.

20. Irvine, P., and Gordon, M., Macromolecules, 13, 761.

21. DeGennes, P.G., Scaling Concepts in Polymer Physics, Cornell University Press, Ithaca, 1979, Chap. III.

22. DesCloizeaux, J., J. Phys. (Les Ulis, Fr.), 1981, 42, 635.

23. Noda, I., Higo, Y., Ueno, N., and Fujimoto, T., Macromolecules, 1984, 17, 1055.

24. DesCloizeaux, J., J. Phys. (Les Ulis, Fr.), 1981, 42, 635.

25. Knoll, A., Schäfer, L., and Witten, T.A., J. Phys. (Les Ulis, Fr.), 1981, 42, 767.

26. Schäfer, L., Macromolecules, 1982, 15, 652.

27. Ohta, T., and Oono, Y., Phys. Lett. A, 1982, 89A, 460.

28. Muthukumar, M., and Edwards, S.F., J. Chem. Phys., 1982, 76, 2720.

29. Schulz, G.V., and Stockmayer, W.H., Makromol. Chem., 1986, 187, 2235.

30. Hager, B.L., Berry, G.C., and Tsai, H.-H., J. Polym. Sci., Polym. Phys. Ed., 1987, 25, 387.

31. Miller, A.R., Proc. Cambridge Philos. Soc., 1939, 38, 109.

32. Guggenheim, E.A., Proc. Roy. Soc., London, A 1944, 183, 203.

33. Kurata, M., Ann. N.Y. Acad. Sci., 1961, 89, 635.

34. Freed, K.F., and Bawendi, M.G., J. Phys. Chem., 1989, 93, 2194.

35. Bellemans, A., and DeVos, E., IUPAC Symposium on Macromolecules, Helsinki, 1972.

36. Madden, W.G., private communication, 1989.

CHAIN STRUCTURE AND SOLVENT QUALITY: KEY FACTORS IN THE THERMOREVERSIBLE GELATION OF SOLUTIONS OF VINYL POLYMERS

H. Berghmans, Ph. Van Den Broecke and S. Thijs
Laboratory for Polymer Research, Katholieke Universiteit Leuven,
Celestijnenlaan, 200 F, B-3030 Leuven, Belgium.

ABSTRACT

The influence of chain structure and solvent quality on the thermoreversible gelation of stereoisomers of polymethylmethacrylate is discussed. In n-butanol, amorphous gels are obtained with the different isomers. Under the correct annealing conditions, the isotactic isomer will also form crystalline gels. In o-xylene and toluene, the syndiotactic PMMA forms gels by a conformational change, followed by an intermolecular association. In the same solvents, the isotactic polymer seems to form gels by crystallization. In solvents like MEK and DMF, the last polymer behaves in the same way as the syndiotactic in the aromatic solvents.

The data are discussed in the more general framework of the different possibilities of physical network formation in relation to the chain structure of vinyl polymers.

INTRODUCTION.

Moderately concentrated solutions of many synthetic and biological polymers solidify on cooling. This phenomenon, known as thermoreversible gelation, results from the formation of some kind of interconnectivity throughout the solution. It is fully reversible as heating restores the original solution[1]-[3]. The term gel originates from polymer chemistry, where these structures are obtained by chemical cross-linking of the molecules to form a permanent, molecular network that can be swollen by a solvent. These gels however are not reversible and the solution can only be restored by decomposing the network.

Another basic difference between chemical and thermoreversible gels is their mechanical behaviour. Chemical gels do not flow and have very characteristic elastic properties, generally not encountered with the thermoreversible ones. These last ones also show an absence of flow when e.g. the test tube in which they are prepared is turned upside down. They nevertheless show these solid-like characteristics only between well determined limits of applied strain, time, etc. These characteristics differ furthermore from one case to the other, making it very difficult to formulate a precise definition.

For that reason, thermoreversible gels must be considered as solutions of moderate concentration, solidified by cooling, with some elastic characteristics within well determined limits. Starting from this general definition, they can be classified according to their formation mechanism resulting from thermal transitions. The most frequently encountered is a liquid-solid (L-S) demixing or crystallization.[2]-[5] Because one is working with solutions, a liquid-liquid (L-L) demixing has not to be excluded as it can interfere with a L-S demixing.[5] Interference with a glass-transition - concentration curve can also be at the origin of a solidification of a solution.[6] On cooling a polymer solution containing chains with a regular structure, a change from a random coil to a helical conformation can also occur.[7] This regular conformation can then be stabilized by intramolecular interactions and by interaction with the solvent.

It is obvious then that many factors have to be well controlled in order to obtain an exact picture of a mechanism of formation and the structure of a thermoreversible gel. This will be illustrated in this paper for the differrent stereoisomers of poly(methylmethacrylate). Isotactic, atactic and different types of syndiotactic isomers are available and an accurate analysis of their tacticity can be obtained from NMR data. These isomers also have rather different properties and consequently are good candidates to illustrate the influence of the experimental factors on their solution behaviour and gelation. The non crystallizable, atactic chain has a glass transiton temperature, T_g, in the vicinity of 105°C. The crystallization from the melt of the isotactic isomer was reported to be very slow, resulting in the formation of a crystalline lattice in which the chains adopt a double helix conformation. [8]. A T_g around 45°C was reported. This transition was found around 120°C in the case of the syndiotactic isomer. Its crystallization from the melt is very difficult. Crystallization on the other hand, induced by solvent vapour, seems to proceed rather easily. It was further shown that the tacticity influences their θ-temperature.

TABLE 1.

Characteristics of the isomers of PMMA

Isomer	$i^*(1)$	$h^*(2)$	$s^*(3)$	$M_w(4)$ $\times 10^{-3}$	$M_n(5)$
isotactic	0.95	0.02	0.03	8.7	4.7
syndiotactic	0.00	0.10	0.90	16.9	9.1
atactic	0.04	0.34	0.62	68.0	14.4

(1), (2) and (3): fractions of isotactic, heterotactic and isotactic triads as obtained from NMR analysis.
(4) weight average molecular weight
(5) number average molecular weight

1.Solution behaviour in a poor solvent: n-butanol

The first phenomenon that is observed when a solution of the different isomers in butanol is cooled to room temperature is a liquid-liquid demixing. This is deduced from the onset of opalescence on cooling or from the exothermic signal in a cooling DSC scan. A typical example of these calorimetric observations is reported in figure 1. The resulting demixing curve is represented in figure 2. The calorimetric observations also allowed for the determination of the dependence T_g on the solvent concentration. This relationship is also represented in figure 2. Both curves intersect at a well defined temperature, corresponding to the temperature at which a rigid gel is formed. This is the consequence of the "freezing" of the concentrated phases that originates from the liquid-liquid demixing.[2][6] Because of the difference in T_g, this intersection point will occur at different temperature and polymer concentration for the different stereoisomers. These values are reported in table 2.

TABLE 2.

Temperature(T_{in}) and polymer concentration(ϕ_2)at the intersection point of the demixing curve and the T_g-concentration curve.

Isomer	T_{in} ('C)	ϕ_2(weight fraction)
Isotactic	2	0.86
Atactic	29	0.70
Syndiotactic	32	0.80

This difference in T_{in} also results in a difference in the gelation temperature on cooling.

In the case of the atactic and syndiotactic isomers, the gels obtained are always amorphous, independent on the experimental conditions. With the isotactic isomer however, the formation of a crystalline structure

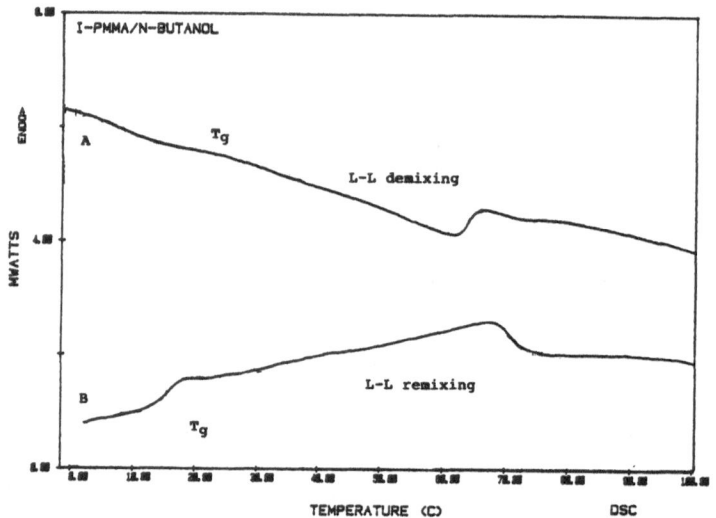

Fig. 1. DSC scans of isotactic polymethylmethacrylate in butanol.
A: cooling; B: heating; scanning rate 5°C/min
concentration: 24% polymer by weight.

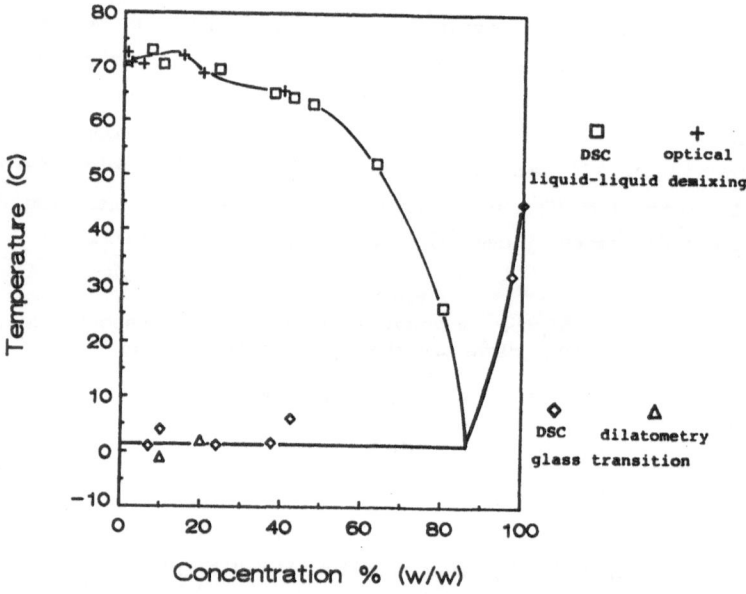

Fig. 2. Temperature-concentration diagram of isotactic polymethylmethacrylate in n-butanol.

is possible and depends on the cooling procedure. When the cooling from a homogeneous solution is performed at e.g. -5°C/min, demixing is followed, at 2°C, by a vitrification of the amorphous domains, resulting in the formation of an amorphous gel. When on the contrary the cooling is stopped at e.g. 43°C, a crystalline structure is obtained after an annealing period of a few days. Its melting point is practically invariant at 70°C, even is the crystallization is performed in a concentration domain outside the demixing domain. It only increases in the very high concentration range, ending at a melting point for the pure polymer sample of 153°C.

2. Gelation in "structure inducing" solvents.

Many solvents are known to induce structure in isotactic and syndiotactic PMMA. And although the same solvents can be used for both isomers, their behaviour is rather different. The syndiotactic isomer forms very rapidly gels in solvents like xylene and toluene.[7] A two step mechanism, composed of an fast intramolecular transition, followed by a much slower intermolecular association, is observed. Evidence was found that the same mechanism, but at a much lower rate, is followed in butanon. When the solvent is removed from these gels, a crystalline diffraction pattern can be obtained, analogous to those reported in the literature.[8] Surprising data however are obtained from the study of the temperature-concentration relationship.(figure 3). The melting point increases only slowly with increasing polymer concentration. The completely dried samples melt at 122°C (maximum of the melting endotherm), just above the glass transition of the polymer.

The behaviour of the isotactic isomer in xyleen is rather different. Crystallization, and consequently gelation, occurs only over a period of days, even at the very high concentrations. The crystalline melting point of the dried sampel is far above T_g and the crystallization from this solvent resembles more the "normally expected" behaviour.

An interesting situation is encountered when solutions of i-PMMA in methyl-ethylketon are cooled to low temperatures.[9] At -35°C, practically transparant gels are already formed with solutions with a concentration as low as 1%. The occurence of a liquid-liquid demixing can be excluded as no whitening of the system on cooling is observed. This very fast gelation is reflected in an exothermic signal in a calorimetric experiment. Melting results in an endothermic peak. When this calorimetric behaviour is studied over the whole concentration range, the analogous surprising data are

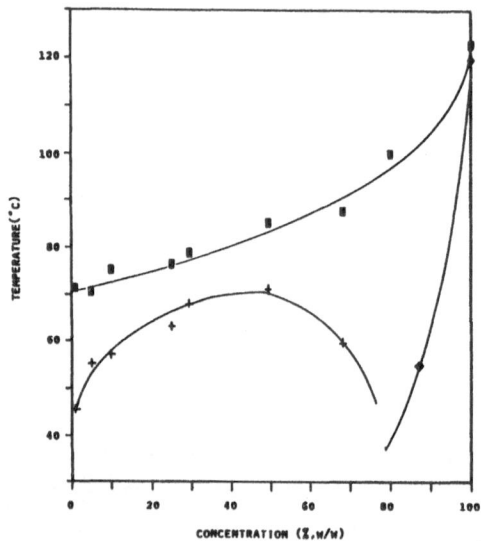

Fig. 3. Temperature-concentration diagram of syndiotactic polymethylmethacrylate in o-xylene.

■ melting point; + gelation temperature;
◆ glass transition; scanning rate: 5°C/min.

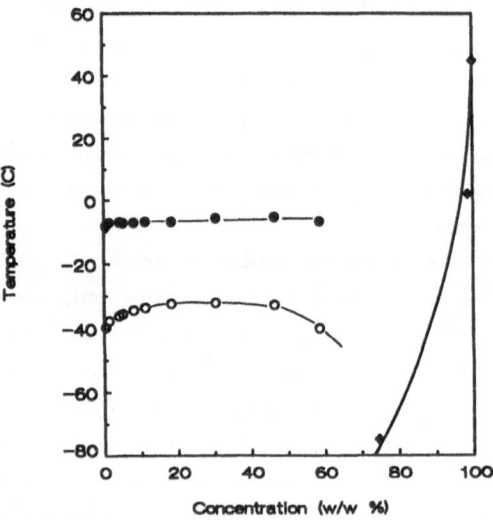

Fig. 4. Temperature-concentration diagram of isotactic polymethylmethacrylate in 2-butanon.

● melting point; ○ gelation temperature
◆ glass transition; scanning rate 5°C/min.

obtained as those reported for s-PPMA in o-xylene. The melting point remains constant over the whole concentration range (figure 4.). The onset of the exothermic signal on cooling as a function of concentration goes through a maximum. This refers to a nucleation controlled phenomenon, slowed down by concentration in the dilute region, and by the approach of Tg in that concentrated region.

This behaviour is also encountered with other solvents like DMF at even lower temperatures. Transparant gels are obtained around -50°C. But the temperature-concentration behaviour is very analogous.

DISCUSSION.

The study of the gelation of solutions of stereoisomers of PMMA leads to interesting conclusions on the formation of thermoreversible gels. It reveals the necessity to investigate the whole temperature-concentration behaviour in order to understand the exact mechanisme of the gelation.

These investigations confirm the mechanism proposed for the formation of amorphous gels with non crystallizable polymers. The gelation temperature depends on their difference in Tg and the position of the L-L demixing curve on the temperature scale.

When working in solvents that can induce the formation of structure in solution, information can be obtained outreaching the phenomenon of thermoreversible gelation. From the behaviour of the syndiotactic isomer in o-xylene and the isotactic isomer in MEK or DMF, it is clear that the expected cryatallization behaviour is not observed. The limited dependence of the melting point on the concentration points to the formation of a structure that is different from the normally expected lamellar crystallites. It seems to form a different phase inside the solution, resulting in a melting behaviour that ignores its surroundings. The rate of formation of these structures is much larger than normally expected for the folded chain crystallization of these polymers. It also seems not to ben influenced by the presence of chain irregularities, making "normal" crystallization very difficult or even impossible.

Up to now it is not yet clear which kind of structure is formed that leads to this specific melting behaviour. The data obtained with s-PMMA suggest that we are dealing with a conformational gelation that is induced by a change in chain conformation, followed by a intermolecular association. This conformation has not to be the same as the one that normally would be encountered in the lamellar crystallites. It has to be stabilize by the environment   e.g. by interaction with the solvent. This was already shown for s-PMMA. [10] The melting behaviour of this "solvent-polymer structure" will be indepent on concentration as it behaves as an independent phase in the solution. The low melting point of the dried samples (close to Tg) could be the result of the melting of these "new" structures. But the transformation to the "normally expected" crystalline morphology with very small dimensions and consequently a very low melting point has not to be excluded. Not enough data however are actually available to draw more definite conclusions.

The analogy between the data obtained with the syndiotacitic and the isotactic isomer also suggest that this mechanism of conformational gelation is not specific for one stereoisomer. The main point is to prevent the "normal" crystallization and to bring the solution into the temperature domain were this very rapid change in conformation is possible. Therefore it will only be possible with very slow crystallizing polymers as i-PMMA or with polymers that have rather irregular chains so that crystallization is strongly reduced or even prevented like s-PMMA. The generallity of this behaviour is strongly supported by the complex phase behaviour of isotactic polystyrene.[3] The formation of lamellar crystallites with a 3_1 helical conformation and of low melting strutures with a extended 12_1 conformation has been reported.

From these observations it is clear that the formation of the well known lamellar crystallites obeing the classical thermodynamic equations, is not the only way by which structure is induced from solution. In presence of a solvent a different structure can be formed resulting from the specific interaction of the solvent with the polymer chain. This allows for the stabilazation of a helical conformations that will further agglomerate to form a physical network or gel. Its morphology and melting behaviour is far from understood. But the fact that it can be observed with polymers that differ by tacticity and even chemical nature, supports the idea that we are dealing with a more general phenomenon that can be observed with many crystallizing vinyl polymers. It further resembles the behaviour of many biological systems like carrageenans, gelatin etc.

REFERENCES

1. H. Berghmans and W. Stoks, in "Integration of fundamental Polymer Science and Technology", Eds. L.A. Kleintjens and P.J. Lemstra, Elsevier Appl.Science, London, 1986, p.218
2. H. Berghmans, in "Integration of fundamental Polymer Science and Technology", Eds. L.A. Kleintjens and P.J. Lemstra, Elsevier Appl.Science, London, 1988, p.296
3. A. Keller, in "Structure-Property relationship of polymeric solids", Ed. A. Hiltner, Plenum Press, 1983, p.25
4. H. Berghmans, N. Overbergh and F. Govaerts, J.Polymer Sci., Phys.Ed., 17, 1251 (1979)
5. W. Stoks, H. Berghmans, P. Moldenaers and J. Mewis, British Polymer J., 204, 361 (1988)
6. J. Arnauts and H. Berghmans, Polymer Communications, 28, 66 (1987)
7. H. Berghmans, A. Donckers, L. Frenay, W. Stoks, F.C. De Schryver, P. Moldenaers and J. Mewis, Polymer, 28, 97 (1987)
8. H. Kusanagi, H. Tadokoro and Y. Chatani, Macromolecules, 9, 531 (1976)
9. Ph. Van den Broeck and H. Berghmans, to be published
10. H. Kusuyama, N. Miyamoto, Y. Chatani and H. Tadokoro, Polymer Comm., 14, 495 (1983)

FLOW INDUCED LIQUID-LIQUID PHASE SEPARATION IN HIGH MOLECULAR
WEIGHT POLYMER SOLUTIONS

P.J. BARHAM
H.H. Wills Physics Lab.,University of Bristol,Tyndall Avenue,
BRISTOL BS8 1TL. U.K.

ABSTRACT

Rheological studies of very high molecular weight
Poly(methylmethacrylate) (PMMA) solutions have revealed three
distinct types of anomolous behaviour. These are interpreted
as being due to: at low rates, the formation of topological
adsorption - entanglement layers; at intermediate rates the
nucleation by adsorption-entanglement layers of a flow-
induced liquid-liquid phase separation, and at high rates a
spontaneous liquid-liquid phase separation. An attempt has
been made to construct the phase diagram for the binary
system PMMA - dimethylphthlate using shear stress, as well as
temperature and solution concentration as thermodynamic
variables.

INTRODUCTION

There have, over the past thirty or so years, been many
reports of anomalous flow behaviour in solutions of very high
molecular weight polymers. It has been shown (1) that such
behaviour is associated with the formation of layers, at
surfaces, which are much thicker than the radius of gyration
of polymer molecules.

Recently Rangel-Nafaile et al. (2) have observed similar
rheological behaviour accompanied by the observation of
turbidity in flowing polystyrene solutions. They have
attributed their observations to a flow induced liquid-liquid
phase separation. Independently, Wolf has come to the same
conclusion (3).

In the present work an attempt is made to distinguish
between flow-induced liquid-liquid phase separation - which
is a pseudo-equilibrium process - and the formation of
'adsorption-entanglement' layers - which is a purely
topological process of entanglement formation stabilised by
flow.

If an alternating (rather than continuous) flow is applied then many fewer <u>different</u> molecules will pass by the surface so fewer configurations will be sampled making entanglements less likely to occur. Further, when the direction of flow changes it is likely to reduce the stability of the entanglements, and may even lead to actual disentanglement, thus making the formation of adsorption-entanglement layers improbable. The effect of an alternating flow field on stress-induced liquid-liquid phase separation should be much less severe. If a constant shear rate is applied, but the sense of rotation periodically reversed, then the resulting shear stress will have a constant magnitude and the first normal stress difference will remain constant (except at the times when the sense of rotation is actually being reversed).

EXPERIMENTAL DETAILS

The polymer was a Polymethylmethacrylate (PMMA) with weight average molecular weight ca 7×10^6. The solvent used was laboratory grade dimethylphthlate (DMP). The rheological measurements were performed using the cone and plate geometry, the cone had a diameter of 50mm and an angle of 0.04 radians.

RESULTS AND DISCUSSION

The results of the alternating steady experiments (see fig. 1) may be summarised as follows in four distinct types of behaviour depending on both strain rate, and amplitude.

1)   Very low rates; - 'normal' behaviour

2)   Low rates: - 'normal' behaviour in small amplitude oscillating flow - shear stress increases to plateau value in continuous flow.

3)   Intermediate rates: - 'normal' behaviour in small amplitude oscillating flow - shear stress increases continuously in continuous flow.

4)   High Rates: - large, erratic, oscillations in shear stress - solutions start to become turbid.

The arguments made in the Introduction suggested that adsorption-entanglement layer formation should be suppressed in oscillating flows, while stress induced liquid-liquid phase segregation should not. Accordingly, it can be deduced that the behaviour in 2 (and possibly also 3) is due to the formation of adsorption-entanglement layers, while that in 4 is caused by flow induced liquid-liquid phase segregation.

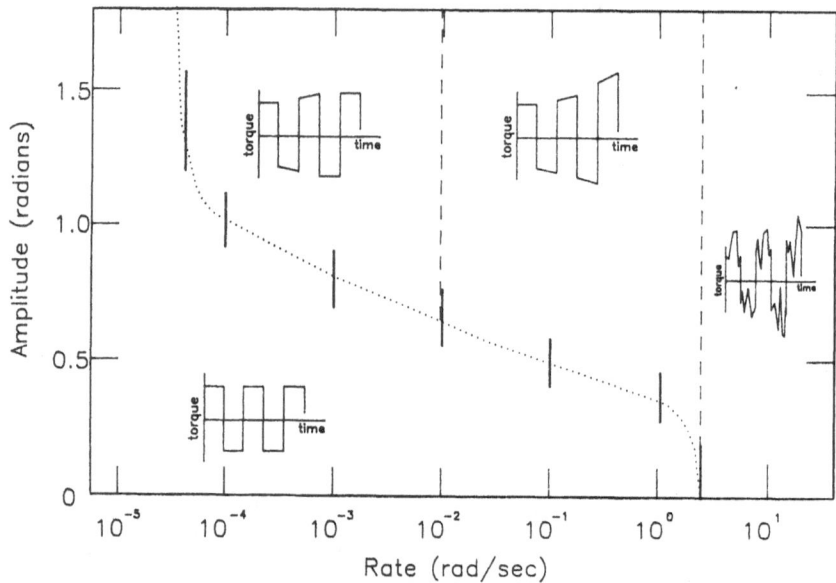

Figure 1.   A diagram illustrating the four types of
behaviour observed in alternating shear with a 2% (w/v)
solution.

    The difference between 2 and 3 suggests the
interpretation that in 2 the surface layer stops growing at a
finite thickness, while in 3 it continues to grow well beyond
that limit.

    A convenient way to find the boundaries between the
various types of behaviour is first to induce the behaviour
as in 4 above by shearing at a high rate and then reducing
the rate in a sudden step and observing the corresponding
torque response. Typical results of this kind of experiment
are shown, for the 2% (w/v) solution, in figure 2.   The
torques at the initial, high, rate (20 rad/s) display the
fluctuations characteristic of type 4 behaviour.   However
when the rate is reduced either of two behaviours are seen at
the lower rate.   If the second rate is low enough then the
torque rapidly decays and reaches a new, steady, value; at
higher second rates the torque level decreases but the torque
continues to show large oscillations.   The rate at which the
transition in the above behaviour occurs is very sharply
defined; in this case it takes place between 9.8×10^{-3} and
1×10^{-2} rad/s.

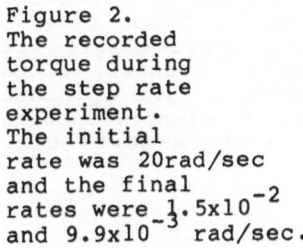

Figure 2.
The recorded
torque during
the step rate
experiment.
The initial
rate was 20rad/sec
and the final
rates were 1.5x10^{-2}
and 9.9x10^{-3} rad/sec.

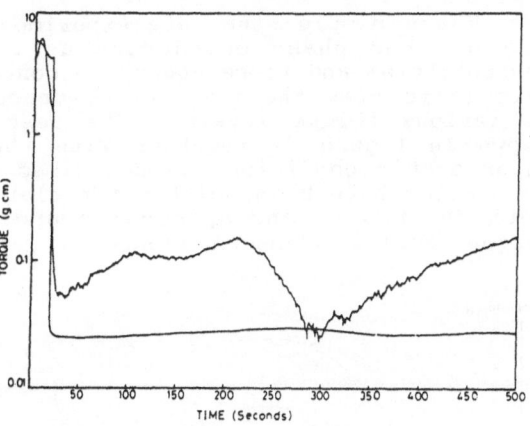

It should be noted that the transition from type 3 behaviour to type 4 on increasing the shear rate occurs at ca. 2 rad/sec, while on decreasing the rate the transition from type 4 to type 2 behaviour occurs at the lower rate of ca. 10^{-2} rad/sec. Furthermore, this transition from type 4 behaviour on reducing the rate corresponds closely with the previously observed transition between behaviours 2 and 3. If we interpret these results in terms of a phase diagram, with shear stress or torque, rather than temperature as the thermodynamic variable, then we can assign the transition between behaviour 3 and 4 on increasing the rate to spontaneous liquid-liquid phase separation (i.e. to the spinodal curve). Similarly we can assign the transition from 4 to 2 on decreasing the rate (and from 2 to 3 on increasing the rate) to the 'equilibrium' co-existence curve. Region 3, by this argument, lies between the phase equilibrium boundary and the spinodal. Accordingly the behaviour in region 3 corresponds to the formation of an adsorption-entanglement layer followed by the growth of this layer due to stress induced liquid-liquid phase segregation.

The flow rate associated with the boundary between 3 and 4 can also be conveniently found using a step rate experiment. If the initial rate is varied and the final rate is kept fixed only slightly above the 2-3 boundary then, as is two types of behaviour are observed at the final rate. When the initial rate is less than the rate at the 3-4 boundary then the torque at the final rate remains more or less constant (i.e. it behaves as if it had no pre-history). However when the initial shearing induces phase-segregation then the torque at the final rate is significantly higher and oscillates in an unstable fashion.

These simple step rate experiments can thus be used to map out the phase boundaries as a function of solution concentration and temperature. A convenient way to represent this is to draw the temperature concentration phase diagram at various torque levels. The resulting phase diagram is shown in figure 3, together with the phase boundary found under static conditions as described earlier. The spinodal boundaries have been omitted for clarity; an example showing both the phase, and spinodal boundaries for a particular torque level is shown in figure 4.

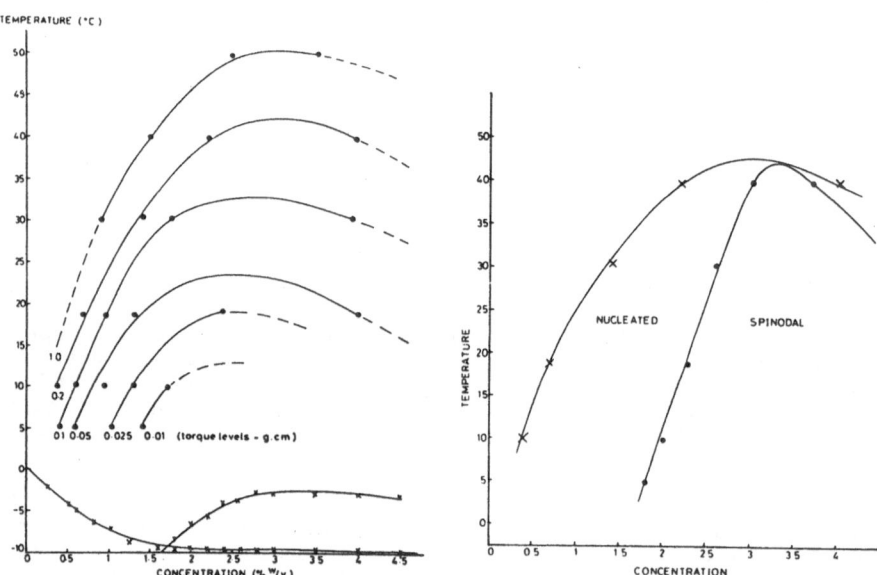

Figure 3. A schematic 'phase diagram' for the PMMA/DMP system at various torque levels. The crosses refer to the static phase diagram.

Figure 4. The proposed phase diagram at a torque level of 0.2gcm showing the 'spinodal' boundary.

References

1.   R.A.M. Hikmet, K.A. Narh, P.J. Barham and A. Keller, Progr. Coll and Polym.Sci., 1986 **264**, 507.

2.   C. Rangel-Nafaile, A.B. Metzner and K.F. Wissbrun: Macromol. 1984, **17**, 1187.

3.   B.A. Wolf: Macromol. 1984, **17**, 615.

LIGHT SCATTERING MEASUREMENTS IN THE SYSTEM POLYSTYRENE-CYCLOHEXANE NEAR THE MISCIBILITY GAP

Th. Philipps and W. Borchard

Angewandte Physikal. Chemie der Universität-GH-Duisburg, FRG

ABSTRACT

The phase separation of polymer-solvent systems is detected by means of dynamic light scattering with a much higher sensitivity than with other methods in use up to now. The spinodal temperature can be determined from the extrapolation of the temperature dependence of the translational diffusion coefficient D at constant composition. The extrapolation of D to zero concentration leads to the mobility of the polymer.

INTRODUCTION

Generally the phase separation of a liquid solution is indicated by a turbidity of the solution. Due to the polydispersity of the polymer component, a polymer-solvent system has to be treated as a multicomponent system. For such a system with an upper critical point it is characteristic that the maximum of the cloud point curve is located above the critical temperature and shifted to lower concentrations /1-4/.

THEORETICAL CONSIDERATIONS

Taking the base molar fraction as generalized concentration variable the relation between the translational diffusion coefficient and the thermodynamic factor $x_2^*(\partial\mu_2/\partial x_2^*)_{T,P}$ reads:

$$D = {}_xu_2\ x_2^*\ \left[\frac{\partial\mu_2}{\partial x_2^*}\right]_{T,P}\ , \tag{1}$$

where ${}_xu_2$ is the mobility, x_2^* the base molar fraction and μ_2 the chemical potential of the polymer /6,8/. In the tempera-ture range below the Θ-temperature the thermodynamic factor decreases. At the spinodal curve the thermodynamic factor vanishes according to the definition of the stability limit. Because the mobility cannot go to infinite values, D decrea-ses to zero at the spinodal curve /3,5,6/. Substitution of the chemical potential in eq.(1) by expression of the polymer solution theories e.g. Flory-Huggins, followed by differenti-ation and extrapolation to zero concentration yields:

$$\lim_{x_2^*\to 0}\ D = {}_xu_{2,0}(T)\ R\ T \equiv D_0 \tag{2}$$

Therefore the spinodal temperatures at constant composition and the temperature dependence of the mobilities of the poly-mer at zero concentration ${}_xu_{2,0}(T)$ can be calculated from measurements of the temperature and concentration dependence of D.

MATERIALS AND METHODS

Three various polystyrene PS samples with narrow molar mass distribution were used for the measurements. Two products are commercial available standards (PS17000: M_w = 17500 g/mol, M_w/M_n = 1.04; PS50000: M_w = 50000 g/mol; M_w/M_n = 1.06). The third PS was polymerized anionically in our group (PSA5: M_w = 18000 g/mol, M_w/M_n = 1.15). Cyclohexane CH p.a. (Merck) was used as solvent, additionally destilled.

The studies were performed with an apparatus for simultaneous measurement of static and dynamic light scattering LS (ALV,

Langen). The apparent diffusion coefficient D_{app} has been calculated from the inverse Laplace transform of the measured correlation function at a scattering angle of 90°.

RESULTS AND DISCUSSION

Static LS Measurements of the System PSA5-CH

The temperatures for these measurements are far above the visually determined cloud point temperatures ⁻CPT. The value of M_w and of the second virial coefficient of the osmotic pressure B do not show any temperature dependence in this range (see tab.1). The values of B are negative. This is due to the vicinity of the miscibility gap of the system PS-CH.

TAB. 1: Results of static LS measurem. of the system PSA5-CH

$T[^{\circ}C]$	$M_w[g/mol]$	$B \cdot 10^5[mol\ cm^3/g^2]$	$\overline{r^2} \cdot 10^{12}[cm^2]$	$\sqrt{\overline{r^2}}\ [nm]$
28	19800 ($\pm$ 610)	-5.7 ($\pm$ 3.9)	7.8 ($\pm$ 0.2)	28
25	20300 ($\pm$ 250)	-6.5 ($\pm$ 3.2)	5.2 ($\pm$ 0.9)	23
22	293000 ($\pm$ ---)	16 ($\pm$ 7.4)	660 ($\pm$ ---)	257

For temperatures just below 22 $^{\circ}C$ the apparent molar mass and the radius of gyration increases about ten times. This can be explained by an aggregation of the PS particles. The experimentell errors of both of these values are very high at this temperature. We found, that this belongs to a temporary change of the scattering intensity of the lower concentrated polymer solutions (1-3% by wt.).

Dynamic LS Measurements at Various PS-CH Systems

Fig.1 shows the D_{app} of PSA5-CH solutions at 22 $^{\circ}C$ with various concentrations. Measurements of the solutions with concentrations above 3% by wt. resulted in only a single D_{app}. For concentrations below 3% by wt. two values of D_{app} were found, differing two orders of magnitude. We explain this behavior by the formation of new phases (microheterogeneous regions) in a very small part of the total volume in the metastable region of the solutions. The observed time dependence for the second D_{app} shows, that the microheterogeneous regions are formed very slowly. It is remarkable, that the appearance of the second diffusion coefficient is detected at temperatures far above the visually determined CPT of the solutions (12.5 ^{o}C-14 ^{o}C). These effects are detected by means of dynamic LS at much higher temperatures than indicated by CPT measurements, because at lower polymer concentrations the volumes of the coexisting phases are too small to be detectable.

The D_{app} of fractionated and unfractionated PS17500-CH solutions with concentrations of 1% by wt. are shown in fig.2. The fractionation was carried out by means of phase separation. With this procedure the fractions with higher molar masses are concentrated in the lower, more concentrated phase. In the PS17500-CH system turbidities were not detected for all temperatures down to the melting point of the solvent (6.5 ^{o}C). In case of the unfractionated PS resp. fractionated PS from the lower phase one diffusion coefficient was detected above 15 ^{o}C resp.25 ^{o}C. At 15 ^{o}C resp.25 ^{o}C and below two D_{app} were observed, which differ again two orders of magnitude. In case of fractionated PS from the upper phase a second D_{app} did not occur. These measurements show, that the appearance of the second D_{app} is caused by a small polydispersity of the polymer. The influence of the molar mass distribution over the cloud point curve is described by Tsuyumoto et al. /7/. Although the PS17500 posses a very narrow molar mass distribution it is possible to detect the existence of a two phase region by means of dynamic LS.

For the system PS50000-CH only one D_{app} is detected in the

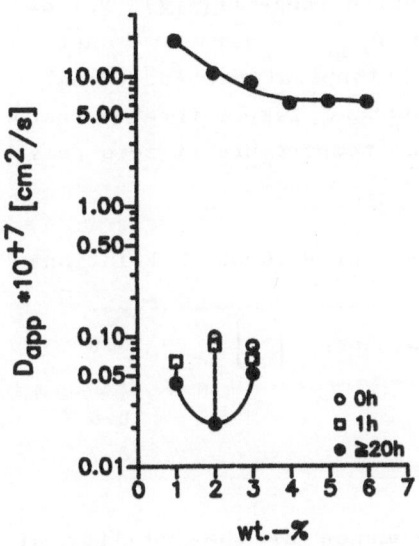

FIG.1: D_{app} of PSA5–CH solut. at 22° C versus concentration

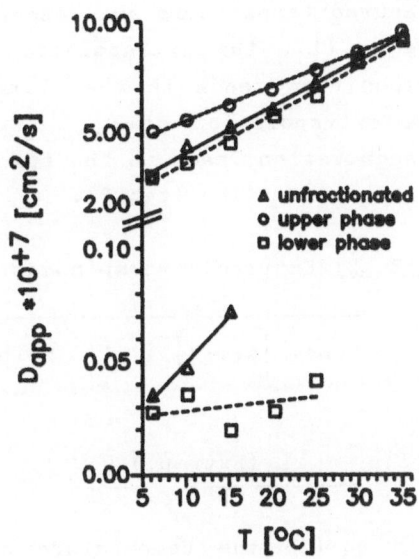

FIG.2: D_{app} of PS17500–CH solut. (1 % by wt.) versus temperature

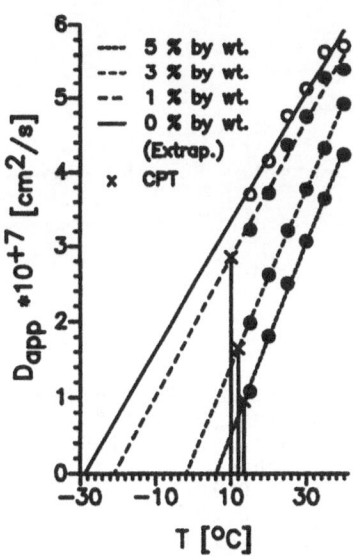

FIG.3: D_{app} of PS50000–CH solutions versus temperature

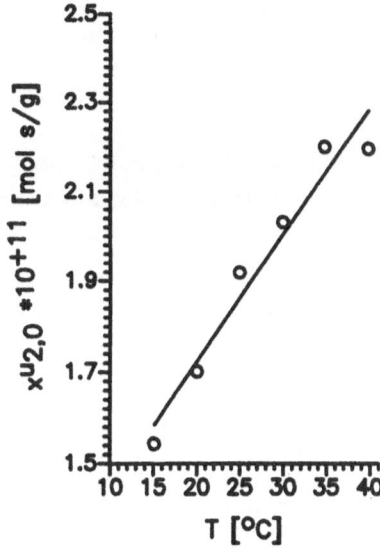

FIG.4: Mobilities of the system PS50000–CH versus temperature

measured temperature and concentration range (fig.3). Following eq.(1), the extrapolation of D_{app} to zero at constant
composition leads to the spinodal temperature. Additionally
the extrapolation of D_{app} at constant temperature to zero
concentration leads to the spinodal temperature at zero polymer concentration (tab.2).

TAB. 2: Extrapolated spinodal temp. of PS50000-CH solutions

conc. $\left[\text{wt.-\%}\right]$:	spinodal temp. $\left[^{\circ}\text{C}\right]$:
1	- 20.9
3	- 1.9
5	6.0
(Extrap.) 0	- 28.7

With eq.(2) the temperature dependence of the mobility at
zero concentration $_xu_{2,0}(T)$ (fig.4) can be calculated. The
values in this paper are - with regard to the difference of
the molar masses - in a good agreement with the values of
other authors /8,9/.

ACKNOWLEDGEMENT

The financial support by the "Deutsche Forschungsgemeinschaft" is greatfully acknowledged.

REFERENCES

/1/ R.Koningsveld, A.J.Staverman; J. Polym. Sci.,1967,16,1775
/2/ R.Koningsveld, A.J.Staverman; J. Polym. Sci.,1968,A-2,325
/3/ G.Rehage, D.Möller, O.Ernst; Makromol. Chemie,1965,88,232
/4/ G.Rehage, D.Möller; J. Poly. Sci.,1967,C 16,1787
/5/ Th.G.Scholte; Europ. Polymer J.,1970,6,1063
/6/ W.Borchard; Ber. d. Bunsenges.,1972,76,224
/7/ M.Tsuyomoto, Y.Einaga, H.Fujita; Polymer,1983,16,229
/8/ G.Rehage, O.Ernst; Kolloid-Z.,1964,197,64
/9/ J.Raczek; Eur. Polymer J.,1983,19,607

Part 2
CHEMISTRY

NETWORK FORMATION IN FREE RADICAL POLYMERIZATION

H. Tobita and A.E. Hamielec
McMaster Institute for Polymer Production Technology
Department of Chemical Engineering
McMaster University
Hamilton, Ontario, Canada L8S 4L7

ABSTRACT

Kinetic models for crosslinking and cyclization in free radical
polymerization of vinyl/divinyl monomers are proposed. In free radical
polymerization each primary polymer molecule experiences a different
history of crosslinking and cyclization, and therefore, a polymer network
possesses "a crosslinkng density distribution". The variance of this
distribution becomes significant when the polymerization conditions
deviate from Flory's simplifying assumptions. The concept of crosslinking
density distribution makes it possible to generalize Flory's theory of
network formation.

INTRODUCTION

In order to build a realistic model for network formation, the specific
reaction scheme of a system must be accounted for. Network formation in
free radical polymerization is a non-equilibrium process, namely, it is
kinetically controlled, and therefore, it is necessary to consider the
history of the generated network structure. As an observing unit for this
change, we are to consider the primary polymer molecule [1]. The primary
polymer molecule is a rather imaginary molecule which would exist if all
crosslinks connected to it were severed, namely, the primary polymer
molecule is a linear polymer. The crosslinking density ρ^f is to be
used to express the degree of crosslinking of the primary polymer molecule.

$$\rho^f = \frac{\text{(number of crosslinked units)}}{\text{(total number of units bound in the polymer chain)}} \qquad (1)$$

where a unit is defined as follows. One vinyl monomer bound in a chain is
equivalent to one unit. One divinyl monomer bound in a chain with a
pendant double bond is equivalent to one unit. When a divinyl monomer
bound in a chain is crosslinked, this is equivalent to two crosslinked
units. A tri-functional branching point (due to chain transfer to
polymer) is equivalent to one crosslinked unit.

From the point of view of physical properties of a polymer network, ρ^f is important. However, to build a kinetic model, it is convenient to consider the crosslinking density ρ which is defined with respect to the number of monomer units.

$$\rho = \frac{(\text{number of crosslinked units})}{(\text{total number of \textbf{monomer units} bound in the polymer chain})} \quad (2)$$

where one divinyl monomer is counted as one whether it is crosslinked or not in the denominator, while the numerator is the same as eq.(1). When the crosslinking density is far smaller than unity, there is no difference between these two definitions of crosslinking density, however, if the mole fraction of divinyl monomer f_{20} is large, one needs to carefully distinguish these definitions. The relationship between these two definitions of crosslinking density will be given in the text.

Flory's theory of network formation [1] basically assumes an equilibrium system. In a kinetically controlled system such as free radical polymerization, Flory's theory is only applicable when the crosslinking densities of all primary polymer molecules born at different times are equal. Generalization of Flory's theory using "crosslinking density distribution" will also be shown.

THEORETICAL

Process of Crosslinking

Let us assume that the crosslinking reaction shown below occurs at conversion ψ and that the primary polymer molecule **A** was formed at conversion Θ ($\Theta < \psi$).

At conversion ψ the primary polymer radical **B** attacks a pendant double bond on the primary polymer molecule **A**, which results in a crosslinkage between two primary polymer molecules. In this case from the point of view of molecule **B**, this crosslinkage is formed during its growth (instantaneous crosslinking density $\rho_i(\psi)$). While from the point of view of molecule **A**, the identical crosslinkage is formed but after it was formed, so that it can be considered as an additional crosslinking ($\rho_a(\Theta,\psi)$). At conversion ψ, the crosslinking density of the primary molecule which was formed at conversion Θ is given by;

$$\rho(\Theta,\psi) = \rho_i(\Theta) + \rho_a(\Theta,\psi) \quad (3)$$

The explicit formation of each type of crosslinking density is given by [2];

$$\frac{\partial \rho_a(\Theta,\psi)}{\partial \psi} = \frac{k_p^*}{k_p(1-\psi)} \quad (4)$$

$$P_i(\Theta) = \int_0^\Theta \frac{\partial P_a(\gamma,\Theta)}{\partial \Theta} \, dy \qquad (5)$$

where k_p^* and k_p are the pseudo-kinetic rate constants [2-4] for crosslinking and propagation respectively, which are given by;

$$k_p^* = k_p^{*0}[F_2(\Theta) - P_a(\Theta,\psi) - P_c(\Theta,\psi)] \qquad (6)$$

$$k_p^{*0} = k_{p13}^* \Phi_1^{\bullet} + k_{p23}^* \Phi_2^{\bullet} + k_{p33}^* \Phi_3^{\bullet} \qquad (7)$$

$$k_p = (k_{11}f_1 + k_{12}f_2)\Phi_1^{\bullet} + (k_{21}f_1 + k_{22}f_2)\Phi_2^{\bullet} + (k_{31}f_1 + k_{32}f_2)\Phi_3^{\bullet} \qquad (8)$$

where

F_i instantaneous mole fraction of monomer i bound in the polymer chain.

P_c cyclization density.

k_{pij}^* kinetic rate constant for crosslinking reaction in which radical of type i reacts with the double bond of type j.

$\Phi_i^{\bullet}$ mole fraction of radical type i.

k_{ij} propagation rate constant in which the radical of type i reacts with the double bond of type j.

f_i mole fraction of monomer of type i.

Subscript 1 is used to designate mono-vinyl monomer, 2 is used for divinyl monomer, and 3 is used for pendant double bonds. Propagation rate constants are defined with respect to the monomer units, not the number of double bonds.

From eqs.(3)-(5), one can calculate the crosslinking density distribution as a function of the birth conversion of the primary polymer molecule. A sample calculation of crosslinking density distribution can be found in [3].

Chain transfer to polymer can similarly be formulated as eq.(4). In this reaction, there is no instantaneous crosslinking, but only additional crosslinkage is formed.

Crosslinking densities given by eqs.(3)-(5) are defined with respect to the number of monomer units (see eq.(2)). But from physical point of view what is important is the crosslinking density defined with respect to the number of units (defined by eq.(1)) especially for the cases where high mole fraction of divinyl monomer is used. The relationship between these two definitions of crosslinking density is given by;

$$\rho^f(\Theta,\psi) = \frac{\rho(\Theta,\psi)}{1 + P_i(\Theta)} \qquad (9)$$

Now let us consider the conditions for which this crosslinking density distribution becomes significant. Flory used the following simplifying assumptions in his theory for vinyl/divinyl copolymerization [1].

(1) The reactivities of all types of double bonds are equal.
(2) All double bonds react independently of one another.

(3) There is no cyclization in finite molecules.
First let us examine these limiting conditions. In this case, the pseudo
-kinetic rate constants for crosslinking and propagation reduce to;

$$k_p^*(\psi) = k_{11}[F_2(\psi)-\rho_a(\Theta,\psi)] \qquad (10)$$

$$k_p(\psi) = k_{11}[1+f_2(\psi)] \qquad (11)$$

Only for this simplified condition does the Mayo-Lewis equation apply for
the calculations of the copolymer composition with reactivity ratios
$r_1=0.5$ and $r_2=2.0$ (details are discussed elsewhere [5]). Applying
eqs.(10) and (11) into eqs.(4) and (5), one obtains analytical solutions
for $\rho_a(\Theta,\psi)$ and $\rho_i(\Theta)$. Therefore, $\rho(\Theta,\psi)$ is given by;

$$\rho(\Theta,\psi) = \frac{2[1-f_2(\Theta)]}{1+f_2(\Theta)} \cdot \frac{f_{20}[1-f_2(\psi)]-(1-f_{20})f_2(\psi)}{(1-f_{20})[1-f_2(\psi)]} \qquad (12)$$

In terms of the number of units,

$$\rho^f(\Theta,\psi) = \frac{2(f_{20}[1-f_2(\psi)]-(1-f_{20})f_2(\psi))}{(1+f_{20})[1-f_2(\psi)]} \qquad (13)$$

Eq.(13) indicates that the crosslinking density defined with respect to
the number of units is solely a function of the present conversion ψ. A
sample calculation is shown in Fig.1. Therefore, for Flory's simplified
condition the variance of crosslinking density is zero.   This
is equivalent to stating that there is no difference between the kinetic
model which considers the history of the generated network structure and
models which assume an equilibrium system.

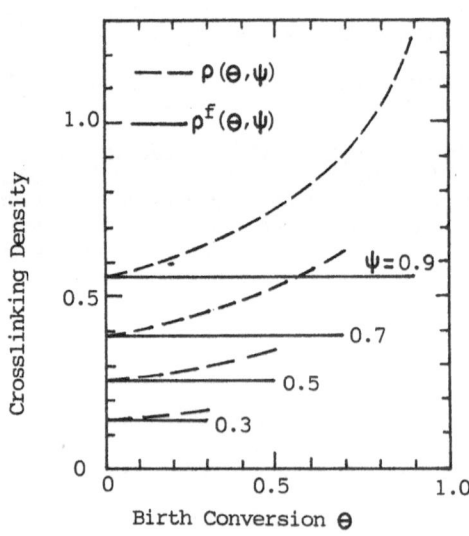

Fig.1 Crosslinking Density
Distribution.

(Flory's Simplifying
Condition.)

$f_{20}= 0.5$

But in general, it is necessary to consider the following non-idealities.
    1) Differences in the reactivities of monomeric double bonds.
    2) The reactivity of pendant double bonds.
    3) The effect of cyclization.
When these effects are significant, the crosslinking density
has a significant variance especially at high conversions in the
post-gelation period. Using appropriate kinetic parameters, our kinetic
model can account for these effects, though we need in addition a kinetic
model for cyclization. In the next section, simple models for cyclization
are derived.

Process of Cyclization

One of the important features of cyclization is that it is controlled, not
by the conventional rate law using average concentrations of functional
groups but by conformational statistics of the sequence of bonds. Strict
treatment of cyclization in a mean-field theory seems to be very difficult
especially for high conversions. In this section, we will show some of
very simple models for cyclization. These simplified models might be
acceptable at this stage in the development of kinetic models for
cyclization.

    In our formalism, it is convenient to divide the cyclization
reactions into two groups, namely, primary and secondary.

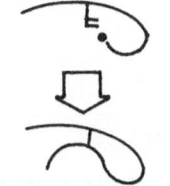

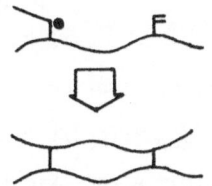

Primary Cyclization   Secondary Cyclization

The primary cyclization is the cycle formed within one primary polymer
molecule, while the secondary cyclization is formed between two or more
primary polymer molecules. The mathematical importance of the difference
between primary and secondary cyclization is that the primary cyclization
is solely a function of the birth conversion ($\rho_{c,p}(\Theta)$), while the
secondary cyclization is a function of both birth conversion and present
conversion ($\rho_{c,s}(\Theta,\psi)$).

    Primary Cyclization: A simple model for the primary cyclization can
be built using "random flight model" [1,6]. If a radical center on a
primary polymer radical is located at the origin of coordinates, the
probability that a randomly selected monomer unit bound on the identical
primary polymer molecule resides in the volume dV at a distance R is given
by;

$$W(\mathbf{R})dV = [3/(2\pi l_s^2 n_s)]^{1.5} \exp[-3R^2/(2 n_s l_s^2)] dV \qquad (14)$$

where l_s is the length of a statistical segment and n_s is the number
of statistical segments in a chain. In order for the primary cyclization
to be formed, R=0.

$$W(0)dV = [3/(2\pi 1_s{}^2 b)]^{1.5} N^{-1.5} dV$$

$$= A N^{-1.5} dV \tag{15}$$

where $n_s = bN$ and N is the number of monomer units. Eq.(15) shows that the smaller cycles have a better chance of formation than the larger ones. The probability of forming a cycle for the primary polymer radical with chain length p is given by;

$$P_{c,r} = \sum_{N=1}^{p} k'_{cp} F_2 N^{-1.5} \tag{16}$$

Therefore, the expectation of the number of cycles formed for a primary polymer molecule with chain length r is given by;

$$E(n_c) = \sum_{p=1}^{r} \sum_{N=1}^{p} k'_{c,p} F_2 N^{-1.5}$$

$$\cong \int_1^r \int_1^p k'_{c,p} F_2 y^{-1.5} dy\, dp$$

$$= 2 k'_{c,p} F_2 (r - 2r^{0.5} + 1) \tag{17}$$

The primary cyclization density ρ_{cp} is given by;

$$\rho_{cp} = E(n_c)/r = 2 k'_{c,p} F_2 (1 - 2/r^{0.5} + 1/r) \tag{18}$$

Since eq.(18) is approximately constant over a sufficient range of chain length except for oligometic chain lengths, as a simple model $\rho_{cp}(\Theta)$ is given by;

$$\rho_{cp}(\Theta) = k_{c,p} F_2(\Theta) \tag{19}$$

The overall cyclization density at present conversion ψ is given by;

$$\overline{\rho}_{cp}(\psi) = k_{c,p} \overline{F}_2(\psi) \tag{20}$$

where $\overline{F}_2$ is the accumulated mole fraction of divinyl monomer bound in the polymer chain. Quite often, the reactivity of divinyl monomer is larger than that of mono-vinyl monomer (since divinyl monomer possesses two double bonds in a monomer unit), so that the primary cyclization density is maximum at the initial stage of reaction.

Secondary Cyclization: The secondary cyclization can be defined clearly in the pre-gelation period. However, it is ambiguous in the post-gelation period, since from a physical point of view it is impossible to distinguish from crosslinking in the post-gelation period. Furthermore, even crosslinking density given by eqs.(4) and (5) should involve secondary cyclization especially in the post-gelation period. Therefore, we are to define the secondary cyclization based on the kinetics of formation. Let us call the formation of junction points which follow eqs.(4) and (5) "crosslinking". The secondary cyclization can be defined as the formation of junction points between primary polymer molecules which does not follow the rate law shown in eqs.(4) and (5). Secondary cyclization may also be regarded as a correction factor in order to connect crosslinking density to other properties.

The secondary cyclization is also dominated by the conformational statistics as it was shown in the modeling of primary cyclization. However, it is convenient to consider the average number of secondary cycles per crosslinking $\eta(\Theta,\psi)$, since in order for the secondary cyclization to be formed it is necessary to have a crosslinkage.

$$\frac{\partial \rho_{cs,a}(\Theta,\psi)}{\partial \psi} = \eta(\Theta,\psi) \frac{\partial \rho_a(\Theta,\psi)}{\partial \psi} \tag{21}$$

$$\rho_{cs,i}(\Theta) = \int_0^\Theta \frac{\partial \rho_{cs,a}(y,\Theta)}{\partial \Theta} \, dy \tag{22}$$

$$\rho_{cs}(\Theta,\psi) = \rho_{cs,i}(\Theta) + \rho_{cs,a}(\Theta,\psi) \tag{23}$$

where $\rho_{cs,a}$ is the additional secondary cyclization and $\rho_{cs,i}$ is the instantaneous secondary cyclization.

In a real system, $\eta(\Theta,\psi)$ should be a very complicated function of the mole fraction of pendant double bonds on the chain, chain length of the primary polymer molecule, molecular conformation, etc. As a first approximation, $\eta(\Theta,\psi)$ can be considered as proportional to the number of pendant double bonds on the primary polymer molecule.

$$\eta(\Theta,\psi) \propto [F_2(\Theta) - \rho_a(\Theta,\psi) - \rho_{cp}(\Theta) - \rho_{cs,a}(\Theta,\psi)] \, r(\Theta) \tag{24}$$

where $r(\Theta)$ is the chain length of the primary polymer molecule which was formed at Θ. For polydisperse systems,

$\eta(\Theta,\psi) \propto$ [average number of pendant double bonds on a primary polymer molecule which was formed at Θ.]

$$\propto \frac{[\text{number of pendant double bonds in P.M.}(\Theta)]}{[\text{number of P.M.}(\Theta)]}$$

$$= k_{cs}[F_2(\Theta) - \rho_a(\Theta,\psi) - \rho_{cp}(\Theta) - \rho_{cs,a}(\Theta,\psi)] \, P_{np}(\Theta) \tag{25}$$

where P.M.(Θ) is used to designate the primary polymer molecules which were formed at conversion Θ, and P_{np} is the number-average chain length of P.M.(Θ). In this approximation the effect of the chain length of the growing primary polymer radical at that moment is neglected.

Since it is reasonable to consider the secondary cycles as effective junction points in terms of elasticity, $\rho_{el}(\Theta,\psi)$ and $\rho_{el}^f(\Theta,\psi)$ are given by;

$$\rho_{el}(\Theta,\psi) = \rho(\Theta,\psi) + \rho_{cs}(\Theta,\psi) \tag{26}$$

$$\rho_{el}^f(\Theta,\psi) = \frac{\rho(\Theta,\psi) + \rho_{cs}(\Theta,\psi)}{1 + \rho_i(\Theta) + \rho_{cp}(\Theta) + \rho_{cs,i}(\Theta)} \tag{27}$$

Sample calculations are shown in Fig.2.

40

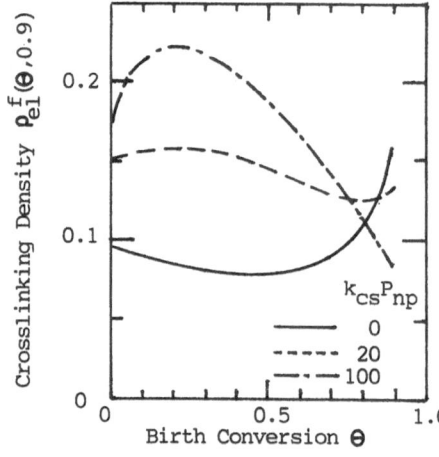

Fig.2 <u>The Effect of Secondary
Cyclization on Crosslinking
Density Distribution</u>.

$f_{20}= 0.1$, $\psi = 0.9$

[Parameters]

$k_{11}=k_{21}=k_{31}= 300$

$k_{12}=k_{22}=k_{32}= 600$

$k^*_{p13}=k^*_{p23}=k^*_{p33}= 100$

Generalized Flory's Theory

Flory's theory of network formation, basically, assumes an equilibrium system, so that generalization is necessary in order for it to apply to a kinetically controlled system such as with free radical polymerization. Quite often the crosslinking density is rather constant at low conversions, and therefore, the effect of crosslinking density distribution may be negligible for pre-gelation period. It is reasonable to approximate the crosslinking density of a primary polymer molecule by the overall crosslinking density ($\bar{\rho}(\psi)$). For example, weight-average chain length $\bar{P}_w$ in the pre-gelation period is given by;

$$\bar{P}_w(\psi) = \frac{\bar{P}_{wp}(\psi)}{1- \bar{\rho}(\psi)\ \bar{P}_{wp}(\psi)} \qquad (28)$$

where $\bar{P}_{wp}(\psi)$ is the accumulated weight-average chain length of the primary polymer molecules. However, for post-gelation period the effect of crosslinking density distribution becomes important. For example, let us consider the weight fraction of sol W_s in the post-gelation period. From the statistical theory by Flory the weight fraction of sol is given by;

$$W_s=\sum_{r=1}^{\infty} w_r\ (1-\rho^f\ W_g)^r \qquad (\rho^f \ll 1) \qquad (29)$$

where w_r is the weight chain length distribution of the primary polymer molecules, and W_g is the weight fraction of gel, i.e., $W_g= 1-W_s$. The parenthesis in eq.(29) is the probability that a randomly selected monomer unit bound in the polymer chain belongs to the sol fraction, and therefore, the meaning of above equation is obvious. Theoretically, the use of eq.(29) is restricted to the low crosslinking density region, however, when crosslinking density is not far smaller than unity, there is practically no sol left in the system, and eq.(29) expresses this behavior well. Practically eq.(29) is applicable for all conversions. Using crosslinking density distribution, eq.(29) can be generalized as follows.

$$W_s(\Theta,\psi) = \sum_{r=1}^{\infty} w_r(\dot\Theta)[1- \rho^f(\Theta,\psi)\ W_g(\Theta,\psi)]^r \qquad (30)$$

Since the primary polymer molecules are linear polymers, we know the functional form of $w_r(\Theta)$.

$$w_r(\Theta) = (\tau(\Theta)+\beta(\Theta))[\tau(\Theta)+(\beta(\Theta)/2)(\tau(\Theta)+\beta(\Theta))(r-1)]r\ \phi^{r+1} \qquad (31)$$

where $\tau = [(\text{rate of termination by disproportionation})+(\text{rate of chain transfer})]/(\text{propagation rate})$, and $\beta = (\text{rate of termination by combination})/(\text{propagation rate})$. $\phi = 1/[\tau(\Theta)+\beta(\Theta)+1]$. Substituting eq.(31) into eq.(30), one obtains;

$$W_s(\Theta,\psi) = A\ G_1\ [T + A\ B\ G_1] \qquad (32)$$

where $T = \tau(\Theta)/[\tau(\Theta)+\beta(\Theta)+ \rho^f(\Theta,\psi)\ W_g(\Theta,\psi)]$

$B = \beta(\Theta)/[\tau(\Theta)+\beta(\Theta)+ \rho^f(\Theta,\psi)\ W_g(\Theta,\psi)]$

$A = T + B$, $\qquad G_1 = 1- \rho^f(\Theta,\psi)\ W_g(\Theta,\psi)$

Equations for other interesting properties can be found in [2].

APPLICATION

We applied generalized Flory's theory to the bulk copolymerization of methyl methacrylate (MMA) and ethylene glycol dimethacrylate (EGDMA) at temperature 70°C. Initial mole fraction of EGDMA was $f_{20} = 5.08 \times 10^{-3}$. Since f_{20} is far smaller than unity, all pseudo-kinetic rate constants for the formation of primary polymer molecules can be approximated by those for homopolymerization of MMA. The effect of diffusion controlled termination in the network polymer is far more significant than that for linear polymers, so that the primary polymer chain length drift should be significant. At present it is unclear how to estimate the decrease in the termination constant theoretically, and therefore, an empirical correlation shown below was applied and all constants were estimated from time-conversion curve.

$$k_p/k_t^{0.5} = 0.129 \quad [1^{0.5}\text{mol}^{-0.5}\text{sec}^{-0.5}] \quad (x<0.143)$$

$$= 0.129\ \exp[8.8(x-0.143)] \quad (0.143< x<0.75)$$

$$= 0.129\ \exp[-28(x-0.75)+8.8(x-0.143)] \quad (x>0.75)$$

where x is the total monomer conversion and k_t is the pseudo-kinetic rate constant for termination reaction.

Details on experimental procedures can be found elsewhere [7]. The comparison with experimental data is shown in Fig.3. The crosslinking density of gel was estimated by swelling experiments in chloroform. Even if all divinyl monomers were used in effective crosslinkages, the final maximum crosslinking density would be $2f_{20} = 1.16 \times 10^{-2}$. Thus, experimental crosslinking densities at high conversions are clearly too high. This may be caused by physical crosslinkages. However, at low conversions the network structure should be quite loose and it may be reasonable to neglect the physical crosslinkages. In the bulk copolymerization of MMA/EGDMA the effect of primary cyclization can be considered to be small [8], and we neglected the primary cyclization.

At present, the kinetics of diffusion-controlled reactions in free radical polymerization are not fully understood at high conversions, but the agreement with gel formation data is satisfactory, and therefore, this model seems applicable for the prediction of the actual gelation phenomena.

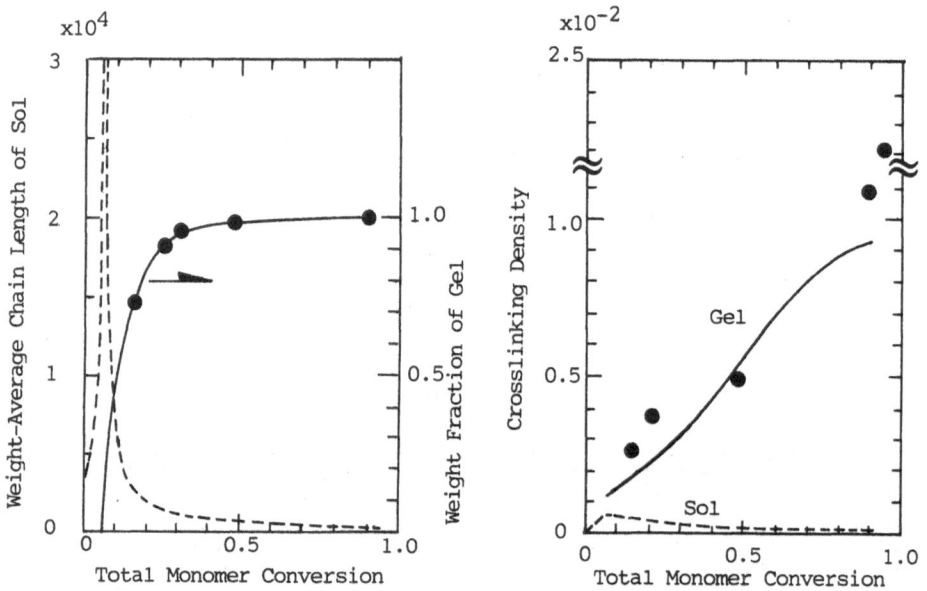

Fig.3 Application of Theory.
(Copolymerization of MMA/EGDMA.)

$$f_{20}= 5.08 \times 10^{-3}, \quad k_p^*/k_p = 0.5, \quad k_{cs} = 0.06$$

REFERENCES

1. Flory, P.J. Principles of Polymer Chemistry; Cornell University Press: Ithaca, NY, 1953
2. Tobita, H.; Hamielec, A.E. ACS Symp. Ser. (in press. A volume on "COMPUTER APPLICATION IN APPLIED POLYMER SCIENCE").
3. Tobita, H.; Hamielec A.E. Makromol. Chem., Macromol. Symp. 1988, 20/21, 501.
4. Tobita, H.; Hamielec, A.E. Macromolecules 1989 (in press.)
5. Tobita, H.; Hamielec, A.E. (will be submitted to Makromol. Chem., Macromol. Symp.)
6. Jacobson, H; Stockmayer, W.H. J. Chem. Phys. 1950, 18, 1600.
7. Li, W.E.; Hamielec, A.E.; Crowe, C.M. Polymer 1989 (in press.)
8. Landin, D.T.; Macosko, C.W. Macromolecules 1988, 21, 846.

PROBLEMS OF MICROMIXING EFFECTS AND REACTOR PERFORMANCE IN FREE RADICAL POLYMERIZATION OF BINARY AND TERNARY SYSTEMS

GUDRUN SCHMIDT-NAAKE, HERMANN SCHMIDT, WINFRIED PIPPEL
Technische Universitaet Dresden, Sektion Chemie,
Mommsenstrasse 13, DDR-8027 Dresden G.D.R.

ABSTRACT

On the basis of experimental data obtained from copolymer systems composed of donor, acceptor and neutral monomers the copolymerization behaviour as a function of the conversion degree was derived. The heterogeneity of the composition was calculated for experiments carried out under discontinuous process conditions (batch reactor). The influence of the reaction conditions on the chemical polydispersity was investigated in connection with continuous process conditions. The effects of incomplete micromixing on free radical copolymerization processes are presented and discussed by model calculations.

INTRODUCTION

The free radical copolymerization of two or more monomers is an important process for the synthesis of polymers with special properties. A general problem in a more component polymerization carried out under non-aceotropic conditions is that with increasing conversion the relative values of the monomeric concentration change, and consequently, in most of the cases, dependent on the process and concentration conditions as well as on the reactivity of the monomers, the composition of the polymer changes, too.

In a range of conversion up to about 5% the copolymer equation describes the relationship between the monomeric and copolymeric composition quite well. However, if conversion ranges of interest to industry are to be considered in order to derive prescriptions for synthesis or appropriate models for process control of the radicalic solution polymerization under discontinuous conditions or in stationary-state operating CSTR type reactors it is useful to apply modified model calculations based on experimental data.

In the production of such sorts of special polymers for which the demand is of the order of only a few kilograms, a discontinuous way of processing is of more interest than a continuous one, requiring more time and costs.

By means of the following model calculations the influence of the processing conditions on the chemical polydispersity of copolymers of the donor (styrene = S), acceptor (maleic anhydride = MA, N-phenylmaleimide = NPI) and neutral monomers (acrylonitrile = ACN, methylmethacrylate = MMA) type shall be investigated. Conversion ranges up to < 60 % will be considered thus excluding diffusion controlled termination steps.

Another problem arrives, especially in bulk or solution polymerization with low solvent concentration when reactants begin to react together prior to becoming mixed uniformly on the molecular scale. In such cases at the input of the reaction vessel, layers (striations) containing both a component with a high and one with a low concentration are formed. The thickness of these striations depends on the viscosity and the energy input of the stirrer. The mass transport between these regions is carried out by diffusion. In these cases inhomogeneities of the initiation reaction can be caused either by concentration gradients when initiator substances are used, or by temperature perfiles in the case of a thermal initiation.

For the modelling of the imperfect micromixing of the monomers A and B and the nonuniform distribution of the initiator a simple layer model, which holds for a striation of thickness independent of time, has been used.

METHODS

For the description of the radical copolymerization, the initiation, propagation and termination are the most important steps. Chain transfer to monomer and solvent can be included into the considerations, if necessary.

The following reaction scheme of binary copolymerization, and respectively a modified scheme for three monomers have been taken as a basis for the model calculations:

Initiator decomposition: $\quad I \xrightarrow{k_i} 2R*$

initiation: $\quad R* + A \xrightarrow{k_{ia}} P_1*$

$\quad R* + B \xrightarrow{k_{ib}} Q_1*$

propagation: $\quad P_j* + A \xrightarrow{k_{paa}} P_{j+1}*$

$\quad P_j* + B \xrightarrow{k_{pab}} Q_{j+1}*$

$$Q_j{*} \;+\; A \;\xrightarrow{k_{pba}}\; P_{j+1}{*}$$

$$Q_j{*} \;+\; B \;\xrightarrow{k_{pbb}}\; Q_{j+1}{*}$$

termination by

recombination:

$$P_j{*} \;+\; P_n{*} \;\xrightarrow{k_{caa}}\; M_{j+n}$$

$$P_j{*} \;+\; Q_n{*} \;\xrightarrow{k_{cab}}\; M_{j+n}$$

$$Q_j{*} \;+\; Q_n{*} \;\xrightarrow{k_{cbb}}\; M_{j+n}$$

disproportion:

$$P_j{*} \;+\; P_n{*} \;\xrightarrow{k_{daa}}\; M_j + M_n$$

$$P_j{*} \;+\; Q_n{*} \;\xrightarrow{k_{dab}}\; M_j + M_n$$

$$Q_j{*} \;+\; Q_n{*} \;\xrightarrow{k_{dbb}}\; M_j + M_n$$

Chain transfer:

to the monomers:

$$P_j{*} \;+\; A \;\xrightarrow{k_{maa}}\; M_j + P_1{*}$$

$$P_j{*} \;+\; B \;\xrightarrow{k_{mab}}\; M_j + Q_1{*}$$

$$Q_j{*} \;+\; A \;\xrightarrow{k_{mba}}\; M_j + P_1{*}$$

$$Q_j{*} \;+\; B \;\xrightarrow{k_{mbb}}\; M_j + Q_1{*}$$

to the solvent:

$$P_j{*} \;+\; L \;\xrightarrow{k_{la}}\; M_j + R{*}$$

$$Q_j{*} \;+\; L \;\xrightarrow{k_{lb}}\; M_j + R{*}$$

Batch reactor processing

In the case of a binary copolymerization the system of
balance equations includes the material balances for [I],
[A], [B], [P*], [Q*], [M]. For the mathematical description
of the molecular weight distribution the method of genera-
ting functions [1] has been applied. Thus, in introducing
the statistical moments for the living polymers P* and Q*
(i.e. the moments of first order as the concentrations of A
and B in P* and Q*, and the moments of second order), and
the dead macromolecules M (moments of first order as the
concentrations of A and B in M, and the respective second

moments) the differential equation system is being expanded to 12, or 21 equations, respectively. From the solution of the established material balance equation system the comonomer- copolymer composition, degree of conversion, number average degree of polymerization and weight average degree can be calculated.

CSTR in stationary state

To describe a free radical co- and terpolymerization reaction in a continuous stirred tank reactor a model corresponding to equation (1) has been derived [2].

$$\frac{dc_i}{dt} = \frac{v}{v_R}(c_i{}^E - c_i{}^A) + r_i(c) = 0 \qquad (1)$$

The source term $r_i(c)$ corresponds to the equation system for the terpolymerization. Restrictions are imposed on the system so that the conversion is not allowed to exceed beyond 60% thus avoiding the gel-effect discussed in literature [3] with the resulting problems to correct the kinetic constants. The model allows a calculation of the stationary composition of the terpolymerization, the monomer composition, the conversion degree and the mean polymerization degrees.

Micromixing model

Model of the copolymerization process in a fixed (static) striation [4]:

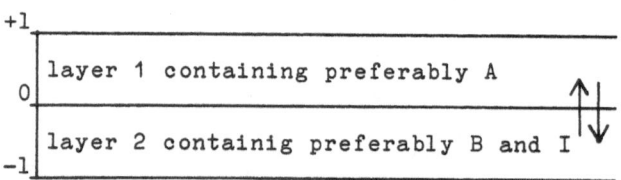

The effect of micromixing is described by a system of diffusion and reaction equations, i.e. a system of partial differential equations with the corresponding initial and border conditions in the form of

$$\frac{\partial c_i}{\partial t} = D\frac{\partial^2 c_i}{\partial x^2} + r_i(c) \qquad (2)$$

For its solution the principle of semidiscretization, i.e. the transformation of the partial into ordinary differential equations by the method of a partial discretization was applied. Thus, in applying a linear implicit multistep procedure according to Curties and Hirschfelder to the stiff equation system the desired final solutions could be derived.

RESULTS AND DISCUSSION

As in a stationary state a well defined equal monomer mix-ture necessarily leads to a constant copolymer composition, the continuous method of processing yields the highest degree of chemical composition homogeneity. So CSTR data ob-tained under stationary conditions are indispensible if prescriptions for synthesis have to be developed.

For AN/S/MA we have carried out a lot of experiments at $100^\circ C$ under continuous conditions. Assuming that $k_{33} = k_{13} = k_{31} = 0$, the results of these experiments could be cal-culated in agreement with the experimental data (see fig.1). Fig. 1 demonstrates that in the stationary state of the continuous operating conditions in the output of the CSTR a fixed comonomer and copolymer combination is achieved.

The conversion dependent copolymer compositions of the different comonomeric mixtures at the reactor input can be represented in a triangular coordinate diagram. Compositions corresponding to the same conversion degree can be connected with another by conversion trajectories.

Fig.2 presents for a AN/S/MA system the conversion dependent alteration of the terpolymer composition for four different monomer mixtures each of them containing 5 mole % MA. Furthermore, fig. 2 represents the copolymer composi-tions obtained in discontinuous experiments (X < 5%) from different monomeric variations each containing 5 mole % of acceptor monomer.

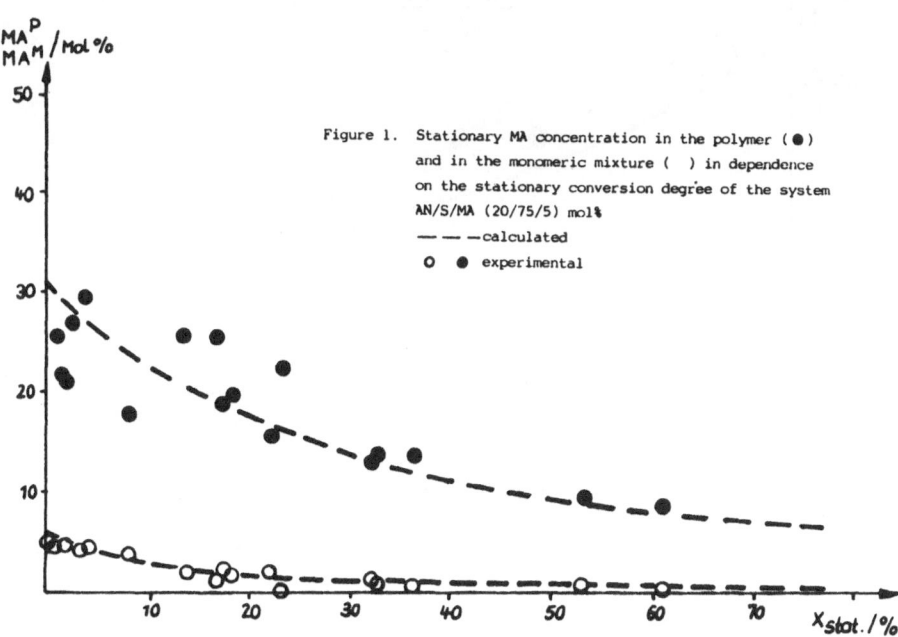

Figure 1. Stationary MA concentration in the polymer (●) and in the monomeric mixture ( ) in dependence on the stationary conversion degree of the system AN/S/MA (20/75/5) mol%
— — —calculated
O ● experimental

48

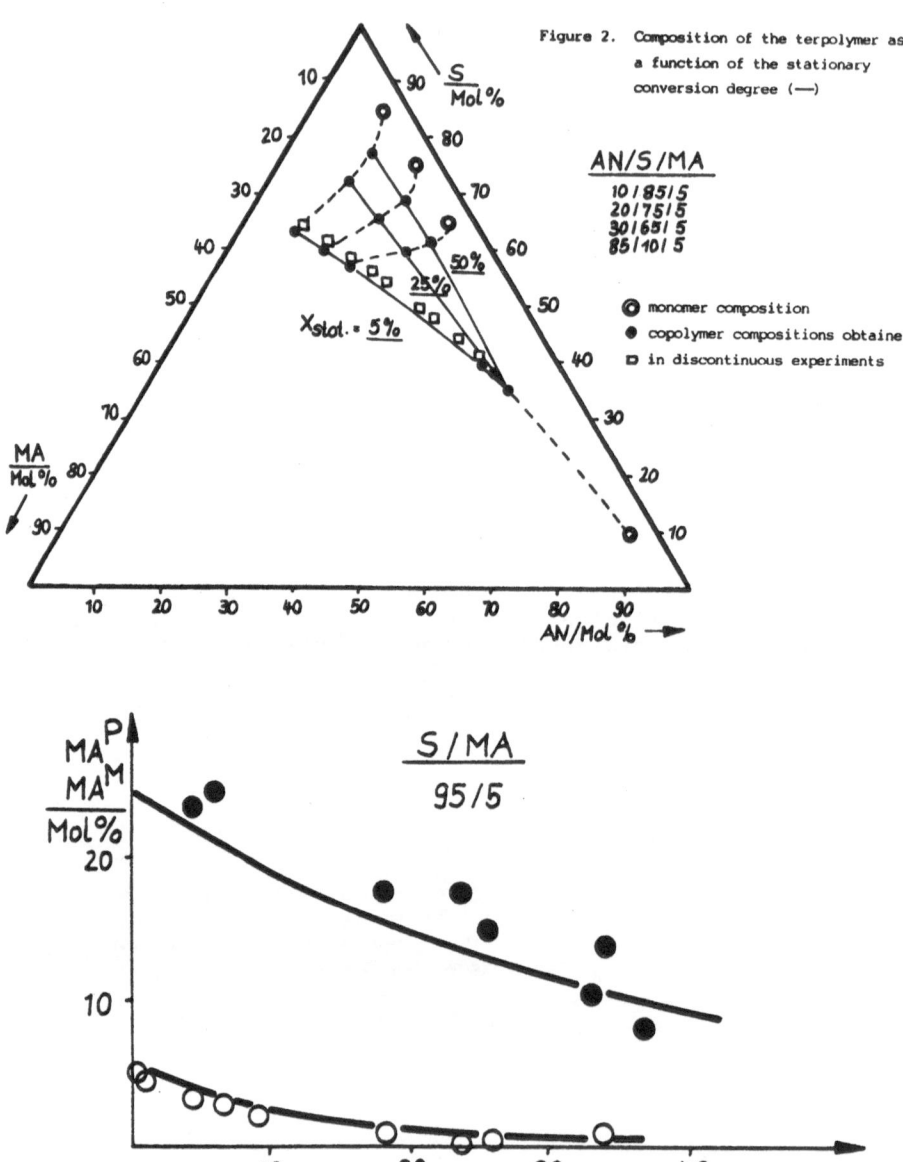

Figure 2. Composition of the terpolymer as a function of the stationary conversion degree (—)

AN/S/MA
10 / 85 / 5
20 / 75 / 5
30 / 65 / 5
85 / 10 / 5

◉ monomer composition
● copolymer compositions obtained
▢ in discontinuous experiments

Figure 3. Stationary MA concentration in the polymeric (●) and the monomeric mixture (○) as a function of the stationary conversion; — calculated.

This representation of the conversion dependent copolymer composition demonstrates that the copolymer composition obtained from discontinuous terpolymerization experiments ($X < 5\%$) can in fact be correlated to the imposed monomer mixtures.

Fig. 3 also demonstrates for the donator-acceptor system S/MA (monomer mixture 95/5) a good agreement between the calculated and experimentally determined conversion dependency of the MA concentration in the polymer and the monomeric mixture for experiments carried out under continuous conditions.

If we want to compare these results with those obtained from discontinuous experiments, it is necessary for all simulation calculations to account for the conversion heterogeneity, i.e. we have to consider the fact that the polymer composition in the course of the polymerization process changes. The discontinuous experiment blends all the compositions over the whole range of conversion.

From the solution of the differential equation system for the batch reactor the monomeric, copolymer and experimentally determined average copolymer compositions are given by the following equations:

$$\text{monomeric:} \quad M_A = f(X) = \frac{[A]_X}{[A]_X + [B]_X} \tag{3}$$

$$\text{copolymer:} \quad P_A = f(X) = \frac{[A]_0 - [A]_X}{([A]_0 - [A]_X) + ([B]_0 - [B]_X)} \tag{4}$$

copolymer average
composition:

$$\overline{P}_A = \frac{1}{X_n} \int_0^{X_n} P_A(x)dx = \frac{1}{X_n} \sum_{i=1}^{i=n} \frac{P_{Ai} + P_{Ai-1}}{2} (X_i - X_{i-1}) \tag{5}$$

Analysing the influence of the operating conditions (i.e. using either a CSTR or a batch reactor) on the chemical polydispersity, the following conclusions can be drawn [5]:

1. Donator-acceptor systems with acceptor contents <20 mole% lead to a drastic change in the copolymer composition as a function of the stationary conversion owing to the strong tendency of the acceptor to become incorporated into the copolymer. Hence, discontinuously derived products will have a high degree of chemical polydispersity; see fig.4, system S/MA.
   Donator-acceptor systems with a high degree of chemical uniformity can be synthesized:
   for A < 20 mole% in a CSTR,
   for A > 20 mole% either in a CSTR or a batch reactor.
2. Neutral-acceptor systems (MMA/Acceptor):
   The preferred incorporation of MMA conduces in the case of an acceptor deficit in the monomeric mixture (A<20

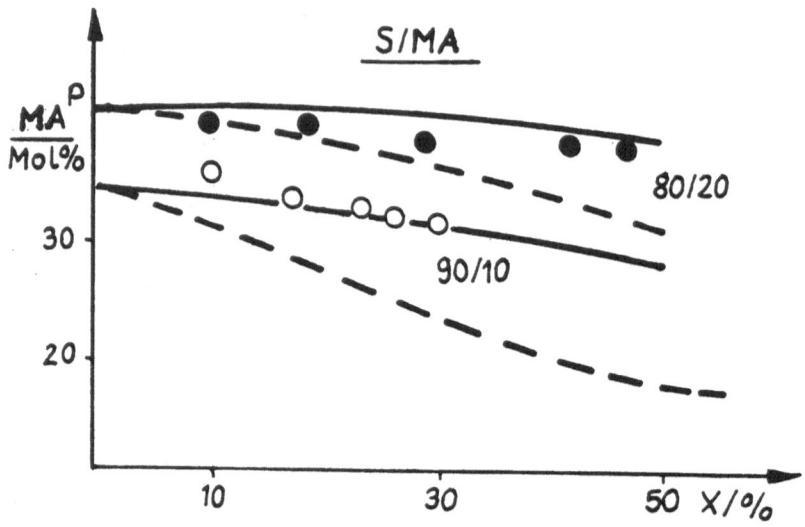

Figure 4. Copolymer composition of S/MA as a function of the conversion for discontinuous and continuous process conditions.

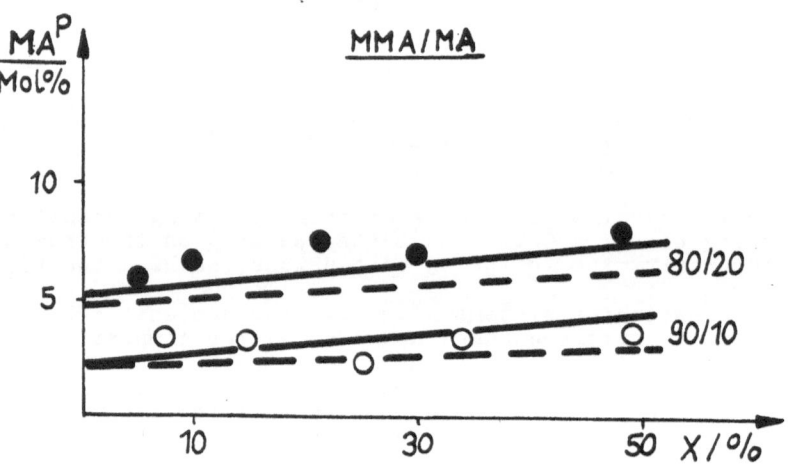

Figure 5. Copolymer compostion of MMA/MA as a function of the conversion for continuous and discontinuous process conditions

mole%) to a nearly conversion independent copolymer composition in a CSTR as well as in a batch reactor (see
fig.5, MMA/MA).
Neutral-acceptor systems with a high degree of chemical
uniformity can be synthesized:
for A < 20 mole% either in a CSTR or a batch reactor
for A > 20 mole% in a CSTR.

Micromixing effects

Finally, the most important results of model calculations
for the simulation of the imperfect micromixing in a
chemical reactor or a demonomerization process shall be
presented.
    Model calculations carried out for a non-stretched
striation demonstrate that the local reduction of the
initiation velocity (decrease of initiator concentration or
temperature) coupled with an imperfect micromixing of the
copolymers (local increase of one of the components) can
cause for small amounts of the reacting system a high molecular (extremely high molecular weights) as well as a high
chemical inhomogeneity (high contents of the enriched component) of the polymer. These mathematically predicted results have been found experimentally in the S/AN bulk polymerization.
    In order to demonstrate this effect, fig.6 shows the
profile of the number average of the molecules A (MMA)
incorporated into the polymer of the system MMA/S dependent
on the time for the defined striation.

CONCLUSION

By means of model calculations the influence of the reaction
conditions on the products has been investigated. From this,
prescriptions for synthesis of copolymers with a minimal
chemical heterogeneity has been derived.
The conversion dependent stationary-state polymer composition observed in the case of continuous process conditions
(with X_{stat}) for terpolymers can be represented in triangle
coordinate systems. This representation demonstrates that
the copolymer compositions obtained from discontinuous
experiments (with X<10%) in deed correlate with the imposed
monomer mixtures.
    It is shown by model calculations that a local decrease
in the initiation velocity (e.g. by a local impoverishment
of the initiator, or drop in temperature, respectively) in
connection with an imperfect micromixing of the comonomers
in the reactor leads to a formation of small amounts of
copolymers with high parts of the enriched component having
extremely high molecular weights.

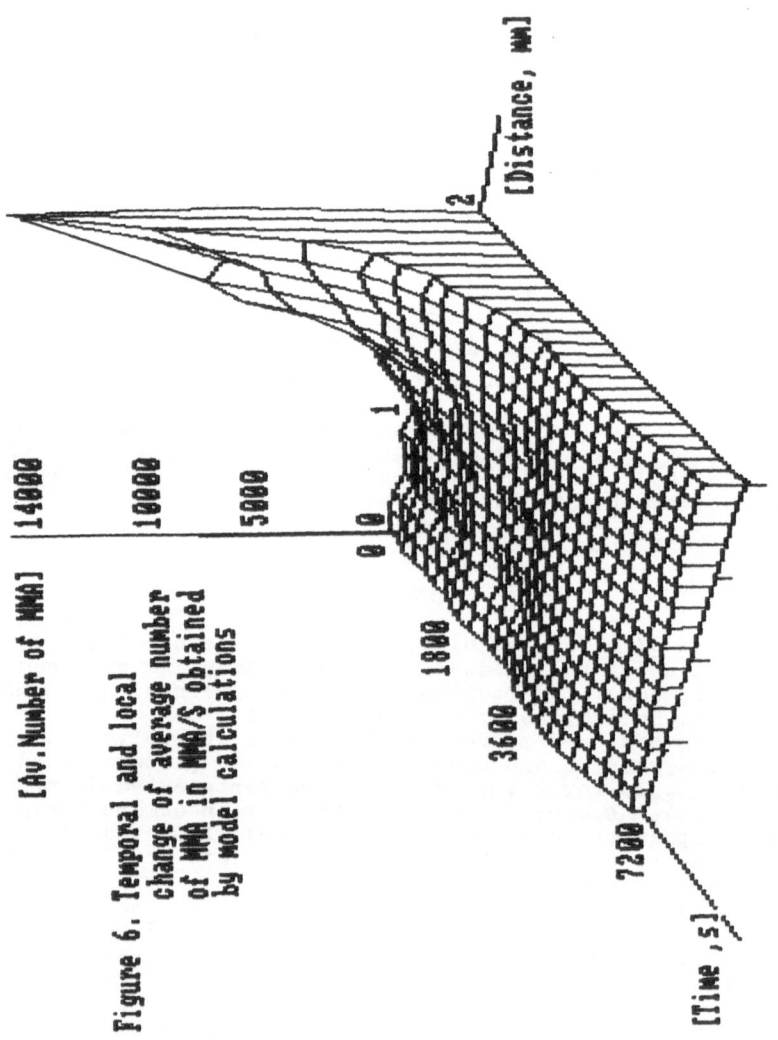

[Av.Number of MMA]

Figure 6. Temporal and local
change of average number
of MMA in MMA/S obtained
by model calculations

REFERENCES

1. Ray, W.H., Mathematical modeling of polymerization reactors. J. Macromol. Sci. - Rev. Macromol. Chem., 1972, C8, pp.1-15.

2. Schmidt-Naake, G., Schmidt H. and Engelmann, U., Zum Copolymerisationsverhalten von ternaeren Donator- Acceptormonomeren bei kontinuierlicher Reaktionsfuehrung. Angew. Makromol. Chemie, submitted.

3. Sharma, D.K. and Soane, D.S., High-conversion diffusion-controlled copolymerization kinetics. Macromolecules, 1988, 21, pp. 700-710.

4. Schmidt-Naake, G., Seifert, P. and Noriega, M., Einfluss der Mikromischung auf die radikalische Polymerisation. Chemische Technik, 1989, submitted.

5. Schmidt-Naake, G., Schmidt, H. and Hamann B., Vergleich diskontinuierlicher und kontinuierlicher Experimente bei der radikalischen Copolymerisation, Acta Polymerica, submitted.

POLYMER DESIGN IN THE ANIONIC POLYMERISATION OF BUTADIENE AND STYRENE

T. GRAAFLAND AND H. VAN BALLEGOOIJEN

Koninklijke/Shell-Laboratorium, Amsterdam (Shell Research B.V.)
P.O. Box 3003, 1003 AA Amsterdam, The Netherlands

ABSTRACT

As a result of the improved understanding of the relation between styrene/butadiene copolymer architecture and the relevant tyre properties there is an increasing demand for tailor-made 'multiple-Tg' SBR copolymers. In this paper it is shown how the 'semi-steady-state batch process' in combination with a simple kinetic model of the S/B copolymerisation can be used to prepare various random SBR block copolymers in which the blocks may differ substantially in styrene and vinyl content.

INTRODUCTION

The anionic solution polymerisation of styrene and butadiene was developed already more than 30 years ago. Thanks to its 'living nature' this process is more flexible than the much older emulsion process, particularly in terms of steering the distribution of the comonomers. Thus, random, tapered and pure block copolymers are accessible. Furthermore, modifiers may be used to influence the ratio of vinyl, cis and trans moieties in the butadiene fraction of the polymer.

In spite of this higher flexibility, a real commercial breakthrough of solution SBR in comparison with emulsion SBR has not been realised up till now. One of the reasons for this may be the fact that until recently only fairly limited knowledge was available on the relation between the polymer architecture and the key properties of tyre treads. However, more and more insight has been gained into the relation between dynamic properties of the tread compound in the laboratory and actual tyre features such as wet grip, rolling resistance and wear resistance [1,2].

With this higher degree of understanding tread compounds are requested containing various structural elements optimised for one or more of the above-mentioned key properties. Such requests may be approached by using tailor-made 'multiple-Tg' SBR (block) copolymers. As discussed above, such polymers can only be made via anionic solution polymerisation. In this contribution we will discuss the technical possibilities to incorporate structurally different S/B elements into a single base polymer by applying the appropriate monomer dosing techniques, with or without a polymerisation modifier.

S/B COPOLYMERISATION BY A SEMI-STEADY-STATE BATCH PROCESS

The preparation of 'multiple-Tg' S/B copolymers may be accomplished by incorporation of two or more 'blocks' which substantially differ in styrene or vinyl content: at block lengths exceeding that of the critical molecular weight separate phases will occur, each showing its own Tg. Preferably, these blocks should have a random styrene distribution, without substantial tapering at the transition.

In a simple batch polymerisation of styrene and butadiene in solution, using a lithium compound as initiator, tapered copolymers are formed with a gradually increasing styrene content along the chain, due to the much higher polymerisation rate constant of butadiene under these conditions. Therefore, several processes have been developed for a random incorporation of styrene, such as (i) a 'steady-state continuous process', (ii) a 'semi-batch high-temperature process' [3], (iii) a 'semi-batch process with butadiene supply' [4] and (iv) the 'semi-steady-state batch process' [5].

The latter process guarantees randomness under a wide variety of process conditions. The reactor is filled with solvent and a certain amount of styrene and butadiene. The polymerisation is then started by addition of butyllithium and a constant supply of both monomers in the ratio to be incorporated in the polymer. In order to maintain a 'steady-state' the concentrations of styrene and butadiene in the reactor have to be such that the rate of polymer formation equals the rate of monomer supply, which in turn implies that the amount of monomers in the reactor is constant and the composition of the feed identical with that of the product.

In the case of the s-BuLi-initiated copolymerisation of styrene and butadiene the above described process requires, in principle, the determination of the four basic propagation rate constants (1)-(4). However, this is very laborious and has led

$$\sim\!\!\sim\!\!\sim S^- + S \xrightarrow{k_{SS}} \sim\!\!\sim\!\!\sim S^- \tag{1}$$

$$\sim\!\!\sim\!\!\sim S^- + B \xrightarrow{k_{SB}} \sim\!\!\sim\!\!\sim B^- \tag{2}$$

$$\sim\!\!\sim\!\!\sim B^- + B \xrightarrow{k_{BB}} \sim\!\!\sim\!\!\sim B^- \tag{3}$$

$$\sim\!\!\sim\!\!\sim B^- + S \xrightarrow{k_{BS}} \sim\!\!\sim\!\!\sim S^- \tag{4}$$

to some ambiguity [6]. Therefore, we have adopted a simplified approach by translating the four basic rate constants into so-called overall rate constants for the disappearance of styrene (k_S) and butadiene (k_B). The depletion of styrene is given by:

$$-\frac{d[S]}{dt} = k_{SS}[S^-]_E \cdot [S] + k_{BS}[B^-]_E \cdot [S] \tag{5}$$

while the butadiene consumption is given by:

$$-\frac{d[B]}{dt} = k_{BB}[B^-]_E \cdot [B] + k_{SB}[S^-]_E \cdot [B] \tag{6}$$

where $[S^-]_E$ and $[B^-]_E$ are the kinetically effective concentrations of styryllithium and butadienyllithium, respectively. In the absence of die-out it is assumed that:

$$[S^-]_E + [B^-]_E = [s-BuLi]_E \tag{7}$$

$$[S^-]_E = x \cdot [s-BuLi]_E \tag{8}$$

$$[B^-]_E = (1-x) \cdot [s-BuLi]_E \tag{9}$$

Substituting (8) and (9) in (5) and (6) results in:

$$-\frac{d[S]}{dt} = k_{SS}x \cdot [s-BuLi]_E \cdot [S] + k_{BS}(1-x) \cdot [s-BuLi]_E \cdot [S] \tag{10}$$

$$-\frac{d[B]}{dt} = k_{BB}(1-x) \cdot [s-BuLi]_E \cdot [B] + k_{SB}x \cdot [s-BuLi]_E \cdot [B] \tag{11}$$

The overall rate constants can be written as:

$$k_S = k_{SS} \cdot x + k_{BS} \cdot (1-x) \tag{12}$$

$$k_B = k_{BB} \cdot (1-x) + k_{SB} \cdot x \tag{13}$$

which reduce (10) and (11) into:

$$-\frac{d[S]}{dt} = k_S \cdot [s-BuLi]_E \cdot [S] \tag{14}$$

$$-\frac{d[B]}{dt} = k_B \cdot [s-BuLi]_E \cdot [B] \tag{15}$$

By determining k_S and k_B under various conditions, (14) and (15) provide a simple model required to adjust the steady-state conditions (the kinetically effective BuLi concentration can be determined e.g. by plotting the total BuLi concentration versus the ratio of observed polymerisation rate of butadiene over the rate at 1 ppm BuLi). It will be obvious that the styryllithium fraction x – and thus k_S and k_B – is dependent on the styrene/butadiene ratio, which makes this model unsuited to describe true batch polymerisations. However, our process ensures an almost constant monomer composition throughout the reaction, justifying application of this model.

By using the 'semi-steady-state batch process' and the S/B copolymerisation kinetic model as outlined above, SBR random block copolymers become technically accessible, as demonstrated by means of a computer simulation in Figure 1a. This graph shows the differential styrene content of a polymer as a function of the monomer conversion. The first block of the polymer contains 65% styrene, while the styrene content in the second block is only 8%. The transition in styrene content between the two blocks can be made very sharp by strongly increasing the rate of monomer supply at the end of the first block. This makes the two blocks

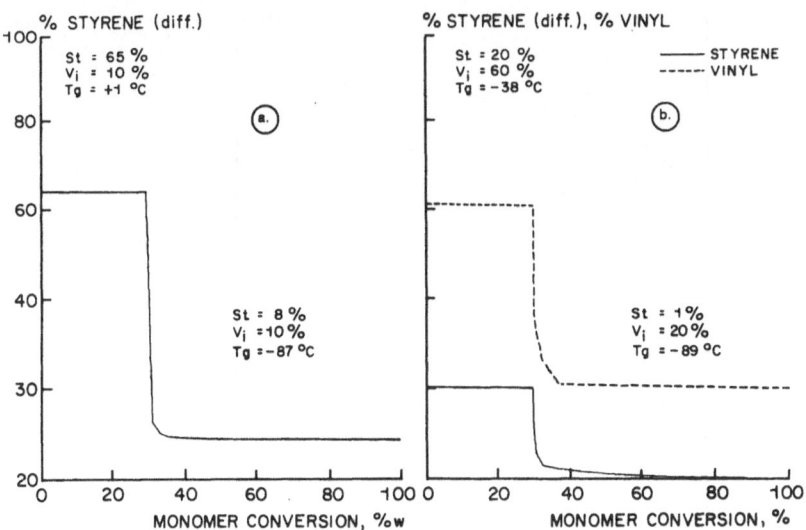

Figure 1. Styrene and vinyl distribution as a function of monomer conversion.

58

incompatible, giving the polymer two glass transition temperatures: at +1 °C and -87 °C. For unmodified copolymers (excluding pure block copolymers) the maximum styrene content in the first block is ca. 70%, while the minimum content in the second block is ca. 4%. Of course, more flexibility is attainable by reversing the order of block preparation but, since these polymers are coupled in practice after 100% conversion, it is necessary to start with the high Tg block in order to obtain optimum tyre tread properties.

EFFECT OF MODIFIERS ON S/B COPOLYMERISATION KINETICS

By adding polar compounds (modifiers) to the alkyllithium initiated S/B copolymerisation, the vinyl content in the butadienyl fraction may be increased [7]. This provides an extra degree of freedom in varying the Tg and imparting certain properties to the polymer. However, next to their modifying behaviour these modifiers also appear to change unequally the overall rate constants k_S and k_B. In most cases k_S increases more strongly than k_B, which means that they tend to become equal. This is exemplified in

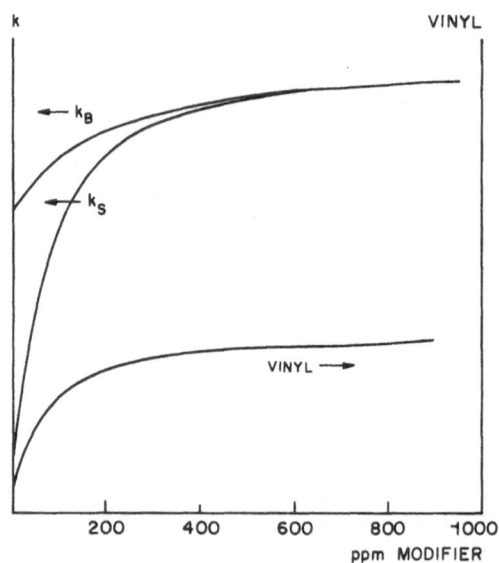

Figure 2. k_S, k_B and vinyl content as a function of the concentration of a typical modifier in cyclohexane.

Figure 2, which shows the effect of a typical modifier on k_S, k_B and the
vinyl content as a function of its concentration in cyclohexane.
Determination of the temperature effect on these parameters (e.g. an
increase of the polymerisation temperature decreases the vinyl content)
allows the preparation of a wide range of random SBR (block) copolymers
varying in vinyl and/or styrene content. Figure 1b gives an extreme example
of the possibilities offered by the semi—steady-state batch process. In
this polymer the first block contains 20 % styrene and 60% vinyl, while the
second block contains only 1% styrene and 20% vinyl. The sharp transition
can be obtained by an increase in temperature and addition of solvent
(dilution of modifier).

CONCLUSIONS

Application of the semi—steady-state batch process and the kinetic model
presented in this paper provides a useful tool to prepare random SBR
(block) copolymers with a wide variety in styrene and vinyl contents. This
offers new possibilities in the development of high-performance, tailor-
made tyre tread compounds.

REFERENCES

1. Bond, R., Morton, G.F. and Krol, L.H., A tailor—made polymer for tyre
   applications. Polymer, 1984, 25, 132–140.

2. Nordsiek, K.H., The "Integral Rubber" Concept – an Approach to an Ideal
   Tire Tread Rubber. Kautsch. u. Gummi Kunststoffe, 1985, 38, 178–185.

3. French patent 1,278,818 (Phillips).

4. French patent 1,369,589 (Firestone).

5. German patent 2,117,995 (Shell).

6. Ohlinger, R. and Bandermann, F., Kinetik der Wachstum—reaktionen der
   Copolymerisation von Butadien und Styrol mit Lithiumorganylen. Macromol.
   Chem., 1980, 181, 1935–47.

7. Antkowiak, T.A., Oberster, A.E., Halasa, A.F. and Tate, D.P.,
   Temperature and Concentration Effects on Polar—Modified Alkyllithium
   Polymerisations and Copolymerisations. J. Polym. Sci., 1972, A–1, 10,
   1319–31.

FREE RADICAL TECHNIQUES OF DERIVING α,ω END FUNCTIONAL
POLYMERS AND BLOCK COPOLYMERS VIA THERMAL INIFERTERS

C.P. REGHUNADHAN NAIR AND G. CLOUET

INSTITUT Charles SADRON (CRM - EAHP)
6 rue Boussingault
67083 Strasbourg-Cedex,FRANCE

ABSTRACT

The initiator, chain transfer and terminator (Iniferter) properties of the
tetraalkyl thiuram disulfides during vinyl polymerization have been
exploited to end-functionalise polymethyl methacrylate (PMMA) and
polystyrene (PS). Typically hydroxyl and amido phosphate groups were
chosen. Polymers bearing the iniferter groups in their backbone served as
polymeric thermal iniferters for vinyl polymerization leading to vinyl
block copolymers. The syntheses of poly(phosphonamide-b-methylmethacrylate)
and poly(phosphonamide-b-styrene) by this technique have been described.
The kinetics of polymerization led to the evaluation of the various kinetic
paramaters pertaining to the initiating, chain transferring and radical
terminating properties of the iniferter groups as a function of their
chemical environment.

INTRODUCTION

Chain-end and chain middle functional polymers are usually designed by the
anionic method by deactivation of the respective living anions using mono,
di or polyelectrophiles[1]. Of late, the interest in this line is oriented
towards the free radical technique because of the several obvious
advantages of this method. The free radical methods usually make us of
functional initiators or chain transfer agents for end functional polymers
or their polymeric analogues for the block copolymers[2,3]. Tetraalkyl
thiuram disulfide functions in three distinctive ways during vinyl
polymerization. Apart from being good chain initiators, they undergo chain
transfer reactions and pose termination of polymer radicals by way of
primary radicals with the result, the resulting polymer chains are
invariably end capped (at either ends) with the iniferter fragments. We
have taken advantage of this property to incorporate certain functional
groups at chain extremes and extended the technique to the synthesis of
block copolymers.

EXPERIMENTAL PART

The two iniferters viz: N,N'-diethyl N,N'-bis(2-hydroxyethyl) thiuram disulfide (DHTD) and N,N'-dimethyl N,N'-bis[2-(N-methyl N-diethoxy phosphinoyl) ethyl] thiuram disulfide (DPTD) were synthesised as described elsewhere[4,5]. α,ω amine terminated poly(phosphonamide) was obtained by reaction between phenyl phosphonic dichloride and excess of piperazine. Reaction of the macrodiamine with CS_2 and I_2 gave the polymeric iniferter (PI)[6].

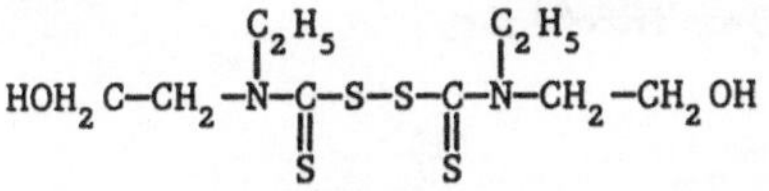

DHTD

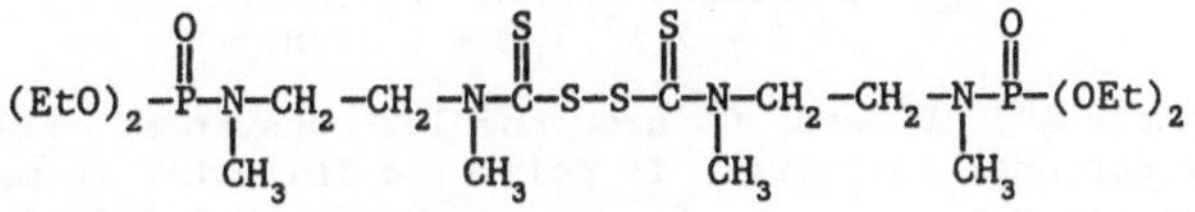

DPTD

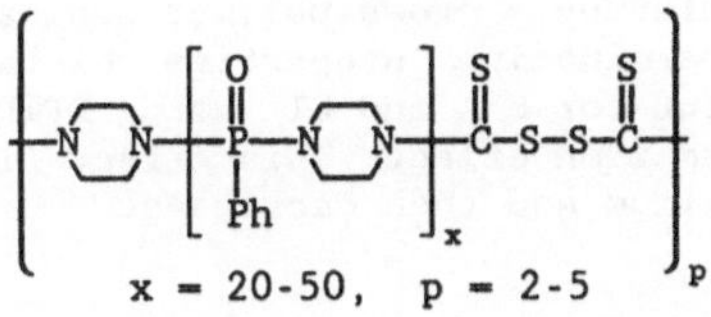

PI

x = 20-50, p = 2-5

Polymerisations were done in bulk in sealed evacuated glass tubes at the required temperature. But fot PI, 2-methoxy ethanol was used as solvent. The polymers were purified by precipitating them in MeOH from solutions in THF or $CHCl_3$. Molecular weights were determined by GPC analyses or from the intrinsic viscosity data.

RESULTS

MMA and ST were polymerized in presence of iniferts and polymeric iniferters to get the end functional polymer or block copolymer as the case may be. The reactions can be depicted as follows:

Scheme 1

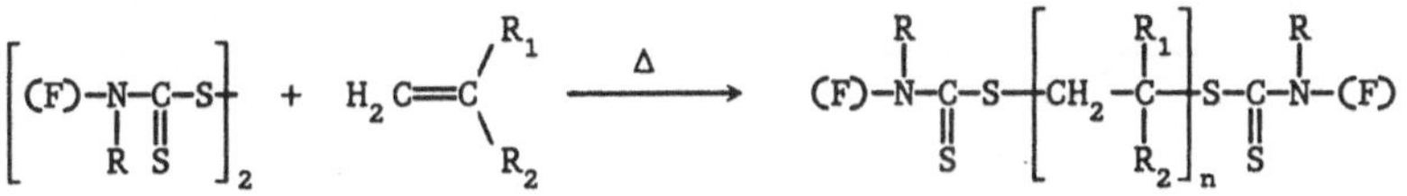

where (F) = Functional group

Scheme 2

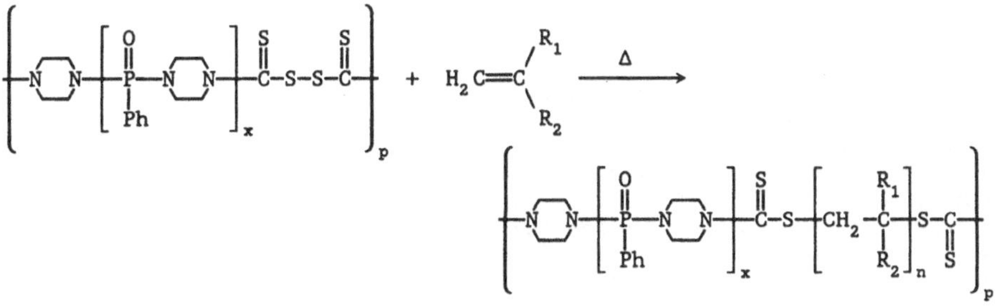

In the case of PI, the block number and the length of the vinyl block of the resulting block copolymer depended on the PI structure, its concentration and the extend of monomer conversion. For functional iniferter as in the case of the PI the polymer molecular weights were governed by the various kinetic parameters of the polymerization reaction.

KINETICS

The rate of polymerization R_p when an iniferter is used to polymerise the vinyl monomer takes the complex form[4]:

$$R_p^2 = \frac{R_{P_o}^2 + A^{*2} I}{1 + 2 B^* I/M + C^* (I/M)^2} \qquad (1)$$

where A^*, B^* and C^* are complex constants; [I] = concentration of the functional iniferter, if polymeric iniferter is used [I]= [PI]. The complex constant A^*, B^* and C^* are measures of initiating, cross polymer radical terminating and mutual primary radical terminating properties of the iniferter. Fig. 1 shows a typical kinetic plot for MMA and ST using DPTD. R_p increases with [I], and then shows a retardation effect. This effect is more prononced for MMA where R_p reaches a maximum and then decreases.

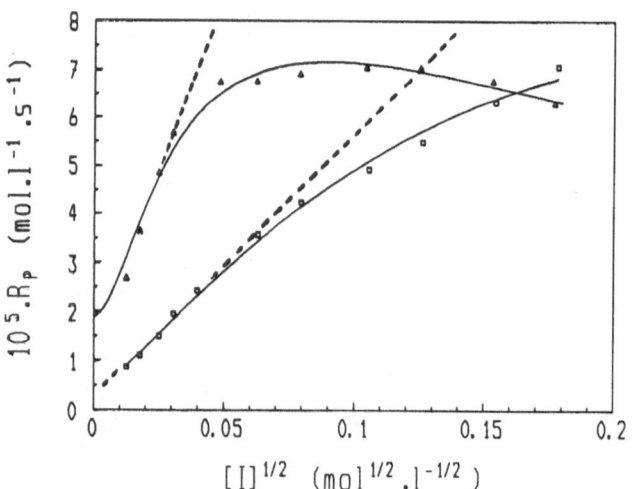

Figure 1. Dependence of R_p on [I], above MMA, below styrene;
(— — —) Theoretical curves in case of simple initiation.

The kinetic parameters calculated by a computer programme and cited in table 1 show that for ST C^* is apparently zero while B^* is considerably lower than for MMA. The initiating capacity of the iniferter is twice more in ST than in MMA. The difference in behaviour for ST is attributed to its high monomer reactivity which enhances the initiation, to the reduced reactivity of its radical towards the thiuramyl radical causing a reduction in primary radical termination and to a relatively decreased mutual termination of the primary radicals. The iniferters were about 1/5th times efficace as initiator in comparison to AIBN. Replacement of the hydroxyl group by the amido phosphate group causes a diminution in the initiating ability of the iniferters towards MMA probably because in the former case specific interaction through H-bonding may be existing between the primary radical and the monomer in the transition state for initiation. Similar H-bonding between the primary radical in the transition state is believed to increase the mutual termination factor (C) for DHTD when compared to DPTD (Tab. 1).

TABLE 1

Various kinetic parameters for the polymerization of methyl methacrylate (MMA) and styrene (ST) at 70°C.

Kinetic constants	MMA			ST	
	DHTD	DPTD	PI	DPTD	PI
$2fk_d.10^6 \ s^{-1}$	4.41	2.11	0.182	3.92	0.71
B	3090	3000	960	170	0
$C.10^{-6}$	5.02	1.40	—	0	0
C_{tr}	0.27	0.41	—	0.70	—

The chain transfer constants for the iniferters were calculated from a modified equation[4].

$$\frac{1}{\bar{P}_n} = \frac{k_t \ R_p}{k_p^2 [M]^2} \left[1 + B^* \ \frac{[I]}{[M]} \right] + C_{tr} \ \frac{[I]}{[M]} \qquad (2)$$

The values (Tab.1) are comparable to those of conventional thiols substantiating the contribution of the chain transfer reaction towards functionalisation.

In the case of polyphosphonamide based polymeric iniferters, the kinetic studies revealed a considerable diminution in its reactivity with respect to initiation and chain termination. In this case the rate expression were simplified, with B^* and C^* in Eq. 1 being zero for ST and MMA respectively[7].

Functionality

Table 2 enlists the polymerization conditions and functionalities of the end-functionalised polymers. Within experimental limits a funtionality of 2 is seen as is anticipated in view of the moderately good initiating ability, high chain transfer constant and a considerable extend of primary radical termination. The hydroxyl terminated polymers can serve as useful prepolymers. The amidophosphate endcapped polymer are intrinsically flame retardant.

Block copolymers

Using a polymeric iniferter shown in scheme 2, for the polymerzation of MMA and ST, the block copolymers of phosphonamide with the vinyl monomers were prepared. In this case, the block copolymer structure depended on the PI-concentration and monomer conversion. The block copolymer formation was evidenced from the increase in molecular weight of the isolated polymer with conversion. This confirms the mechanism of polymer growth by the successive insertion of the vinyl blocks between the S—S linkage of the poly(thiuram disulfide). Calculations showed that at 10% monomer conversion, multiblock copolymers with 3 to 4 A-B blocks were easily formed. An increase in average vinyl blocklength with increase in monomer conversions can be seen. A concomitant diminution in phosphonamide content confirms the postulated mechanism of chain growth. A better functionalisation was seen for styrene. The detailed results can be found in Ref 7. The P-functionalised polymers serve as flame resistant polymers.

TABLE 2
Functionalities of the end functional polymer

Reaction system	[I] mol.1^{-1}	Temp °C	$\overline{M}n.10^{-4}$ (GPC)	% Sulfur (found)	Functionality
	0.061	95	2.10	0.60	2.0
	0.061	85	2.73	0.55	2.3
MMA,DHTD	0.092	70	2.96	0.55	2.5
	0.031	95	3.24	0.43	2.2
	0.061	70	3.50	0.41	2.2
	0.061	90	2.50	0.60	2.3
MMA,DPTD	0.123	90	1.70	0.82	2.2
	0.246	90	1.10	1.04	1.8
	0.032	70	2.20	0.70	2.3
ST,DPTD	0.160	70	0.67	1.88	2.0
	0.250	70	0.47	2.42	1.8
	0.320	70	0.42	2.94	1.9

CONCLUSION

Tetraalkylthiuram disulfide based functional iniferters are useful for conferring α,ω functionality to vinyl polymers. When polythiuram disulfides are used multiblock copolymers are easily formed enabling synthesis of block copolymers of the type condensation-addition or addition-addition type. The P-functionalisation by this technique is a means to impart flame retardancy to intrinsically inflammable vinyl polymers.

REFERENCES

1. M. SZWARC, Nature, 178, 1168 (1956)
2. S.F. REED, J. Polym. Sci., A-1, 9, 2029 (1971)
3. H. INOUE, A. UEDA and S. NAYAI,
   J. Polym. Sci., Chem. Ed., 26, 1077 (1988)
4. C.P.R. NAIR, G. CLOUET and P. CHAUMONT,
   J. Polym. Sci., Chem. Ed., 27, (1989)
5. C.P.R. NAIR and G. CLOUET, Makromol. Chem., 190, (1989)
6. C.P.R. NAIR, G. CLOUET and J. BROSSAS,
   J. Makromol. Sci., A-25(9), 1089 (1989)
7. C.P.R. NAIR and G. CLOUET, Polymer, 29, 1909 (1988)

BRANCHING KINETICS OF EPOXY POLYMERIZATION OF 1,4-BUTANEDIOL
DIGLYCIDYL ETHER WITH CIS-1,2-CYCLOHEXANEDICARBOXYLIC ANHYDRIDE

BENJAMIN CHU AND CHI WU
Department of Chemistry
State University of New York at Stony Brook
Long Island, New York 11794-3400, USA

ABSTRACT

The copolymerization of epoxy (1,4-butanediol diglycidyl ether (DGEBA) with
anhydride (cis-1,2-cyclohexanedicarboxylic anhydride (CH) in the presence
of benzyl dimethyl amine (CA) as a catalyst produces a branched epoxy
polymer. We show that the branching kinetics of the copolymerization
reaction and the molecular weight of the branched polymers formed during
the copolymerization reaction before the gelation threshold can be
approximated by using Smoluchowski's coagulation equation.

INTRODUCTION

Epoxy polymers are complex systems. The epoxy polymers are highly branched
before the gelation threshold. After the gel point, the epoxy polymers
form permanently cross-linked polymer networks. The molecular weight (M),
the molecular weight distribution (MWD) and the degree of branching (DB)
changes with reaction time (t), as well as experimental conditions.
Experimentally, determinations of such macromolecular parameters are not
easy because routine methodologies for molecular weight determinations of
highly branched epoxy polymers do not exist. Size exclusion chromatography
becomes difficult to use, especially when the degree of branching changes
with reaction time. After the gelation threshold, the epoxy polymer
product becomes insoluble because of network formation throughout the
macroscopic domain. In this article, we wish

I. to outline the methodologies for the determination of macromolecular parameters, such as M, MWD, and DB, of highly branched

epoxy polymers before the gelation threshold, and

II. to test theoretical kinetic models based on the experimentally determined macromolecular parameters.

METHODOLOGIES FOR DETERMINATION OF MACROMOLECULAR PARAMETERS AND POLYMER NETWORK FORMATION

Macromolecular parameters can be determined by using a combination of techniques including,

(1) chemical analysis to determine the extent of the polymerization reaction,

(2) laser light scattering (LLS) to determine the weight-average molecular weight (M_w), MWD, the second virial coefficients, the radius of gyration (R_g) and the hydrodynamic radius (R_h) with R_g/R_h being related to the degree of branching.

(3) small angle x-ray scattering (SAXS) with synchrotron radiation [1] to determine the random structure of the epoxy polymer formed, in terms of the concept of fractal geometry [2],

(4) light scattering intensity and angular distribution measurements to observe the structure of epoxy polymer in situ beyond the gelation threshold.

The methodology has been implemented to study the curing reaction of 1,4-butanediol diglycidyl ether (DGEB) with cis-1,2-cyclohexane-dicarboxylic anhydride (CH), in the presence of a catalyst, benzyl dimethyl amine (CA) [2-4].

For epoxy polymer studies, SAXS is the proper technique to determine R_g of the branched epoxy polymer during initial stages of the epoxy polymerization process,

$$\lim_{C \to 0} \frac{M_w HC}{R_{vv}(K)} \overset{(LLS)}{\left(= \lim_{C \to 0} \frac{M_w H'C}{I(K)}\right)} \overset{(SAXS)}{\simeq} 1 + (R_g/3)K^2 R_g = 1 + (R_g/3)\chi \qquad (1)$$

where H and H' are optical constants for light and x-rays, $R_{vv}(K)$ is the excess Rayleigh ratio using vertically polarized incident and scattered light, K is the scattering vector, and I(K) is the excess SAXS intensity. In plots of $\lim_{C \to 0} M_w HC/R(\theta)$ or $\lim_{C \to 0} M_w H'C/I$ versus $\chi(=K^2 R_g)$, as shown in Figure 1, we see that LLS measurements, as denoted by filled symbols are clearly appropriate for R_g values of a few hundred Å. It becomes increasingly more difficult at smaller R_g values because of the small KR_g ($\lesssim 1$) ranges accessible to LLS. The filled and hollow diamonds demonstrate an overlap of two independent scattering techniques (LLS and SAXS) on an absolute determination of R_g.

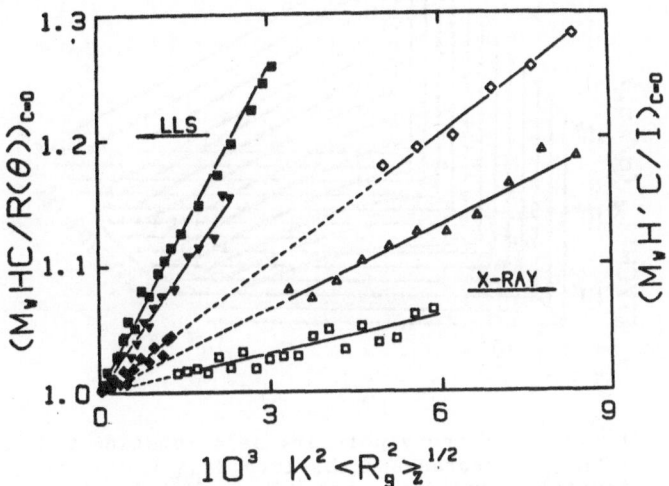

Figure 1. Plots of $M_w HC/R_{vv}$ (for light scattering) and $M_w R'C/I$ (for SAXS) as a function of $K^2 \langle R_g^2 \rangle_z^{1/2}$. According to Eq. (1), $\lim_{c\to 0} \frac{M_w HC}{R_{vv}}$ ($= \lim_{c\to 0} \frac{M_w H'C}{I}$) $= 1 + \frac{R_g}{3} K^2 R_g$. Thus, the slope is equal to $R_g/3$. The plots demonstrate overlapping regions of the two scattering techniques. Laser light scattering is denoted by filled symbols, while SAXS is denoted by hollow symbols. [Fig. 4 of ref. 3]

By dissolving the branched epoxy polymer in methyl ethyl ketone (MEK), SAXS and LLS measurements can determine M_w, MWD, R_g and R_h over a large range of the polymerization reaction.

Figure 2 shows static structure factors $S(K)$ [$= I/CH'$] from SAXS as a function of % CH conversion. In plotting the scattering curves, we have reduced the intensities to its absolute value. Thus, the y($= \lim_{c\to 0} I/CH'$) axis has units of g/mol and at $K = 0$, denotes the M_w of the epoxy polymer as a function of % CH conversion. Such an approach is feasible up to the gelation point, beyond which the epoxy polymer can no longer be dissolved as individual macromolecules in MEK. At high % CH conversion, the epoxy polymer has reached fairly high molecular weights (~10^5 g/mol). Thus, SAXS measures mainly the fractal geometry of the branched epoxy polymers in solution according to $I(K) \sim K^{-d_K}$ with d_K being a fractal dimension.

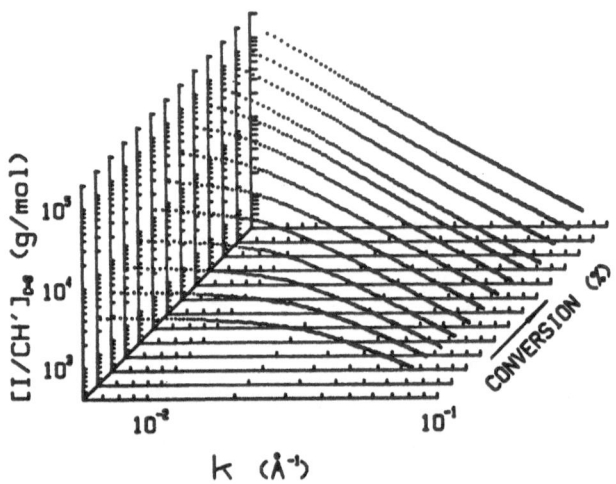

Figure 2. Structure of epoxy polymers as a function % CH conversion. $[I/CH']_{C=O}$ represents absolute SAXS intensity at infinite dilution in units of g/mol. Thus, $\lim_{K \to 0} [I/CH']_{C=O} = 1/M_w$. The scattering curves are numbered with increasing % CH conversion. Properties of the 13 samples represent epoxy polymers during different stages of the curing process. [Fig. 6 of ref. 4]

BRANCHING KINETICS OF EPOXY POLYMERIZATION

The copolymerization of DGEB and CH in the presence of CA is dominated by alternative linkages between DGEB and CH, i.e., the reactions between the epoxy resins and between the anhydrides are suppressed [5]. Highly branched epoxy copolymers are formed in the copolymerization reaction because the sum of the reaction functionalities of DGEB (two epoxy rings which can form four chemical bonds) and of CH (one anhydride ring which can form two chemical bonds) is six, which is larger than the gelation criterion, i.e., a minimum total of five functionalities is needed to form a branching point. At the initial reaction stage, most of the polymers formed are linear. As the reaction proceeds, the degree of branching of the branched epoxy copolymer increases.

By assuming that: (1) the reactivity is independent of polymer chain length, and (2) rings do not form, we are able to present a simplified model which can be identified with the Smoluchowski coagulation equation for the branching kinetics in the copolymerization process.

Figure 3 shows how M_w changes with p, using p_o = 0.065, $\bar{n}(p_o)$ = 1, and DGEB:CH:CA = 1:2:0.001, with p, p_o and $\bar{n}$ being the extent of CH conversion, the initial extent and the average number of active sites, respectively.

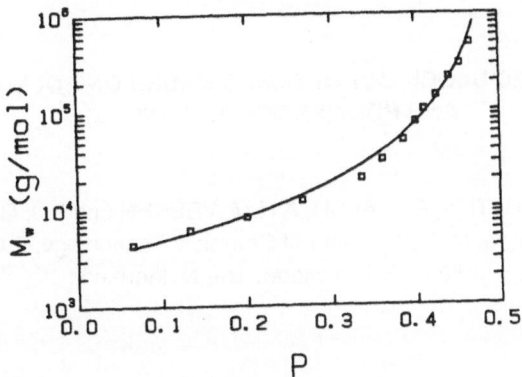

Figure 3. Plot of weight-average molecular weight M_w versus p. Solid line represents the calculated values. Hollow squares show the real experimental data with the same ratio of DGEB:CH:CA at 80°C. [Wu, Chu, Stell, J. Chem. Phys., to be published]

REFERENCES

1. Chu, B., Wu, D.-Q. and Wu, C., A Kratky block-collimation small-angle x-ray diffractometer for synchrotron radiation. Rev. Sci. Instrum., 1987, 58, 1158-1163.

2. Chu, B., Wu, C., Wu, D.-Q. and Phillips, J., Fractal geometry in branched epoxy polymer kinetics. Macromolecules, 1987, 20, 2642-2644.

3. Chu, B. and Wu, C., Structure and dynamics of epoxy polymers. Macromolecules, 1988, 21, 1729-1735.

4. Wu, C., Zuo, J. and Chu, B., Molecular weight distribution of a branched epoxy polymer: 1,4-butanediol diglycidyl ether with cis-1,2-cyclohexane-dicarboxylic anhydride. Macromolecules, 1989, in press.

5. Fischer, R. F., Polyesters from epoxides and anhydrides. J. Polym. Sci., 1960, 44, 155-172.

6. Smoluchowski, M. V., Drei vortrage uber diffusion, Brownsche bewegung und koagulation von kolloidteilchen. Physik. Z., 1916, 17, 585-599.

7. Scott, W. T., Analytic studies of cloud droplet coalescence I. J. Atmos. Sci., 1968, 25, 54-65.

8. Lu, B., The evolution of the cluster size distribution in a coagulation system. J. Stat. Phys., 1987, 49, 669-684.

SEGMENTED BLOCK COPOLYMERS BASED ON POLYAMIDE-4,6 AND POLY(PROPYLENE OXIDE)

P.F. VAN HUTTEN, E. WALCH, A.H.M. VEEKEN and R.J. GAYMANS
University of Twente, Department of Chemical Technology, P.O. Box 217,
7500 AE Enschede, The Netherlands

ABSTRACT

Segmented block copolymers were obtained through copolymerization of nylon-4,6 and poly(propylene oxide) (PPOx). An amine-terminated PPOx ("soft") segment of molar mass 430 was reacted with nylon-4,6 salt in an autoclave (30 minutes, 210 °C, 7 bar). In this reaction a prepolymer was prepared, which was postcondensed afterwards. In ethanol, a considerable amount of material could be extracted from the solid copolymers of low polyamide content, i.e. those with short polyamide ("hard", crystallizable) segments. Torsional tests showed an extended rubber plateau of fairly constant modulus. Whereas PPOx has a T_g of -60°C, the T_g's of the copolymers were higher as a result of dissolution of uncrystallized polyamide segments in the PPOx phase. The high melting point of nylon-4,6 was approached only by copolymer of high polyamide content.

INTRODUCTION

The idea behind Thermoplastic Elastomers (TPE) is to extend the region of rubbery behaviour which characterizes semi-crystalline homopolymers between T_g and T_m, in order to obtain materials which are truly rubbery from room temperature (or below) up to a fairly high temperature (say 250 °C) [1]. Since crystallites serve as cross-links in such an elastomer, it can be melt-processed.
Thermoplastic homopolymers usually satisfy the relation $T_g/T_m \approx 0.6$ (T in K), and in order to extend the rubber range one must construct a block copolymer, the chains of which consist of two types of blocks: "soft" (low T_g) ones and "hard", crystallizing ones. In the solid state such a copolymer shows a phase-separated morphology: hard, crystalline domains are dispersed in an elastomeric matrix. In copolymers obtained through condensation polymerization block lengths have to be small, and

so-called multi-block or "segmented" TPE are obtained. Examples: polyurethane, poly(siloxane urea), co-polyester, poly(ether ester), and poly(ether amide).

We have chosen polyamide-4,6 (T_m= ca. 290 °C), which has been extensively studied in our group [2], to be the hard component in a number of different TPE. In previous research the TPE of nylon-4,6 and poly(tetramethylene oxide), PTMO, soft segments was investigated [3]. TPE's based on nylon-4,6 and poly(propylene oxide), PPOx, will be the subject of this paper.

EXPERIMENTAL

Materials and copolymer synthesis

The synthesis of N46-PPOx copolymers is a two step procedure: first a prepolymer is prepared, which is postcondensed afterwards in order to increase the molar mass.

The prepolymer syntheses were carried out in a 1.4 ltr autoclave (Juvo 142) equipped with anchor stirrer and heating device. Starting materials were PPOx–diamine (Jeffamine D-400 from Texaco Chemical Company, molar mass 430) and a stoechiometric amount of adipic acid in order to have the right functionality for copolymerization with the nylon-4,6 blocks. Jeffamine is an oligoether capped with an amino group at both ends of the chain:

$$H_2N\text{-}CH(CH_3)\text{-}CH_2\text{-}[\text{-}O\text{-}CH_2\text{-}CH(CH_3)\text{-}]_n\text{-}NH_2$$

Nylon-4,6 was added in the form of the salt, which was prepared from adipic acid and 1,4-diaminobutane as described previously [2]. In order to reduce the tendency to phasing, m-cresol was added to the reaction mixture in amounts of 1 ml per gram of reactants. The reaction was carried out at 210 °C for 30 minutes, in which a maximum pressure of about 7 bar was attained. At the end of the reaction period the pressure was released and the temperature was raised to 260 °C in order to have water and m-cresol evaporate from the mixture (1 hour).

Postcondensation took place in a rotating glass vessel situated in an oven. The finely divided prepolymer (milled autoclave product) was kept in a stream of dry nitrogen at 260 °C for 24 hours.

Characterization

DSC measurements were carried out with a DuPont 990 Thermal Analyzer. An indium standard was used for calibration. The procedure was as follows: 1st heating run at 20 °C/min up to 20 °C above T_m; equilibration during 5 minutes; cooling run at 10 °C/min down to 150 °C below T_m; 2nd heating run at 20 °C/min.

Torsional moduli were measured with a computer controlled Myrenne pendulum. The samples were compression moulded bars, dried in vacuum at 110 °C for 16 - 18 hours prior to testing. The test ran at 1 Hz, while the sample was continuously heated from -150 °C at a rate of 1.8 °C/min.

TABLE 1

Characterization of as-prepared N46-PPOx copolymers

sample	w_{PA}	[–COOH] (meq/g)	[–NH$_2$] (meq/g)	$\overline{M}_n$ (10^3 g/mol)	η_{inh}^* (dl/g)	mass fr. extracted
PA.26	0.26	-	-	-	0.35	1.0
PA.40	0.40	0.147	0.034	11.1	0.47	0.99
PA.47	0.47	-	-	-	0.54	0.74
PA.56	0.56	-	-	-	0.73	0.24
PA.75	0.75	0.102	0.011	17.7	0.76	0.04

* 1 g/dl solution in m-cresol at 25°C.

RESULTS AND DISCUSSION

Remarks on polyamide content

The polyamide content w_{PA} (mass fraction) is defined such that each amide bond is taken into account: $w_{PA} + w_{PEth} = 1$. w_{PEth} refers to the polyether part of the PPOx-diamine chains; the –NH$_2$ end groups are excluded. These amine end groups are thus included in the value of w_{PA}. For pure nylon-4,6 , w_{PA} equals 1.

End group analysis, solution viscosity, extraction

Potentiometric titration of end groups shows that excess acid is present in the samples investigated (Table 1). The evaporation of diaminobutane may have played a role [2]. This imbalance reduces the number average molar mass. Low molar mass is furthermore indicated by the values of the inherent viscosity.

Pure PPOx is soluble in water and in ethanol. The poly(ether amide) prepared from PPOx-diamine and adipic acid (sample PA.26) proved to be completely soluble too. It is therefore expected that, in the case of N46-PPOx copolymers, those chains which are not anchored in the polyamide crystallites will be extractable in ethanol. The hard block content or the length of the hard blocks in these chains may be insufficient for taking part in crystallization.

Compression moulded samples were subjected to Soxhlet extraction with ethanol for 16 hours. In similar experiments on N46-PTMO copolymers, the extract proved to be of high polyether content [3]. At a polyamide content of 0.4, the N46-PPOx copolymer is almost fully extractable. This implies that most of the polyamide is extracted as well. Obviously, the elastomer network is not fully developed.

Torsional testing, differential scanning calorimetry

Fig.1 shows the storage and loss moduli of the dry N46-PPOx copolymers. The glass transition lies around -15 °C for low polyamide contents (Table 2). This is ca. 40 °C higher than the T_g of PPOx homopolymer [4]; it is close, however, to the T_g of the liquid copolymer obtained from PPOx-diamine and adipic acid (sample PA.26).

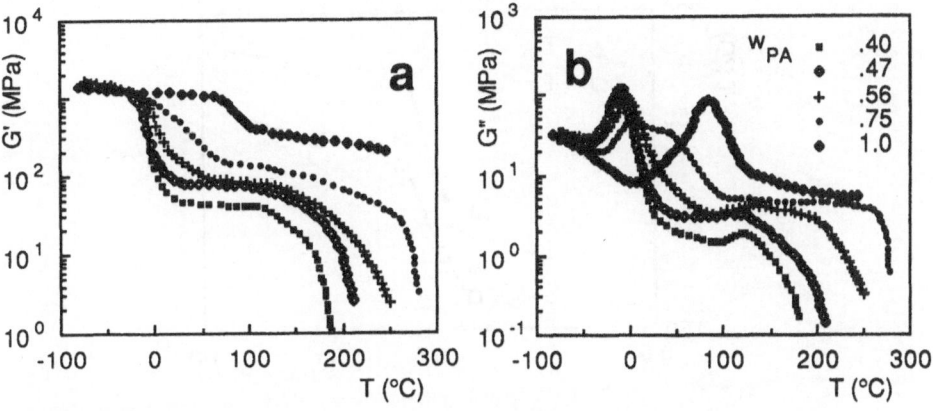

Figure 1. Torsional moduli of N46-PPOx copolymers of varing polyamide content;
a: storage modulus G'; b: loss modulus G".

TABLE 2

Data from torsional testing and DSC for N46-PPOx copolymers

sample	T_g (°C)	$G'_{20°C}$ (MPa)	$G'_{100°C}$ (MPa)	T_{flow} (°C)	T_c (°C)	ΔH_m^* (J/g)	$\Delta H_{m,PA}$ (J/g)	T_m^* (°C)
PA.26	-20	-	-	-20	-	-	-	-
PA.40	-12	50	40	170	165	1.9	4.8	180, 205
PA.47	-12	85	70	200	186	3.2	6.8	185, 214
PA.56	-3	170	80	220	241, 228	14	25	273
PA.75	2, 38	550	140	270	241	43	58	274
Nylon-4,6	82	1100	**400	>270	256	84	84	285

* values obtained in the second heating run.
** G' at 100 °C is close to 300 MPa when extrapolated from the plateau range.

With increasing polyamide content T_g is displaced to higher temperatures, until at 75% polyamide two separate transitions can be distinguished. The temperature shift results from mixing of non-crystallizable polyamide blocks with the polyether blocks, so as to form a single amorphous phase. The peak splitting at higher polyamide content indicates partial segregation, but pure phases (consisting of one type of block only) are not formed.

Melting points found in the second DSC-heating run (Fig. 2) are 10 - 15 °C lower than those recorded during the first heating run. Table 2 lists the overall heat of fusion, ΔH_m, as well as $\Delta H_{m,PA}$, which pertains to the nylon-4,6 mass fraction.

For the copolymers of low polyamide content, the values of both melting point and heat of fusion are low in comparison with nylon-4,6. In these softer copolymers crystallization is hampered, since the average number of repeat units in the hard blocks is much lower than 5, the value for high-melting nylon-4,6 crystallites.

74

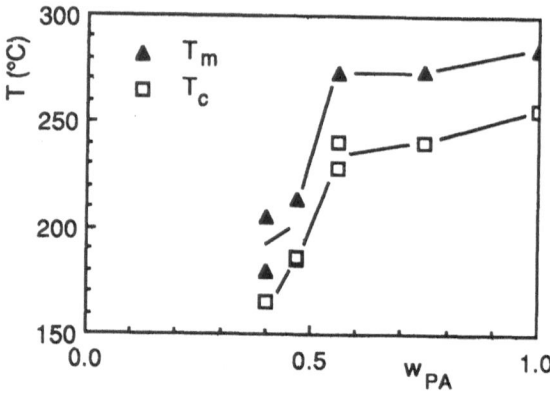

Figure 2. DSC results for N46-PPOx copolymers: □: crystallization temperature;
▲: melting temperature found in the second heating run.

CONCLUSIONS

This work has shown the feasibility of preparing thermoplastic elastomers based on
the high-melting polyamide-4,6 and a poly(propylene oxide) soft segment.
The restricted length of the PPOx chains (molar mass 430) between amide bonds
leads to T_g values which are high with respect to pure PPOx. Low heats of fusion
imply the presence of a considerable amount of uncrystallized polyamide blocks.
Due to their restricted length, these polyamide segments do not segregate from the
polyether phase but rather mix with the latter. The lower bound of the rubber region
of this TPE thus rises with increasing polyamide content. Resorting to a somewhat
higher molar mass, e.g.1000, of the PPOx-diamine oligomer may improve on this,
promoting crystallization and reducing compatibility in the amorphous phase.

REFERENCES

1. Thermoplastic Elastomers, eds. Legge, N.R., Holden, G. and Schroeder,
   H.E., Hanser Publishers, Munich, 1987.
2. Gaymans, R.J., van Utteren, T.E.C., van den Berg, J.W.A. and Schuyer, J.,
   Preparation and some properties of Nylon 46, J. Polym. Sci. Polym. Chem.
   Ed., 1977, 15, 537-45.
3. Gaymans, R.J., Schwering, P. and De Haan, J.L., Nylon 46–Polytetra-
   methylene oxide segmented block copolymers, submitted to Polymer, 1989.
4. Yui, N., Tanaka, J., Sanui, K. and Ogata, N., Polyether-segmented
   polyamides as a new designed antithrombogenic material: Microstructure of
   poly(propylene oxide)-segmented nylon 610, Makromol. Chem., 1984, 185,
   2259-67.

STUDY OF THE POLYURETHANE REACTIONS BY DIFFERENTIAL SCANNING CALORIMETRY

TATJANA MALAVAŠIČ, IRENA ANŽUR, UČI OSREDKAR
Boris Kidrič Institute of Chemistry
Hajdrihova 19, 61000 Ljubljana, Yugoslavia

ABSTRACT

The courses of reaction between hexamethylene diisocyanate or isophorone diisocyanate and 1,4-butanediol, 1,5-pentanediol or triethylene glycol were measured by differential scanning calorimetry. Two types of reaction mixtures, either with excess of hydroxy or with excess of isocyanate componente, were prepared. From DSC curves the enthalpies of reaction, the activation energies and the reaction orders were calculated. For the above systems enthalpies of reaction between -75 and -80 kJ/eq. of urethane bond, activation energies between 74 and 130 kJ/mol and reaction orders between 1.17 and 1.63 were found.

INTRODUCTION

The kinetics of the polyurethane reactions between different aromatic or aliphatic isocyanates and monomeric or polymeric hydroxyl compounds in presence or absence of catalysts have been studied by many authors (1 - 11). Due to complexity of systems, to variety of possible reactions and methods of measurements different mechanisms of reaction were proposed and different reaction orders were obtained. To be able to easier describe the complicated urethane systems in many studies model reactions with monofunctional monomers were used.

To obtain more data about polyurethane reactions we studied the courses of polymerization of simple polyurethane bifunctional reaction systems. For measurements differential scanning calorimetry (DSC) was used. To avoid difficulties in interpretation of DSC curves on account of melting enthalpies and phases separation and to avoid a low reaction temperature liquid and less reactive monomers were chosen.

MATERIALS AND METHODS

Materials

Diols: 1,4-butanediol (B), 99 % (Aldrich) and 1,5-pentanediol (P), 98 % (Merck) were dried over molecular sieves 4 A, triethylene glycol (T), anhydrous, (Fluka) was used as received.

Diisocyanates: 1,6-hexametylene diisocyanate (HDI) and isophorone diisocyanate (IPDI), (Hüls), were used as received.

Method

The enthalpy of polymerization was measured in a differential scanning calorimeter DSC-7, Perkin Elmer.

Diol and diisocyanate, which are immiscible at room temperature, were weighed directly in a standard pan for volatile samples under dry nitrogen atmosphere with the excess of one of the two monomers. To mix the components the sealed pan was put for one minute in a ultra sonic device. The pan was then placed in the calorimeter and heat with a rate of 5 degrees/minute in the temperature range of 20 to 200 °C. The excess of one of the two monomers was needed since the size of samples was small (10-20 mg). Therefore it was not possible to weigh them in equivalent ratios.

The enthalpies of polymerization were calculated from the areas under the DSC curves and the kinetic of the reaction using the Perkin-Elmer Kinetics Software.

RESULTS AND DISCUSSION

The urethane formation is described by the equation

$$RNCO + B \underset{k_2'}{\overset{k_1'}{\rightleftharpoons}} [RNCO:B] \xrightarrow[R'OH]{k_3'} RNHCOOR' \qquad (1)$$

Where B can be any compound with a free electron pair. In the case of the non-catalyzed reaction the complex with alcohol is formed.

For the non-catalized reaction between aliphatic isocyanates and alcohols the third order of reaction following the overall rate equation

$$\frac{dx}{dt} = k_1 (a-x)(b-x)^2 + k_2 x(a-x) (b-x) \qquad (2)$$

was suggested by Sato (3). a and b are the initial concentrations of isocyanate and alcohol, x the concentration of the urethane and k_1 and k_2 the rate constants.

Due to two possible isomers and two isocyanate groups of different reactivity the reaction for IPDI is even more complicated (8). Nevertheless after simplification the equation becomes of second order.

With the excess of one monomer the kinetic equation becomes of pseudo first order with respect to isocyanate concentration and in the dependence on the proposed mechanism of first (8) or intermediate between 1 and 2 order (1) or of second order (3) with respect to alcohol concentration.

With DSC the generation of the enthalpy of polyurethane formation, which is proportional to polyurethane concentration, was measured. The conversion (x) is assumed to obey the following rate equation

$$\frac{dx}{dt} = k(1-x)^n \qquad (3)$$

with Arrhenius temperature dependence. The "Kinetics Software" uses the linear form of the last two relations

$$\ln(\frac{dT}{dt}\frac{dx}{dT}) = \ln Z \; - \; \frac{E_a}{RT} \; + \; n \; \ln(1-x) \qquad (4)$$

The use of the multilinear regression enables the evaluation of n, Z and E_a just from one DSC measurement.

On Figure 1 the courses of reaction of B, P and T with HDI (H on Figure) with an excess of diols are shown. The courses differ in the intensity of the heat flow at the maximal rate of reaction and in the temperature interval of reaction. For all other studied combinations (excess of isocyanate, combinations with IPDI) similar curves were obtained. However, the curves of reaction mixtures with IPDI are not of a such regular shape, what can be ascribed to different reactivites of its isocyanate groups.

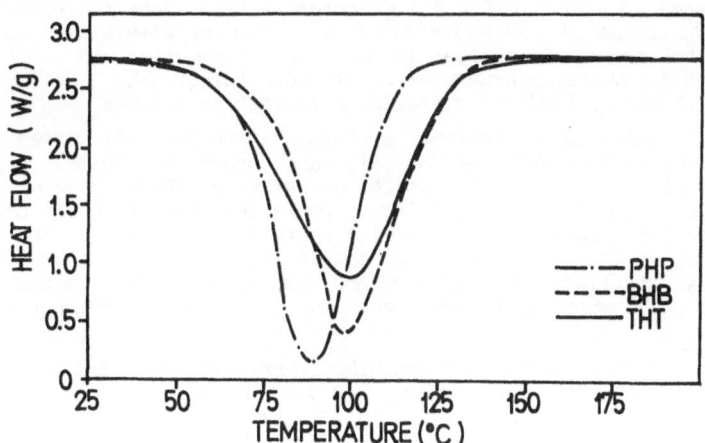

FIGURE 1: Courses of reaction of B, P, and T with HDI (excess of diol)

In Table 1 the temperatures of maximal reaction rates (as interval for paralell measurements), the Δ H on equivalent of urethane bonds, the orders of reaction n and the activation energies E_a are given.

TABLE 1

Enthalpies of reaction (ΔH), temperatures of maximal reaction rates, orders of reaction with respect to polyurethane formation and activation energies for reaction between B, P, or T and HDI(H) or IPDI (I) with excess of diol or diisocyanate

System	ΔH (kJ/eq.)*	$T_{max.}$	n**	E_a (kJ/mol)**
B-H-B	-75.5	85-123	1.63±0.03	116.6±1.88
H-B-H	-80.4	108-132	1.48±0.04	109.4±2.15
P-H-P	-75.6	89- 96	1.59±0.04	130.6±2.49
H-P-H	-78.3	108-115	1.40±0.03	94.6±1.88
T-H-T	-79.0	99-100	1.17±0.01	74.2±0.29
B-I-B	-76.0	121-124	1.58±0.59	96.7±2.47
I-B-I	-74.6	126-130	1.64±0.03	118.4±2.17
P-I-P	-75.2	116-122	1.61±0.08	115.6±4.48
I-P-I	-74.9	116-122	2.51±0.14	182.4±9.47
T-I-T	-76.7	87- 90	1.81±0.04	163.4±3.72

*average of all measurements
**kinetic calculated on the base of most regular curve

The ΔH seems to be equal for all mesaured systems since the values found are in the range of -74.6±3.8 % kJ/eq. which is within the limits of experimental error. The values we found are comparable with values for ΔH found by outher authors -77.5 to -105 kJ/eq. (4,6), -73.6 kJ/eq. (11), -60.3 kJ/eq. (10) for different polyurethane systems.

The reaction order for the urethane formation, calculated on the basis of the concentration of the deficient monomer was in most cases not 1 as expected, but near 1.5 which was already found by some authors (10,11). Exceptions were system T-H-T, where the found reaction order was near 1 and the systems I-P-I and T-I-T where the order was 2. The minimal differences in the reaction orders between systems with excess diol, in which side reactions are minimal and the systems with excess of the isocyanate, in which side reactions are to be expected, are somewhat surprising.

The calculated activation energies varry for different systems and are with exception of the value 74.2 kJ/mol for the system T-H-T in comparison to literature data (41 to 69 kJ/mol (6), 42 kJ/mol (7), 29.3 to 64.8 kJ/mol (11), 48.1 to 54.8 kJ/mol (4)) for different urethane systems to high.

CONCLUSIONS

For less reactive and at room temperature liquid monomers it was possible using DSC to obtain many informations on the courses and kinetics of the reaction only with few measurements. Probably due to side reactions, specially to reaction with water, which is very difficult to eliminate completely from the reaction system, the reproducibility of the measurements was not always satisfactory. Additional troubles were caused by the immiscibility of the monomers at room temperature and by the small size of the samples which implicated the excess of one monomer.

The △ H and the reaction orders were comparable to values given in the literature, while our values for the activation energies were much higher. Anyhow the results are promising and with more experiments it will be possible to obtain more informations on polyurethane reactions.

REFERENCES

1. Baker, J.W. and Gaunt, J., J. Chem. Soc., 1947, 9-18, 19 -24, 27-31

2. Dyer, E., Taylor, H.A., Mason, S.J., and Samson, J., J.Am. Chem. Soc., 1947, 71, 4106-09

3. Sato, H., J. Am. Chem. Soc., 1960, 82, 3893-97

4. Lowering E.G., Laidler, K.J., Can. J. Chem., 1962, 40, 31-36

5. Anzuino, G., Pirro, A., Rossi, O., and Friz, L.P., J. Polym. Sci., 1975, 13, 1657-66

6. Hager, S.L., MacRury, T.B., Gerkin, R.M., and Critchfield, F.E., Urethane Block Polymers, Kinetic of Formation and Phase Development. In Urethane Chemistry and Applications, ed. K.N. Edwards, ACS Symposium Series 172, American Chemical Society, Washington, 1981

7. Feger, C., Molis, S.E., Hsu, S.L. and MacKnight, W.J., Macromolecules, 1984, 17, 1830-34

8. Cunliffe, A.V., Davis A., Farey, M., and Wright, J., Polymer, 1985, 26, 301-06

9. Kelemen-Haller, A., Farkas, F., Thermochim. Acta, 1985, 92, 297-300

10. Lipshitz, S. and Macosko, C.W., J. Appl. Polym. Sci., 1977, 21, 2029-39

11. Marciano, J.H., Rojas, A.J., Williams, R.J.J., Polymer, 1982, 23, 1489-92

CROSSLINKED POLYPROPYLENE AND RELATED MATERIALS

I. Chodák

Polymer Institute CCR, Slovak Academy of Sciences
842 36 Bratislava, Czechoslovakia

ABSTRACT

A survey of polypropylene crosslinking initiated by thermal decomposition of peroxide is given. The role of coagents of the crosslinking is demonstrated and mainly consists in inhibition of macroradical fragmentation. The properties of crosslinked polypropylene as well as of crosslinked polypropylene-polyethylene blends are shown to depend strongly on the effectivity of crosslinking.

INTRODUCTION

Polypropylene macroradicals formed by thermal decomposition of peroxides can decay via recombination, fragmentation, or disproportionation. Recombination leads to crosslinking, while disproportionation causes a decrease of the crosslinking efficiency. Fragmentation leads to a more severe drop of the crosslinking efficiency but also to considerable degradation and to inferior ultimate properties of the material. Since a substantial portion of macroradicals decays by fragmentation in polypropylene, the crosslinking can be reached under extreme conditions only, e. g. at high concentration of an efficient initiator (1) or with high doses of the high-energy radiation. (2).

CROSSLINKING OF POLYPROPYLENE BY PEROXIDES

When the crosslinking is initiated by decomposition of organic peroxides, it is crucial to consider how the reaction conditions can influence the ratio of recombination to fragmentation. Obviously, the gel content will increase with raising the
peroxide concentration. Less evident is that also the efficiency of crosslinking rises (3), since e. g. in polyethylene the
efficiency does not change with peroxide concentration.
The reason for the increase is the higher stationary level of
macroradicals. Consequently, the second-order recombination
prevails over monomolecular fragmentation. This phenomenon is
demonstrated in Fig. 1, showing straight line dependence of the
gel point (the highest peroxide concentration when no gel is
formed) on the halftime of decomposition of various peroxides
at 170 ^{o}C, which is reciprocal to the rate of peroxide decomposition.

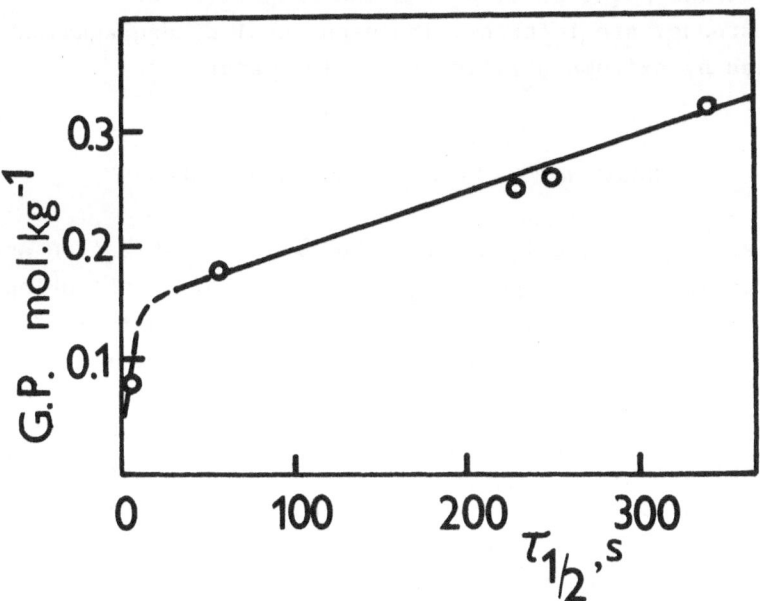

Figure 1.: The dependence of gel point (G.P.) on the halftime
of peroxide decomposition ($\tau_{1/2}$)

The amount of gel formed is affected by temperature of cross-linking (3). The dependence shows a maximum in the vicinity of the melting temperature of polypropylene. At lower temperature, the temperature increase leads to faster peroxide decomposition as well as to an increase of the mobility of the polymer segments bearing the radical centre. Both effects contribute to higher efficiency of the crosslinking. Fragmentation prevails at higher temperature because of much higher activation energy of chain scission when compared with the recombination of macroradicals, the result being a decrease of the gel content. The temperature dependence of gel formation is also qualitatively similar when the crosslinking is initiated by uv irradiation (4). A clearly identified maximum in the range between 35 and 60 ^{o}C indicates that the efficiency of crosslinking is influenced by the segment mobility.

We may conclude that polypropylene can be crosslinked to a rather high gel content using organic peroxides. However, the concentration of the peroxide is too high, and unacceptable from the practical point of view, and the properties of the cross-linked product are inferior. The high level of degradation is indicated by extreme brittleness of the material.

CROSSLINKING IN THE PRESENCE OF COAGENT

A remarkable increase in the ratio between recombination and fragmentation can be reached using coagents of crosslinking applied together with the initiator. The addition to the macro-radical is the main mechanism of their action, stabilizing the macroradical against the fragmentation without decay of the ra-dical centre. Efficient coagents possess two or more active functional groups in the molecule reacting independently with different macroradicals. The crosslink is formed actually by coagent molecule. Polyfunctional monomers are commonly used as coagents (5). At the higher concentration of the monomer various side reactions, mostly homopolymerization, take place. This affects the overall efficiency. In addition to polyfunctional

monomers, other compounds are known, e.g. acetylene (6), chlor-
anil (7), sulfur (8), and a mixture of sulfur with terephthaloyl
chloride (9). None of these was applied in industry either because of low efficiency or due to various technological complications.

When studying the polypropylene crosslinking, we investigated several potential coagents, such as thiourea and its derivatives, various biphenols, and aromatic compounds with various
substituents (e.g. p-nitrophenol, p-nitroaniline). Very different and mostly low efficiency was found with these coagents.
Hydroquinone and p-benzoquinone were found to have highest efficiency. These coagents are superior when compared to any published data on other compounds. The system with benzoquinone and
some polyfunctional monomers are compared in Table 1.

The system with benzoquinone evidently initiates much
higher gel formation than the systems with polyfunctional monomers at the same or even lower concentrations of peroxide.

TABLE 1

Gel formed with initiating system dicumyl peroxide - coagent
(PETA - pentaerythritol tetraallyl ether, DAM - diallyl maleate,
BG - benzoquinone).

Peroxide mass. %	Coagent mass. %	Gel, %		
		PETA	DAM	BQ
2	1	22	-	89
3	2	55	-	93
5	1	68	-	92
5	2.5	-	45	96

The amount of gel is affected significantly by the concentration of both components of the initiating system (Fig. 2).
The gel content rises with the increase of the peroxide concentration at any coagent amount. On the other hand, the plot of
the gel content vs. benzoquinone concentration goes through a
maximum. This course is explained according to a scheme

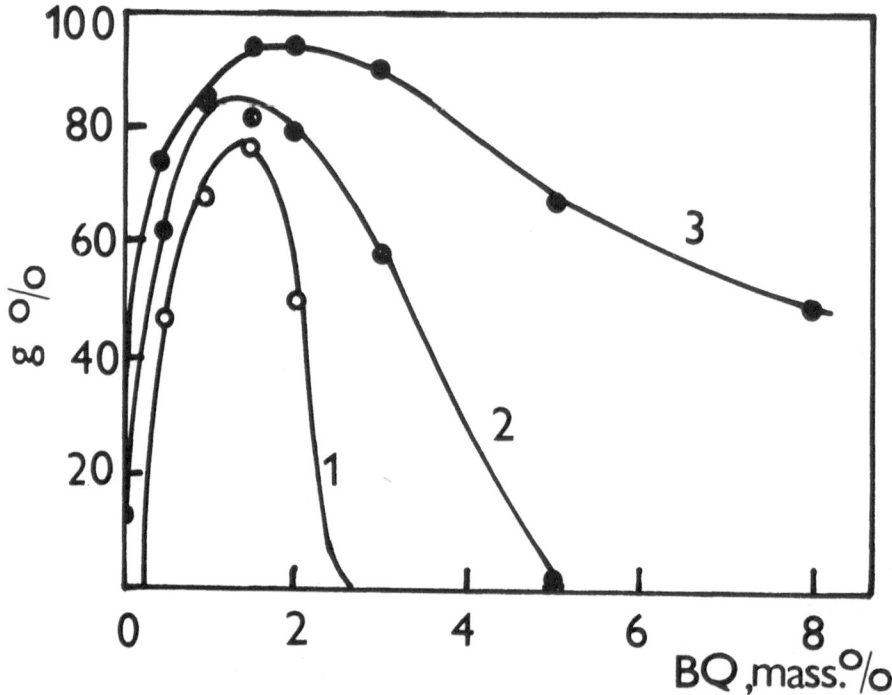

Figure 2.: Gel content of crosslinked polypropylene in dependence on benzoquinone concentration for different tert. butyl perbenzoate concentrations (1 mass. % - 1, 2 % - 2, 4 % - 3).

$$ROOR \longrightarrow 2\ RO^{\cdot}\ (R^{\cdot}) \qquad (1)$$

$$PH + RO^{\cdot} \longrightarrow ROH + P^{\cdot} \qquad (2)$$

$$P^{\cdot} \longrightarrow P_1{=} + P_2^{\cdot} \qquad (3)$$

$$2\ P^{\cdot} \longrightarrow P\text{-}P \qquad (4)$$

$$P^{\cdot} + Q \longrightarrow P\text{-}Q^{\cdot} \qquad (5)$$

$$P\text{-}Q^{\cdot} + P^{\cdot} \longrightarrow P\text{-}Q\text{-}P \qquad (6)$$

$$Q + RO^{\cdot}\ (R^{\cdot}) \longrightarrow Products \qquad (7)$$

namely by the reaction (7) which represents several paralell or subsequent reactions between the coagent and the low molecular radicals in the system. The reaction (7) is more significant at higher than optimal coagent concentration. The decay of macroradicals via reactions (5) and (6) prevails at optimal ratio of the components. Reactions (3) and (4) proceed when no coagent is present. The scheme is supported by the fact that coagent is

bound into the network formed as well as by complete suppression
of crosslinking at high benzoquinone concentration.

PROPERTIES OF CROSSLINKED POLYPROPYLENE

The properties of crosslinked polypropylene depend on the gel
content. E.g. the dependence of the melting temperature and of
the heat of fusion on benzoquinone concentration is approximately reciprocal to the dependence of the gel content vs. coagent
amount shown in Fig. 2. Apparently, higher content of benzoquinone protects polypropylene against radical modification of
any kind. Consequently, the properties are in this case close
to those of the original unmodified polypropylene.

The most important practical reason for polyolefin crosslinking is the increase of the temperature resistance of the
polymer so that it keeps the shape at the temperature over T_m.
The temperature resistance of the crosslinked polypropylene is
compared with the original polymer in Table 2. The crosslinked
polymer is much more temperature resistant at temperature above
T_m at the relatively high load applied so that the short-term
resistance is evident even at 315 ^{o}C.

The high temperature resistance is also connected with
thermooxidative stability of the polymer. Two interesting features were observed when investigating this topic. First, the stability of crosslinked unstabilized polypropylene was surprising-

TABLE 2

Time of penetration (s) of a needle with area of 1 mm^2 in a depth
of 1 mm in original (PP) and crosslinked (XLPP) polypropylene

| Load | 150 ^{o}C | | 205 ^{o}C | | 260 ^{o}C | | 315 ^{o}C | |
kg	PP	XL	PP	XL	PP	XL	PP	XL
2.5	600	600	11	600	3	600	2	40
5.0	600	600	3	600	1	21	1	5
10.0	600	600	1	600	1	2	1	2
20.0	600	600	1	40	1	2	1	1

ly high as compared with the unmodified stabilized polypropylene. Second, the plot of the induction period of oxidation vs.
peroxide concentration shows a maximum (Fig. 3). The maximal

oxidative stability is observed with 0.5 % of hydroquinone and
1.0 % of peroxide. At this composition the coagent is overdosed
and no gel is formed. Simplified explanation can be given
according to the reaction scheme. The maximal gel content is
reached at the optimal ratio of components when mostly reactions
1, 2, 5, and 6 proceed. If there is a shortage of peroxide re-
garding the coagent amount, reaction 5 takes place but there is
not enough macroradicals to accomplish also next step 6. No crosslinking therefore occurs but rather stable free radicals of se-
miquinoid nature are bound to the polymer. The radicals have
much higher stabilizing activity than the original hydroquinone.
If the system contains more peroxide, the radicals decay in subsequent reactions with macroradicals leading to an increase of
gel content but also to a decrease of thermooxidative stability.

The crosslinking affects also the impact strength of polypropylene (10). At the temperature under -30 ^{o}C almost no

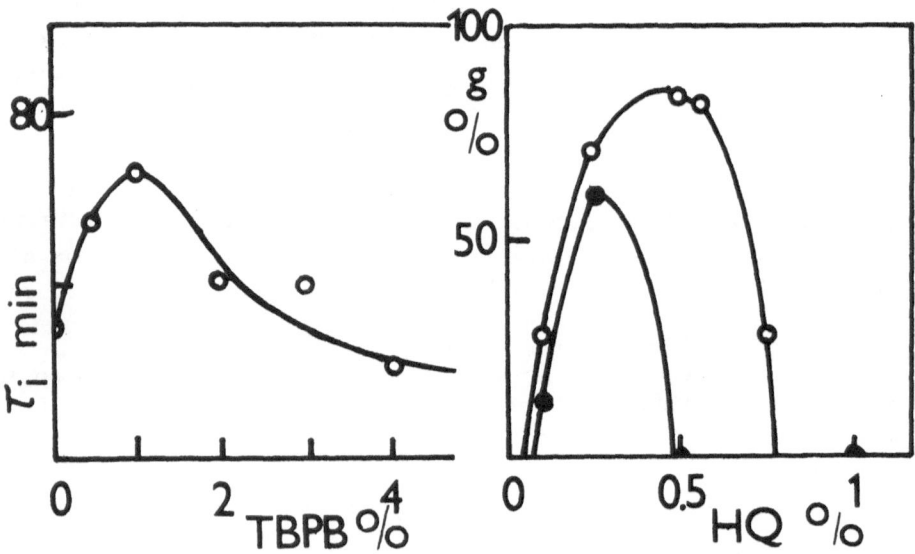

Figure 3.: Induction period of the oxidation of polypropylene
crosslinked with 0.5 mass. % of hydroquinone vs. peroxide concentration (left) and gel formation in polypropylene vs. hydroquinone concentration with 1 or 2 mass. % of tert. butyl perbenzoate (right).

difference between crosslinked and original polypropylene was observed, but at -10 $^{\circ}$C and above the impact resistance increases remarkably with the crosslinking degree.

CROSSLINKING OF POLYPROPYLENE-POLYETHYLENE BLENDS

The crosslinking can positively affect the properties of the blends containing polypropylene. The properties of polypropylene-polyethylene mixture are inferior in most cases because of the low compatibility of the components. Interaction beteeen the phases can be substantially modified by crosslinking and the properties can be improved. Crosslinked blends of PP/LDPE have much higher impact strength than uncrosslinked ones, especially above -10 $^{\circ}$C. Of course, the effect is only evident if the degradation during the crosslinking is negligible, i.e. the crosslinking proceeds in the presence of an efficient coagent. Table 3 shows that the values of impact strength W_i at -20 $^{\circ}$C are rather low for uncrosslinked blends and a minimum occurs when the polyethylene content is 25 - 50 mass. %.
To achieve the high impact strength, the crosslinking to high degree is crucial (system A) since at the lower crosslinking degree (system C) the W_i values are almost the same as in uncrosslinked blend. The crosslinking without coagent leads to

TABLE 3

The values of impact strength W_i of uncrosslinked (O) and crosslinked (XL) blends of polypropylene-polyethylene. Crosslinking system A: peroxide 3 mass. %, hydroquinone 0.5 %, B: peroxide 3 %, C: peroxide 1 %, hydroquinone 0.25 %.

LDPE mass %	PP mass %	system	gel %	O	W_i, kJ x m^{-2} S_x	XL	S_x
100	0	A	75.5	75	14	>110	
75	25	A	81.7	73	30	>110	
50	50	A	69.9	13	6	>110	
25	75	A	60.3	15	5	>110	
0	100	A	67.4	21	10	27	4
50	50	B	40.6	13	6	3	2
50	50	C	23.1	13	6	13	8

much lower W_i in spite of rather high gel formation. This can
be explained by the degradation of polypropylene leading to
brittle material.

In regard to the mechanical ultimate properties, the tensile strength is an additive value and does not depend on crosslinking unlike the values of elongation at break, where great
difference was observed, the crosslinked blends having much
higher elongation at break than uncrosslinked mixtures.

According to these results we can expect the crosslinking
degree to depend on both the composition of the blend and the
composition of the crosslinking system. Fig. 4 shows the dependence of gel on the ratio of PP to PE. When the process is initiated by peroxide in the absence of a coagent, the gel amount
decreases linearly with the increase of PP content in accord
with the expected increase of fragmentation. In the presence
of the coagent the gel amount is almost independent of the blend
composition.

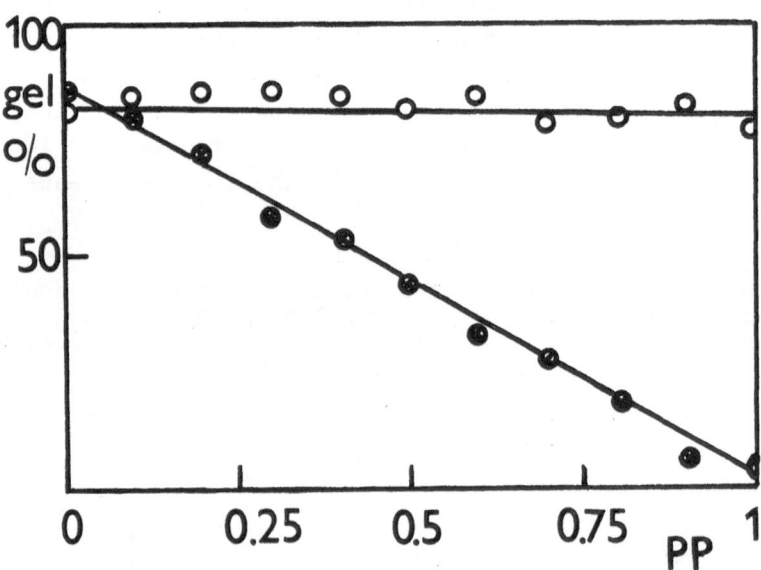

Figure 4.: Gel formation vs. PP content in PP/LDPE blend
crosslinked with 3 mass. % of tert. butyl perbenzoate (●) and
3 % of TBPB + 0.75 % of hydroquinone (O).

89

The crosslinking also influences the melting and crystallization temperature of both components. Due to immiscibility,
DSC scan shows both phase transitions separately for polyethylene and for polypropylene. The melting temperature of PP does
not change with composition but T_m of the PE part decreases
slightly. The decrease is more pronounced in crosslinked blend.
Apparently, the PE crystals formed in the presence of PP crystals are less perfect. A more significant decrease of T_{mPE} in
crosslinked blends indicates the interaction between PE and PP
molecules as a result of crosslinking. Polypropylene chains
linked with polyethylene interfere with PE crystallization.
On the other hand, the PP crystals are not influenced by the
presence of PE since the PE melt is squeezed out from the
growing PP crystals. The effect of crossbonded PE/PP chains on
the crystallization of PP is eliminated in the same way. It
should be emphasized that an evident decrease of T_m of both
polymers takes place as a result of crosslinking. This phenomenon is well known from previous papers (e. g. 11).

The above-mentioned results show that the crosslinking of
polypropylene in the presence of an efficient coagent can give
materials with some interesting properties and with the potential practical application.

REFERENCES

1. Borsig, E., Fiedlerová, A., and Lazár, M., J. Macromol. Sci.
   Chem. 1981, 16, 513.
2. Starenkii, A.G., Lavrentovich, Y.I., and Kabakchi, A.M.,
   Vysokomol. soed. 1970, 12, 2476.
3. Chodák, I., and Lazár, M., Angew. Makromol. Chem.,1982,106,153
4. Chodák, I., and Zimányová, E., Eur. Polymer J., 1984, 20, 81
5. Chodák, I., Fabiánová, K., Borsig, E., and Lazár, M.,
   Angew. Makromol. Chem., 1978, 69, 107.
6. Kitamaru, R., and Hyom, S.H., J. Polymer Sci., Macromol.
   Reviews, 1979, 14, 207.
7. Pat. US 3 285 883.
8. Čapla, M., Borsig, E., and Lazár, M., Angew. Makromol. Chem.
   1985, 133, 53.
9. Pat. US 3 285 885.
10. Chodák, I., Plastics Proc. Appl., 1985, 11, 14 (in slovak).
11. Narkis, M., Raiter, I., Shkolnik, S., Siegmann, A., and
    Eyerer, P., J. Macromol. Sci.-Phys. 1987, B26, 37.

THE ELUCIDATION OF THE CROSSLINKING AND SCISSION PROCESSES RELATIONS DURING PHOTOCHEMICAL MODIFICATION OF POLYOLEFINES

PAVEL ZAMOTAEV
Oil Chemistry Department of Physical Organic Chemistry
and Coal Chemistry Institute,
Academy of Science of the Ukr.SSR
Kharkovskoe shausse,50, Kiev 252160, USSR

ABSTRACT

Low density polyethylene, linear polyethylene and polypropylene films were photochemically crosslinked in the presence of different initiators. Their gel content, equilibrium swelling degree and stress-strain behaviour in the swollen state were determined. It was found that the correlations between crosslinking and chain scission reactions depend on the photoinitiators nature. Solid-state crosslinking causes trapping of entanglements with the following contribution to the effective network density determined by solvent swelling or stress-strain behaviour up to the melting point.

INTRODUCTION

The polyolefines crosslinking improves its form stability above the normal melting point, its environmental stress cracking resistance, its creep, tensile yield stress and some others physical properties. Irradiation and peroxides crosslinking methods are mainly used nowadays for convertion thermoplastic polyolefines into thermosetting materials. But there is one more method, photochemical, that has some advantages for the films crosslinking. It includes the addition of some ingredients (photoinitiators) to polyolefines which can photodecompose or photoreduce in polymer with macroradical formation. These macroradicals can take part in crosslinking and chain scission reactions. These reactions interrelations determine the level of critical UV irradiation doses, required to obtain the gel point and the magnitude of maximum attainable gel content. There are numerous investigations in which the ratio of scission to crosslinking events (p_o/q_o) for irradiation and peroxides crosslinking of polyolefines were determined. But for

photochemical crosslinkig this problem is still in the air.
The present work is devoted to comparison of crosslinkability
of low density polyethylene (LDPE), linear polyethylene
(LLDPE) and polypropylene (PP) under UV irradiation in the
presence of S_2Cl_2, SO_2Cl_2, xanthone (XN) or 2-ethyl-9,10-
anthraquinone (EAQ) as a photoinitiators.

MATERIALS AND METHODS

The films used in present study were prepared from LDPE with
number average (Mn) and weight average (Mw) molecular weights
of $17 \cdot 10^3$ and $110 \cdot 10^3$ kg/kmol, LLDPE with Mn=$16 \cdot 10^3$ and Mw=
$59 \cdot 10^3$ kg/kmol or PP with Mn=$86 \cdot 10^3$ and Mw=$55 \cdot 10^4$ kg/kmol.
Photoinitiators were inserted in the films from gaseous (S_2Cl_2)
or SO_2Cl_2) or liquid phases (EAQ,XN) and during compression
molding (EAQ,XN). The lamp used for irradiation was DRT-
1000W. Extraction of sol-fraction was performed in boiling
p-xylene. The stress-strain measurements were carried out in
p-xylene at 95°C using different loads.

RESULTS AND DISCUSSION

Analyses of the solubility behaviour of crosslinked polyolefines has remained the principal technique used to elucidate
the ratio between chain scission and crosslinking processes.
On the basis of Charlesby and Pinner theory, generally
accepted and used nowadays, we may determine the p_o/q_o ratio
from Inokuties' equation (1):

$$g_{max} = \frac{1}{2}(1 - p_o/q_o + (1 + 2p_o/q_o)^{1/2}) \qquad (1)$$

where g_{max} - maximum attainable gel content. Complete extraction of crosslinked polymer with known initial molecular
distribution function makes it possible to determine the
effective network chain density ($\mathcal{V}^*$) (2). It is possible to
check the validity of the previouse theories using the same
networks in the experiments evaluating their stress-strain
behaviour and equilibrium swelling degree. The data of the
stress-strain measurements could be well described by means
of the semi-empirical Mooney-Rivlin equation, modified for
swollen systems (3):

$$\frac{\mathcal{O} \, q_1^{-1/3}}{\lambda - \lambda^{-2} \, q/q_1} = 2C_1 + 2C_2 \lambda^{-1} \qquad (2)$$

where $\mathcal{O}$ is the equilibrium retractive force of the network
per unit cross-section of the dry unstrained elastomer, λ,
is the elongation ratio reffered to the unstrained swollen
length, q and q_1 are the initial degree of swelling and the
elongated sample degree of swelling, $2C_1$ and $2C_2$ are the
Mooney-Rivlin constants. According to Hermans, the term q/q_1
is approximately equal to $\lambda^{1/2}$ (3). Using this approximation
the Mooney-Rivling plot for the different photocrosslinked

samples were obtained. These plots yield negative values of $2C_2$ constants which lay in the range $(-9 \div -3)\cdot10^{-3}$ MPa. The negative $2C_2$ values were obtained for the swollen gels of peroxide crosslinked polyethylene (4) and some elastomers (5). Since these results indicate on some faults of the selected model, additional measurements of the swelling degree dependence on the elongation ratio were performed. These tests featured that the q of photocrosslinked samples is proportional to λ ,that corresponds to the stress-swelling behaviour of some elastomers (6). Therefore the approximation $q/q_1 = \lambda$ in equation (2) will be used. The resulting Mooney-Rivlin plots for swollen extracted photocrosslinked gels are presented in figure 1 and are characterised by $2C_2$ cnstants which aspire to nought.

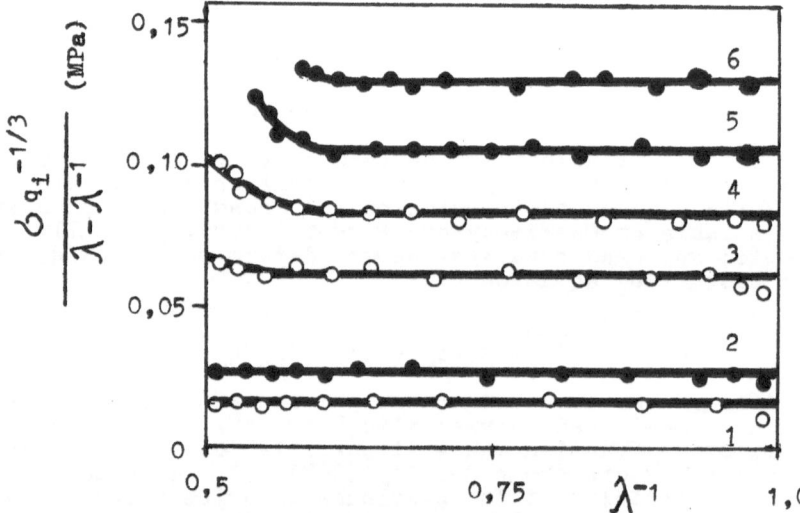

Figure 1. Stress-strain isotherms for the swollen extracted
LDPE networks plotted according to the eq.2 (No are
determined in tabl.1)

This shows the coincidence of the elasticity behaviour of extracted networks and ideal phantom network. For a phantom network ν^* can be calculated by means of the relation (7):

$$\nu^* = \frac{\ln(1-q_1) + q_1 + \chi q_1{}^2}{(2/\phi - 1)\, V_1\, q_i{}^{-1/3}} \qquad (3)$$

where χ is interaction parameter, V_1 - the molar volume of the solvent, ϕ is network funktionality equals 4. At the same time we can use $2C_1$ constant for determination of ν^* (3-5,7):

$$\nu^* = 2C_1 q_i{}^{2/3}((1- 2/\phi)RT)^{-1} \qquad (4)$$

where R is the gasconstant, T is absolute temperature. The derived $2C_1$ constants, q_1 and ν^* obtained by different methods are compiled in tables 1,2. This tables show also the rapture stress (σ^+) and elongation for break for swollen samples(λ^+)

TABLE 1

Mechanical properties of swollen photocrosslinked by SO_2Cl_2 LDPE gels after extraction and comparison of estimates of ν^* obtained by different methods

No	$[SO_2Cl_2]$ (%)	$10^2 \cdot 2C_1$ (MPa)	q_1	g (%)	$10^2 \cdot \sigma^+$ (MPa)	λ^+	$10^2 \nu^*$ (mol/l) dry film		
							Calcd from $2C_1$	Calcd from q_1	Calcd from g
1	1,8	1,5	10,8	47,3	7,9	2,70	2,2	2,0	2,6
2	3,9	2,7	8,0	67,9	10,1	2,08	4,8	4,6	4,8
3	7,7	6,2	5,6	78,3	17,8	2,01	9,6	9,0	7,2
4	11,3	8,3	4,7	84,0	19,3	1,85	12,9	12,7	1o,1
5	14,1	10,6	4,3	89,0	20,9	1,72	16,5	15,5	14,5
6	18,7	13,5	3,7	91,6	23,4	1,62	19,6	20,3	16,9

TABLE 2

Mechanical properties of swollen crosslinked films, different methods estimates of ν^* and p_o/q_o terms for different photo-crosslinking processes

	$10^2 \cdot 2C_1$ (MPa)	q_1	g (%)	$\dfrac{p_o}{q_o}$	$10^2 \cdot \sigma^+$ (MPa)	λ^+	$10^2 \nu^*$ (mol/l) dry film		
							Calcd from $2C_1$	Calcd from q_1	Calcd from g
					LDPE				
a	23,2	3,0	91,6	0,67	20,4	1,35	31,6	32,1	16,9
b	19,5	3,5	89,1	0,80	8,3	1,15	29,6	24,4	14,1
d	25,3	2,6	92,8	0,55	25,5	1,43	31,7	40,4	16,5
e	22,8	3,0	90,3	0,73	12,1	1,20	31,5	34,9	15,3
f	20,1	3,4	92,8	0,55	29,4	1,60	29,7	26,0	16,5
g	7,0	5,1	86,8	0,88	2,1	1,09	13,6	14,1	17,3
h	8,3	4,3	88,3	0,84	5,2	1,21	14.4	17,4	17,1
i	6,2	4,7	88,0	0,90	2,0	1,10	11,5	15,1	17,4
j	4,6	5,6	88,3	0,84	4,1	1,28	9,6	11,9	17,1
					LLDPE				
b	13,4	3,4	72,0	1,50	7,9	1,2o	19,9	26,1	12,8
c	12,5	3,7	74,0	1,44	14,2	1,42	18,5	22,2	13,3
g	4,5	7,2	64,0	1,81	3,5	1,20	11,0	8,1	10,3
h	5,1	6,6	68,5	1,70	3,9	1,23	11,8	9,o	10,7
					PP				
a	6,2	4,8	93,4	0,4θ	5,3	1,23	11,6	14,7	9,4

b	5,7	5,1	94,2	0,40	5,7	1,28	11,1	14,4	9,0
c	4,1	6,6	93,4	0,4o	5,4	1,41	9,5	9,0	8,8

photocrosslinked by SO_2Cl_2(a,b,c),S_2Cl_2(d,e,f),EAQ(g),XN (h,i,j),prior(a,d,g,h) or after recrystallization (b,e,i) or extraction (c,f,j)

The extracted photocrosslinked samples show good coincidence of ν^* obtained by different methods. It verifies the validity of the p_o/q_o definition method. The p_o/q_o term depends on the nature of polyolefines and photoinitiators that were used. The gratest divergence of p_o/q_o values for S_2Cl_2, SO_2Cl_2, EAQ and XN (0,40;0,53;2,90;3,16) were obtained in PP. The ν^* values after $2C_1$ or q_1 for unextracted films are larger than for extracted. It may be due for the contribution of trapped entanglements. Last contribution to the $2C_1$ and q_1 values are decreasing after recrystallization and are aspiring to nought after sol extraction.

REFERENCES

1. Inokuti M., Gel formation in polymers resulting from simultaneous crosslinking and scission. J.Chem.Phys.,1963, 38, 2999-3005.

2. Saito O., Statistitcheskay teoriy poperechnogo scivaniy. In Radiatzionay Chemia Macromolecool, ed.M.Dole, Atomisdut, Moskow, 1978, pp.205-38.

3. de Boer J. and Pennings A.J. The effect of pendant chains on the elasticity behaviour of polyethylene networks. Colloid and Polymer Sci., 1983, 261, 750 - 6.

4. de Boer J. and Pennings A.J., Polyethylene networks crosslinked in solution: preparation, elastic behaviour, and oriented crystallization.2. Macromolecules, 1977,10,981-9.

5. Hoffmann M. Determination of network structure by extraction and random degradation. 3. Makromol.Chem., 1983, 183, 2237-56.

6. Volynski A.L. Mechanical behaviour of crosslinked natural rubbers under deformation in liquid media. Vysokomol.soed. 1988, 30B, 859-62 .

7. Quesiel J.P. and Mark J.B., Characterization of elastomeric networks structures using the effects of swelling on stress-strain isotherms and the extents of swelling at thermodynamic equilibrium. Polym.Bull., 1983, 10,119-25.

Part 3
EMULSION POLYMERIZATION

INTRODUCTION TO
EMULSION POLYMERIZATION

Mohamed S. El-Aasser
Department of Chemical Engineering
Emulsion Polymers Institute
Center for Polymer Science and Engineering
Lehigh University
Bethlehem, PA 18015 U.S.A.

ABSTRACT

Emulsion polymerization is a heterogeneous free-radical process in which the kinetic events take place simultaneously in more than one phase. The following four mechanisms for particle nucleation have been discussed: micellar, homogeneous, coagulative, and in miniemulsion droplets. The roles of surfactants in determining the latex particle number are outlined.

INTRODUCTION

Emulsion polymerization is a heterogeneous, free-radical polymerization process which has wide industrial application in the production of polymer colloids or latexes of several different types of polymers: polybutadiene and butadiene-styrene copolymers, poly(vinyl acetate) and vinyl acetate copolymers, acrylate ester copolymers, poly(vinyl chloride) and vinyl chloride copolymers, vinylidene chloride copolymers, polyethylene and ethylene copolymers, polytetrafluoroethylene, polyacrylamide, and acrylamide copolymers. These latexes are used in a wide variety of applications: synthetic rubber, floor coatings, paints, adhesives, binders for non-woven fabrics, high impact polymers, latex foam, additives for construction materials such as cement and concrete, flocculants and rheological modifiers. Latexes are also used in numerous biomedical applications: such as diagnostic tests, immunoassays, biological cell-labeling, (identification and separation), and drug delivery systems. Small quantities of monodisperse polymer colloids are used as size calibration standards and find extensive use to test theories in physics, colloids, and rheological studies.

Emulsion polymerizations are usually carried out using one of the following three types of processes: batch polymerization, in which all ingredients are added at the start of the reaction; semi-batch (sometimes called semi-continuous), in which one or more of the ingredients usually the monomer either neat or in emulsion, is added continuously or in increments; and continuous, in which all ingredients are added continuously and product latex is continuously removed. The semi-batch emulsion polymerization is the most common process used industrially to produce commercial latexes. This is due to its flexibility and the ability to control the heat transfer in the polymerization reactor as well as the copolymer composition of the latex particles by controlling the monomer feed streams.

The purpose of this introductory article is to review the mechanisms and kinetics of the emulsion polymerization, and the role of the surfactants in particle nucleation. Several books are available on the subject and should be consulted (1-10).

CHEMISTRY OF FREE-RADICAL POLYMERIZATION

Bulk Polymerization

The four major reactions in free radical polymerization are initiation, propagation, termination and transfer.

The initiation takes place when a primary radical formed by decomposition of the initiator adds a monomer molecule to form a monomer radical. Examples of simple thermal initiators are benzoyl peroxide and potassium persulfate. Examples of redox initiator systems, usually used for low temperature emulsion polymerization, are potassium persulfate/sodium bisulfite/iron II or hydrogen peroxide/iron II (Fenton's reagent). The propagation involves successive addition of monomer molecules to monomer radicals to form a long polymer chain ending in a monomer radical. Termination takes place when two polymer radicals combine to form one polymer molecule, or disproportionate to form two polymer molecules. The transfer involve reaction of a growing polymer radical with a large variety of molecules by abstracting a hydrogen or a halogen atom from another compound to terminate the polymer radical and form a new radical, which adds monomer molecules to grow another polymer chain.

The rate of free radical bulk polymerization is given by equation 1.

$$R_p - k_p \ [R_i/k_t]^{1/2} \ [M] \tag{1}$$

where R_i is rate of initiation, k_p is the propagation rate constant, k_t is the termination rate constant,

The kinetic chain length ν, defined as the number of monomer units used up per active chain, is given by equation 2.

$$\nu - k_p \ (1/R_i \ k_t)^{1/2} \ [M] \tag{2}$$

The degree of polymerization $\overline{X}_n - \nu$ for polymer chains formed by disproportionation termination reaction. The degree of polymerization $\overline{X}_n - 2\nu$ for polymer chains formed by combination termination reactions.

Emulsion Polymerization

In conventional emulsion polymerization, a water-immiscible monomer is emulsified in an aqueous continuous medium using an oil-in-water emulsifier type, and polymerized using water soluble or oil-soluble initiator. The initial monomer emulsion is comprised of droplets in the size range of 1-10μm. The final latex system is comprised of colloidal dispersion of polymer particles in water with particle size usually in the submicron range.

The basic recipe for heterogeneous emulsion polymerization contains, in addition to the monomer and initiator, water as the continuous phase and a surfactant. The polymerization in this system can take place simultaneously in one or more of the following sites: the continuous phase, the monomer droplets or the monomer-swollen polymer particles. Consequently, all the kinetic events outlined above may occur in more than one of these three phases simultaneously. As a result the partition of monomers, free radicals and chain transfer agents between the various phases during the course of polymerization play a significant role in determining the polymerization profile as well as the properties of the polymer product. One of the most important characteristics of heterogeneous polymerization, in contrast to bulk or solution polymerization, is that the free-radicals grow in relative isolation. The degree of isolation of the free-radicals depends on the degree of subdivision of the reaction sites and the flux rate of free-radicals. Hence this process is well suited to prepare high molecular weight polymers at relatively high rates of polymerization.

MECHANISMS AND KINETICS OF EMULSION POLYMERIZATION

The emulsion polymerization reaction can be divided into two stages: particle nucleation and particle growth; the two stages occur simultaneously. The particle nucleation stage is more important because it determines the number of particles present in the system and thus the rate of polymerization. The particle nucleation stage begins by initiation of polymerization of the monomer, which can physically be located as solubilized in micelles, as monomer dissolved in the continuous phase, as monomer emulsified in droplets, and as monomer absorbed in the adsorbed surfactant layer. The particle growth is usually considered to take place by polymerization of monomer in the monomer-swollen particles i.e. by propagation. However, there is strong evidence for particle growth by flocculation phenomena.

The following is a brief qualitative review of the mechanisms and kinetics of emulsion polymerization. The cited literature articles will provide the reader with more detailed quantitative treatments.

Micellar Nucleation

Harkins mechanism considered that the major source of particle nucleation was the monomer-swollen surfactant micelles (11). According to Harkins, radicals generated in the aqueous phase enter monomer-swollen micelles and initiate polymerization to form monomer-swollen polymer particles nuclei. The nuclei grow by polymerization of monomer supplied to the monomer-swollen particles by diffusion from the monomer droplets through the aqueous phase. The surfactant molecules required to stabilize the growing particles are supplied from the uninitiated micelles. Usually one of every 100-1000 micelles capture a radical and become a polymer particle. The particle nucleation stage ends with the disappearance of the micelles. The major locus of polymerization was postulated to be the monomer-swollen polymer particles.

Micellar nucleation mechanism is generally applied to monomers which are sparingly soluble in water, in the concentration range of 0.34-15 mM, if the emulsifier is present in concentrations above the cmc. These include emulsion polymerization of monomers such as n-octyl acrylate, dimethyl- styrene, vinyl toluene, n-hexyl acrylate, styrene, n-butyl acrylate, chloroprene, and butadiene.

Smith and Ewart (12) developed a quantitative model to describe the particle growth based on Harkins mechanism. Their population balance equations for particles containing $\underline{n}$ free radicals were solved for three limiting cases based on the value of n: case 1 for n<<1; case 2 for n = 0.5; and case 3 for n>>1. Case 2 is the most popular, and generally applies for styrene emulsion polymerization where the monomer is sparingly soluble in water.

The polymerization rate according to Case 2 is given by:

$$R_p = k_p \, [M] \, (N/2) \tag{3}$$

where the number of particles per unit volume N is given by:

$$N = k \, (\rho_i/\mu)^{0.4} \, (a_s S)^{0.6} \tag{4}$$

where ρ_i is the rate of production of free radicals in the aqueous phase, μ is the rate of growth of the volume of a latex particle, a_s is the cross section area of a surfactant molecule, and S is the surfactant concentration. The value of the constant k was given as 0.37 for the case where both micelles and monomer-swollen-particles compete for the free-radicals; and 0.53 for the case where free radicals enter micelles only until they disappear.

The number average degree of polymerization is given by equation 5.

$$\overline{X} = k \, [M] \, (N/R_i) \tag{5}$$

Equations 4 & 5 show that both the rate of polymerization and the number-average degree of polymerization are proportional to the number of the particles. This is contrasted with the inverse variation of R_p and $\overline{X}$ in the bulk polymerization given by equations 1 and 2. The simultaneous increase in the polymerization rate and molecular weight with the increase in the number of polymer particles is known as "emulsion polymerization kinetics". This kinetics require that the free-radicals must be segregated, and the number of polymerization loci must be within a few orders of magnitude of the number of the free-radicals.

The Smith-Ewart theory has been useful in predicting the dependency of the number of particles on the initiator and emulsifier concentrations for few systems (13, 14). However, many deviations were reported between the experimental results and the theoretical predictions, particularly for emulsion polymerizations of monomers with water solubilities higher than that of styrene. Several modifications of the Smith-Ewart theory were attempted in order to account for these discrepancies by incorporating the radical exit from particles, aqueous-phase termination, and the possibility of radical re-entry (15-18).

Homogeneous Nucleation

According to the homogeneous nucleation, charged free radicals generated in the aqueous phase react with soluble monomer to form soluble oligomeric radicals. The oligomeric radicals grow by further addition of monomer units until they exceed their solubility limit in the aqueous phase and precipitate from solution. The precipitated oligomeric radicals form spherical particles and adsorb surfactant molecules to form "primary" particles. The oligomeric radicals formed therafter may precipitate to form more primary particles or may be captured by already formed particles. Primary polymer particles then swell with monomer and grow by propagation. Primary particles may also flocculate with themselves or with growing particles, depending on the effectiveness of the surfactant as a stabilizer. Consequently, particle growth according to homogeneous nucleation mechanism may take place by both propagation and flocculation. The number and size of the latex particles are determined by the amount of the surfactant and its effectiveness in stabilizing the primary particles and the growing particles.

The particle nucleation of monomers with relatively high water-solubility, higher than 290 mM, is generally considered to proceed by homogeneous nucleation (19,20). These are exemplified by monomers such as vinyl acetate, ethylene, methyl acrylate, acrylonitrile, and acrolein. For monomers such as, vinyl chloride, ethyl acrylate, methyl methacrylate, and vinyl chloride, with water solubility in the range of 66-170 mM, both micellar and homogeneous nucleation mechanisms have been proposed in separate occasions, but most consider homogeneous nucleation the more appropriate mechanism. Homogeneous nucleation was also proposed as the primary mechanism for particle formation in surfactant-free emulsion polymerization systems (21-23). The stabilization of the primary particles and growing particles is due to electrostatic stabilization mechanism as a result of the presence of charged initiator fragments on the particle surface, e.g. negatively-charged sulfate ions.

Fitch and coworkers (24,25) developed a quantitative treatment of homogeneous nucleation, where the rate of particle nucleation is given in terms of the rate of radical generation in the aqueous phase R_i, the rate of capture of oligomeric radicals by existing particles R_c, and the rate of flocculation of polymer particles R_f.

$$dN/dt = R_i - R_c - R_f \qquad (6)$$

At their early stage of polymerization, the rate of formation of primary particle is essentially equal to the rate of radical generation R_i. After particles have been formed, a steady state is reached between initiation and capture of the radicals and flocculation of particles. If the time to reach the steady state is short, e.g. because of high rate of radical generation, then all the particles will grow at the same average rate and end up at about the same size. This leads to narrow particle size distribution. On the other hand if the time to reach steady state is long, e.g. because of slow initiation, the particles formed early will be larger than those formed at a later stage. This leads to a broad particle size distribution.

Ugelstad pointed out that the radical capture by collision mechanism as suggested by Fitch (24), assuming that the capture rate is proportional to the particle surface area, leads to lower value of R_c and thus a higher particle number (26). Instead, Ugelstad (27,28), and others (25,29) proposed a radical capture mechanism based on diffusion theory, where R_c is proportional to the particle radius.

Coagulative Nucleation

Recent work by Napper and Gilbert's group in Australia proposed that the particle nucleation involves at least two mechanistic steps (30). The first step is the formation of the "precursor" particles, most likely by homogeneous nucleation. The second step is the formation of "mature" latex particles (of certain volume) by aggregation of the precursor particles. The coagulative mechanism assumes that the "primary precursors" are extremely small in size (approximately 3nm-in radius). As a result of their small size, the precursors exhibit poor colloidal stability to coagulation, and their monomer concentrations are much less than in normal mature latex particles. Consequently, they grow in volume much faster by coagulative rather than by propagational processes. Ultimately, the aggregational and propagational processes of the primary and higher precursors lead to "mature", colloidally stable latex particles which are swollen with monomer fully. The formation of more mature particles leads to an increase in the probability of their heterocoagulation with the newly-generated precursor particles, so that the rate of production of new particles progressively declines and ultimately ceases at the end of the nucleation stage. This mechanism was formulated based on particle size distribution measurements during the initial stage of styrene emulsion polymerization, both in the presence and absence emulsifier (31-34). Their results of the particle size distribution (plotted as a function of particle volume, rather than radius), showed positive skewness; indicating that most of the "mature" latex particles are produced late during the nucleation stage. This is contrasted with the negative skewness which characterize the particle size distributions for particles generation by micellar entry or simple homogeneous nucleation mechanisms. In this case, larger number of particles generated in the early stage of the nucleation period will grow to larger volumes at the end of the nucleation period. Where as fewer particles are generated late during the nucleation period, and thus grow to smaller size particles.

Nucleation in Monomer Droplets

The probability of particle nucleation in monomer droplets was dismissed in conventional polymerization based on the unfavorably too small surface area, and fewer number, of the monomer droplets to compete effectively with the monomer-swollen micelles for the free radicals, as shown in Table I. This is due to the relatively large size monomer droplets (1-10μm in diameter), compared to the size of the monomer-swollen micelles (generally 10-30nm in-diameter). However, despite this unfavorable statistical probabilities some monomer droplets capture radicals and polymerize to form large-size microscopic particle, which can easily be seen by examination of the final latex using optical microscopy (35).

The monomer droplets could become significant locus for particle nucleation and polymerization if their surface area were large, i.e. if the droplet size could be made small. This is the basis for polymerization in miniemulsion and microemulsion monomer systems. The reduction

TABLE I

RELATIONSHIP BETWEEN MICELLES AND MONOMER DROPLETS

	Number Density (No./g emulsion)	Surface Area (cm^2/g emulsion)
Micelle	10^{17}	3×10^6
Monomer Droplet	10^{6-9}	$3 \times 10^{2-3}$
Miniemulsion Droplet	5×10^{14}	2×10^5
Microemulsion Droplet	2×10^{18}	8×10^5

in the monomer droplet sizes (generally 100-500nm in-diameter for miniemulsions, and 5-40nm in-diameter for microemulsions) results in increased both the number and surface area of the droplets by several orders of magnitude, relative to conventional emulsion droplets,(see Table I). This results in an effective competition of initiation in monomer droplets with other particle nucleation mechanisms such as miceller and aqueous phase. Indeed, Ugelstad, El-Aasser, and Vanderhoff demonstrated experimentally that monomer droplets could become the principal locus for particle formation in styrene miniemulsions systems (36).

Miniemulsion Polymerization

Monomer miniemulsions are stable oil-in-water emulsion systems, with droplet size of 100-500nm in-diameter. Mixed surfactant systems comprising ionic surfactant in the presence of co-surfactants such as fatty alcohol (36,37), or alkane (38,39) are used to prepare these miniemulsions. Latexes are prepared by initiation of polymerization in the monomer miniemulsion droplets using water-soluble or oil-soluble initiators. The final particle size of the resulting latex is in the same range as the initial droplet size. Thus, the miniemulsion polymerization process is characterized by the introduction of the monomer to the polymerization reactor in a high degree of subdivision, with droplet size in the submicron-range. This results in a drastic change in the profile of the polymerization kinetics for the miniemulsion polymerization of styrene (37,40), and miniemulsion copolymerization of vinyl acetate-methyl acrylate (39,41-43), vinyl acetate-methyl acrylate (44) and styrene-methyl methacrylate (45). These studies showed that there are several major differences in the kinetic features between miniemulsion and conventional emulsion polymerization, which are summarized as follows.

1. The submicron monomer droplets are the main locus of particle nucleation in the miniemulsion process.

2. The radical entry into monomer miniemulsion droplets is a process of low efficiency, thus the particle nucleation stage is unusually long. It has been postulated that the intermolecular complexes of the surfactant/additive, or liquid crystals formed at the oil/water interface provide a physical barrier which prevents oligomeric free radical from entering the monomer droplets. As a result of this low initiator capture efficiency, not all droplets can be initiated, and the fraction becoming particles is determined by the level of initiator. This leads to a relatively strong dependence of the particle number on the initiator concentration.

3. The miniemulsion latex has a very broad particle size distribution (% standard deviation in the range of 10-20), regardless of the initiator concentration (0.133-2.660 mM for potassium persulfate, and 0.2-4.0 mM for azobis methyl butyronitrile) or polymerization temperature (50-70°C). This is contrasted with the common experience encountered in conventional emulsion polymerization where the breadth of the particle size distribution is inversely proportional to initiator concentration and polymerization temperature.

4. The miniemulsion polymerization does not exhibit the interval-II characteristic of a conventional emulsion polymerization of styrene, i.e., a constant rate of polymerization. Instead, when the particle nucleation stage ends (marked by the disappearance of all monomer droplets), the polymerization rate begins to decrease immediately due to the decrease in monomer concentration in the monomer-swollen particles.

5. For the miniemulsion copolymerization of vinyl acetate-n-butyl acrylate monomer mixtures, using hexadecane as the co-surfactant along with sodium hexadecyl sulfate as the surfactant in preparing the miniemulsions, the copolymer composition during the initial 70% conversion was found to be rich in vinyl acetate monomer units for the miniemulsion process compared to the conventional process. The dynamic mechanical properties of the copolymer films showed less mixing between the poly(butyl acrylate)-rich core and the poly(vinyl acetate) rich shell in the miniemulsion latex films compared to the conventional latex films.

6. During the copolymerization process the miniemulsions gave lower overall polymerization rates, and larger particle size compared to conventional polymerizations, however, the polymerization rate per particle was similar in both cases.

7. The roles of the cosurfactant in miniemulsion polymerization are as follows. First, the cosurfactant allows the formation and stabilization of submicron monomer droplets, whose surface adsorb most of the surfactant. These submicron monomer droplets compete effectively for radical capture over other nucleation mechanisms, and thus become the main locus of particle generation. Second, its presence in the monomer-swollen polymer particles increase their swelling capacity for monomer. Third, its presence in the uninitiated monomer droplets reduces the equilibrium concentration of monomers in the polymer particles. The last two effects, which seem to be in opposition to each other, are actually responsible for the differences in instantaneous comonomer composition within the monomer-swollen polymer particles, and thus the observed

differences in the instantaneous copolymer composition and the microstructure of the miniemulsion latexes compared to the conventional latexes.

A complete mathematical model based on thermodynamic and kinetic parameters was developed to describe the role of the hexadecane co-surfactant on the monomer distribution between the various phases, the polymerization rate, and the copolymer composition during the course of the miniemulsion copolymerization process (43). It should be emphasized that the incorporation of hexadecane in the bulk monomer does not affect the kinetics of the polymerization, unless a miniemulsification process is used to form and stabilize the submicron droplets (42).

Potential Advantages of Miniemulsion Polymerization

The major potential advantages of miniemulsion polymerization process in latex technology can be summarized as follows. First, it represents a novel method of introducing the monomer to the polymerization reactor with a high degree of subdivision in the submicron size range. The result of polymerization is particle size distribution which is relatively broad, with about the same average particle size as the initial droplets. Thus, it provides an approach for controlling the particle size distribution. Second, due to the broad particle size distribution, some times bimodal or trimodal (36), it provides an approach for making high solids latexes, without resorting to scheduled additions of surfactants during the course of the polymerization process. Third, the presence of the hexadecane additive and its effect on the comonomer distribution at the site of polymerization in a miniemulsion copolymerization process provides an approach for controlling the microstructure of the polymer particles and the instantaneous copolymer composition. Thus, conceivably addition of different types and concentra- tions, or mixtures, of co-surfactants in the initial process for making the miniemulsions of the various monomers before initiation of polymerization may result in the required polymer microstructure in the final latex particles. Fourth, the results of the different dependency of polymerization rates and particles numbers on various polymerization parameters, such as initiator concentration, emulsifier concentration, and temperature, in miniemulsion process compared to the conventional process may provide yet another control strategy tool.

Microemulsion Polymerization

Microemulsion polymerization of monomer-in-water systems represent yet another degree of further subdivision of the monomer droplets, in this case the starting droplet size is in the range of 5-40 nm in-diameter. Polymerizations of styrene or methyl methacrylate in microemulsion systems were carried out using water-soluble (46-48), oil-soluble (46-50) chemical initiators or ultraviolet light initiation (51). A recent study investigated the effects of initiator concen-tration, polymerization temperature, and monomer concentration in styrene microemulsion polymerization on the kinetics, particle size distributions and molecular weight distributions (46). The polymerization profile showed some similarity to miniemulsion polymerization of styrene. The polymerization showed a long particle nucleation stage characterized by an increase in the polymerization rate until a maximum, which is reached at 20-25% conversion, compared to 30% conversion in miniemulsion processes and 2-15% in conventional process. The increased polymerization rate in the microemulsion relative to miniemulsion is due to the increased number of polymerization loci, as a result of radicals entry into microemulsion droplets. Interval I ends when all microemulsion droplets have disappeared either by becoming monomer-swollen polymer particles or, if not initiated, by diffusion of their monomers to active polymer particles. Interval II is marked by a decrease in the polymerization rate due to the decrease of monomer concentration inside the monomer-swollen polymer particles by propagation reactions. There was no constant rate period in the styrene microemulsion polymerization. The final microemulsion latex particles were small in size (20-30nm in-diameter), with broad distribution. The weight average and number average molecular weight of the polymer were high ($1-2 \times 10^6$), which imply that each microemulsion latex particle consists of two to three polystyrene molecules. A particle nucleation mechanism was proposed for microemulsion polymerization, similar to that of miniemulsion polymerization, which is dominated by initiation of polymerization by radical capture into the preformed microemulsion droplets.

THE ROLE OF SURFACTANTS IN EMULSION POLYMERIZATION

Surfactants play major roles in the emulsion polymerization process and in the formulation and application of latexes. In emulsion polymerization, the roles of surfactants are numerous; (i) solubiliza-tion of highly water-insoluble monomers, (ii) determines the mechanism of particle nucleation, (iii) determines the number of particles nucleated and thus the rate of polymerization, (iv) maintains colloidal stability during the particle growth stage, and (v) controls average particle size and the size distribution of the final latex system. The surfactants are also essential for stabilization of the latexes during post polymerization processes, such as stripping to remove residual monomers, formulation, storage, shipping and applications.

The underlying fundamental property of surfactant molecules in all these functions is their adsorption at the various interfaces,of the latex systems. The adsorption isotherms of surfactant molecules on the monomer droplets or the polymer particles are influenced by the following factors: The interaction between the particle surface and the surfactant molecules; the mutual interaction between surfactant molecules in the adsorbed layer; and the interaction between the ions in the bulk solution with the adsorbed layer. This section outlines the specific role of surfactants in determining the particle number in each one of the nucleation mechanisms described in the previous section.

Effect of Surfactant on Number of Latex Particles

Surfactants play a major role in determining the number of latex particles formed during the nucleation stage of the emulsion polymerization process. Each one of the nucleation mechanisms outlined in the previous section dictates a certain role for the surfactant during this stage. It should be noted that in reality more than one nucleation mechanism may be operating simultaneously in emulsion polymerization. Consequently, the actual role of the surfactant is more complex than outlined by each mechanism.

Several practical reasons were cited (52) which warrant exercising maximum control on the particle number during the course of an emulsion polymerization; these include management of heat transfer during the course of polymerization, polymer properties such as molecular weight and copolymer composition, latex stability, and reproducibility in latex properties. The profile of the particles number during the entire polymerization process depends on the number of particles generated during the nucleation stage as well their stability against flocculation. Once again, the main factors affecting the particle number density are the surfactant type and concentration, and the polymerization temperature. Equations 3 and 5, show that both polymerization rate and polymer molecular weight are proportional to the number of particles, i.e. inversely proportional to the particle size. A decrease in latex particle size increases the rate of polymerization. Since reactors in an industrial setting usually are operated under conditions of maximum cooling, an increase in the polymerization rate in the presence of excess monomer will cause a maximum temperature rise and consequently an increase in polymerization rate beyond what is permitted by the rate of heat removal. In addition, a decrease in particle size can sometimes lead to undesirable and uncontrolled increase in polymer molecular weight. On the other hand, an increase in particle size causes a decrease in the polymerization rate and a decrease in molecular weight. A decrease in polymerization rate will cause the reaction temperature to drop and as a result a monomer build-up in the reactor. This in turn leads to a dangerous run-away reaction and ultimately undesired copolymer composition and latex properties.

Micellar Nucleation. According to Harkins picture of emulsion polymerization, the role of surfactants is to provide the micelles, which when swollen with monomer become the main locii of particle nucleation upon entry of water-borne free radicals. During the particle growth stage, both particle volume and surface area increase, the role of the surfactant becomes critical in maintaining the colloidal stability of the particles. According to Smith and Ewart theory, the number of polymer latex particle formed is proportional to the surfactant concentration to the power 0.6. Indeed several studies confirmed that this relationship holds at least for sparingly water soluble monomer such as styrene (14,26,53).

The role of micelles in particle nucleation is still a controversial issue, since there are no direct way to to proof (or disproof) their function when present(30). Roe (21) indicated that micelles are not essential for the development of the relationship between particle number and the emulsifier concentration as given by Smith and Ewart. Indeed, he was able to derive the same equations for styrene polymerizations assuming initiation in the aqueous phase and provided that sufficient surfactant molecules are available for stabilization. Sutterlin (54) showed that for persulfate-initiated styrene emulsion polymerization, the 0.6 power dependence of number of particles on the surfactant concentration holds only at a surfactant concentration considerably higher than the cmc, whereas at lower concentration the exponent varies drastically within a relatively narrow surfactant concentration. It has been suggested that the increase in the number of particles just below the the cmc may be due to the presence of micelles, as a result of aggregation of surfactant molecules, or as a result of solution polymerization of styrene to form surface-active sulfate oligomer; both are capable of solubilizing monomer and serving as a locus for particle nucleation (55). The formation of micelles in aqueous surfactant solutions below the cmc has been established experimentally by measurements of the partial specific volumes (56).

Ugelstad et al. (26) studied the effect of the surfactant concentration on the number of particles in seeded and unseeded emulsion polymerizations of styrene and methyl methacrylate systems. At surfactant concentrations below the cmc, the presence of seed particles drastically decreased the number of new particles generated for both monomer systems. Above the cmc, on the other hand, the curves for the number of new particles formed in seeded and unseeded polymerizations coincided; both showing an increase in the number with increasing the surfactant concentration. These results were taken to indicate that when present, micelles play a key role in particle nucleation. Also, the results of Dunn (57,58) on homologous surfactants and Piirma (59,60) on mixed surfactants support the importance of micelles in particle nucleation of styrene polymerization.

Homogeneous Nucleation. For initiation in the aqueous phase, the main function of the surfactant is to stabilize the primary particles which are formed by precipitation of the oligomeric radicals from the aqueous phase. During the particle growth stage, both particle volume and surface area increase, the role of the surfactant becomes critical in maintaining the colloidal stability of the particles.

In homogeneous nucleation, the steady state particle number, which is reached after the nucleation period, is determined by the initiation rate in the aqueous phase modified by the rate of capture of the free radicals and oligomers by the particles, and the coagulation rate of the latex particles. The surfactant type and concentration influence both the nucleation period and the steady state number density of particles, through their effectiveness is stabilizing the primary particles and the latex particles against flocculation. The role of the surfactant is to stabilize greater number of particles and shorten the nucleation period by stabilizing particles at smaller size so that their degree of coalescence is reduced. The result is reflected in smaller average size of the latex particles and narrower particle size distributions. Fitch presented various practical scenarios for controlling the particle size distribution (narrow, broad, or bimodal), based on qualitative description of the influence of the surfactant concentration on the nucleation period and the particle coagulation (61).

Dunn et al (62) applied the Derjaguin, Landau, Verwey and Overbeek, (DLVO), theory of colloid stability (63,64), to the problem of aggregation during the course of emulsion polymerization. The adsorption of the surfactant on the particle surface causes an increase in the surface charge density, and thus a decrease in the coagulation rate of the particles. In the absence of surfactant, the primary particles coagulated at about the Smoluchowski's fastest rate. Upon coagulation and by capturing more charged oligomeric radicals, they become more stable. The addition of surfactant, cause particle stability at earlier stage in the polymerization reaction, by adsorption of the surfactant molecules on the particles, so that nucleation stops at a lower conversion level. They also noted that nucleation can occur throughout the entire polymerization reaction, however, the total number of the particles remains constant. This is due to the instability of the primary particles which are formed at the later stages of the polymerization, and their coagulation with the larger particles already present. Fitch indicated that bimodal particle size distribution can be obtained by addition of surfactant at some later stage of the polymerization, which can stabilize the new primary particles as they are formed (61).

In addition to surfactants, stabilization of the primary particles may also result from the initiator fragments, particularly if they are ionized, such as sulfate end-groups from persulfate initiators. In the emulsion polymerization of methyl methacrylate, Fitch et al (24), showed that the particle number was proportional to the surfactant concentration to the power 1.1 when persulfate-bisulfite-iron initiator was used. On the other hand, this dependency was 3.9 when the initiator system was hydrogen peroxide-iron, which indicates greater dependence upon the surfactant concentration in the nonionic system of initiator. This indicates that the sulfate end groups are more effective stabilizer than the hydroxyl groups, and contribute to the stabilization of the growing chain and the nucleated particles.

Sutterlin investigated systematically the influence of sodium lauryl sulfate concentration on the number of particles in two series of emulsion polymerizations of several acrylate and methacrylate monomers (54). He demonstrated that the relationship between the number of particles and surfactant concentration is not described by a single exponent. At a surfactant concentration below the cmc, the number of particles for both series of polymerizations was found to display a strong function of the equilibrium water-solubility of the monomer. The number of particles increased with increasing the monomer water-solubility, which is taken as an indication of the dominant role of homogeneous nucleation mechanism for particle formation. At a surfactant concentration above the cmc, the dependency of the number of latex particles on the surfactant concentration for a given monomer was described by a single exponent; its value was found to decrease with increasing the monomer solubility in water. For example, the value of the exponent decreased from 0.8 for ethyl acrylate(water solubility 0.54 mM), to 0.16 for methyl acrylate (water solubility 616 mM). The decrease in the exponent with increasing the monomer water solubility (or monomer polarity), was taken as an indication of increased particle agglomeration tendency. This increased agglomeration was attributed to the decreased packing density of surfactant molecules at the particle surface with increasing monomer hydrophilicity (65,66).

Coagulative Nucleation. The role of surfactants in this mechanism is to modify the coagulation rate coefficient between the precursor particles. The adsorption of surfactant molecules on the particles increase their surface charge density, which in turn increases the repulsive potential energy and reduces the coagulation rate coefficient. Adsorption isotherms of the surfactants were used to determine the surface concentration of the surfactants on the particles. Considerations were also given for the added surface charge density due to charged initiator fragments on the particles.

The rate of particle formation according to this theory is given (34) in terms of the the the rate of homogeneous nucleation and formation of primary precursors, based on Hansen-Ugelstad-Fitch-Tsai theory (26), the kinetics of coagulation among precursor particles, based on Smoluchowski-Mullert-Fuchs theory, with the coagulation rate coefficient being calculated based on DLVO theory modified to encompass heterocoagulation (64,67-69), and propagational growth.

According to coagulative theory, the effect of the surfactant concentration on the number of particles is explained qualitatively in terms of the coagulation rate coefficients B_{ij} which is given by:

$$B_{ij} = (4/3) \ (k_B \ T/\eta \ W_{ij}) \ (1 + r_i/r_j)^2 \ / \ (2 \ r_i/r_j) \qquad (7)$$

where k_B is the Boltzman's constant, T is the temperature, η is the viscosity of the medium, r_j is the radius of the particle, and W_{ij} is the Fuchs stability ratio, which is related to the height of the potential energy barrier that must be surmounted for coagulation to occur between the particles. The height of the potential energy barrier is related to the surface charge density of the particles, according to the DLVO theory. There are two contributions to the particle surface charge density: due to the ionic initiator fragments and the adsorbed surfactant molecules. At low surfactant concentration, most of the particle surface charge density is due to initiator fragments, and is independent of the surfactant concentration. Consequently, the coagulation rate constant is high and independent of the surfactant concentration, which results in a relatively low number of particles. As the surfactant concentration is increased, the surface charge density increases rapidly, and the coagulation rate coefficient decreases, with the result of more precursors grow by propagation to mature particles. With further increase in the surfactant concentration, a point is reached where the particle surface is now saturated and the surface charge density become independent of the surfactant concentration. Consequently, the coagula- tion rate coefficient and the particle number become independent of the surfactant concentration.

Nucleation in Miniemulsion Monomer Droplets. The role of the surfactant in this nucleation mechanism is the formation and stabilization of submicron size monomer droplets. The mixed emulsifier system used to prepare these submicron monomer emulsion droplets usually results in increased surface concentration of the surfactant molecules through some complex association or liquid crystals formation. This interfacial layer seems to retard the radical capture efficiency, as suggested by the relatively long nucleation stage (37,40,41,46).

The presence of liquid crystals in aqueous solutions of cetyl alcohol-sodium lauryl sulfate mixture has been proven using birefringence measurements and microscopy examinations under cross polarized light (70). An emulsification mechanism in these systems has been postulated to involve unidirectional swelling of these liquid crystals upon addition of the oil phase, which is diffused through the aqueous phase and become localized in the hydrophobic regions of these liquid crystals (71). At a certain concentration of the oil phase, the swollen liquid crystals break up forming very fine droplets, which grow by further diffusion of oil and/or by collisions. The presence of mixed emulsifier system at the oil/water interface cause stabilization of the emulsion droplets by a combination of electrostatic and steric mechanisms.

The mechanism of emulsification in the presence of hexadecane a co-surfactant, was not yet identified. However, Ugelstad (72) suggested that the stabilization of miniemulsion could be explained by the Higuchi and Misra concepts (73); i.e. stabilization by retardation of the interdroplet diffusion of the oil phase due to the presence of the relatively water-insoluble hexadecane (10^{-8}M), inside the droplets.

For miniemulsion polymerization, the fact that the monomer droplets are in the submicron range and most of the surfactant molecules already exist on the surface of the droplets prior to initiation of polymerization, usually results in a less dependence of the latex particle number on the surfactant concentration. In a comparative study on conventional and miniemulsion copolymerization of vinyl acetate-n-butyl acrylate, the number of particles in the miniemulsion case showed 0.23 power dependence on the sodium hexadecyl sulfate concentration, where as the conventional process showed a 0.68 dependence (41).

REFERENCES

1. El-Aasser, M.S. and Fitch, R.M., Eds., Future Directions in Polymer Colloids, NATO ASI, Series E, Applied Sciences No. 138; Nijhoff, The Hague, 1987.

2. Poehlein, G.W., Ottewill, R.H., and Goodwin, J.W., Eds., Science and Technology Of Polymer Colloids, (2 Volumes), NATO ASI, Series E, Applied Sciences No. 67, Nijhoff, Dordrecht, 1987.

3. Piirma, I., Ed., Emulsion Polymerization, Academic Press, New York, 1982.

4. Bassett, D.R. and Hamielec, A.E., Eds., Emulsion Polymers and Emulsion Polymerization, ACS Symp. Ser. No. 165, Washington, D.C., 1981.

5. El-Aasser, M.S. and Vanderhoff, J.W., Eds., Emulsion Polymerization of Vinyl Acetate, Applied Science, London, 1981.

6. Fitch, R.M., Ed., Polymer Colloids II, Plenum, New York, 1980.

7. Blackely, D.C., Emulsion Polymerization, Applied Science, London, 1975.

8. Bovey, F.A., Kolthoff, I.M., Medalia, A.L. and Meehan, E.J., Emulsion Polymerization, Interscience, New York, 1955.

9. V.I. Eliseeva, S.S. Ivanchev, S.I. Kuchanov, and A.V. Lebedev; Emulsion Polymerization and its Application in Industry, Consultants Bureau, New York, 1981.

10. Calvert, K.O., Ed., Polymer Latices and Their Applications, Macmillan Publishing Co., New York, 1982.

11. Harkins, W.D., _J. Am. Chem. Soc._, 1947, **69**, 1428; _J. Polym. Sci._, 1947, **5**, 217.

12. Smith, W.V. and Ewart, R.H., _J. Chem. Phys._, 1948, 16(6), 592.

13. Gerrens, H., _Ber Bunsenges. Phys. Chem._, 1963, **67**, 741.

14. Vanderhoff, J.W., in _Vinyl Polymerization II_, Ham, G.E., Ed., Dekker, New York, 1969, p. 210.

15. Parts, A.G., Moore, G.E. and Watterson, J.G., _Makromol. Chem._, 1965, **89**, 156.

16. Gardon, J.L., _J. Polym. Sci._, _Polym. Chem._, 1968, 6, 623;643;665;687;2853;2859; and 1971, **9**, 2763.

17. Harada, M., Nomura, M., Eguchi, W. and Nagata, S., _J. Appl. Polym. Sci._, 1972, **16**, 811.

18. Ugelstad, J., Mork, P.C. and Aasen, J.O., _J. Polym. Sci._, _Polym. Chem._, 1967, **5**, 2281.

19. Jacobi, B., _Angew. Chem._, 1952, **64**, 539.

20. Priest, W.J., _J. Phys. Chem._, 1952, **56**, 1077.

21. Roe, C.P., _Ind. Eng. Chem._, 1968, **60**, 20.

22. Kotera, A., Furusawa, K. and Takeda, Y., _Kolloid A. u. Z. Polymere_, 1970, **239**, 677; 1970, **240**, 837.

23. Goodwin, J.W., Hearn, J., Ho, C.C. and Ottewill, R.H., _Colloid Polym. Sci._, 1974, **252**, 464.

24. Fitch, R.M. and Tsai, C.H., in _Polymer Colloids_, Fitch, R.M. Ed., Plenum, New York, 1980, pp. 73, 103.

25. Fitch, R.M. and Shih, L-B, _Prog. Colloid Polym. Sci._, 1975, **56**, 1.

26. Hansen, F.K. and Ugelstad, J., in _Emulsion Polymerization_, Piirma, I., Ed., Academic Press, 1982, p. 51.

27. Ugelstad, J. and Hansen, F.K., _Rubber Chem. Technology_, 1976, **49**, 536.

28. Hansen, F.K. and Ugelstad, J., _J. Polym. Sci._, _Polym. Chem._, 1978, **16**, 1953.

29. Barrett, K.E.J., _Dispersion Polymerization in Organic Media_, Wiley, New York, 1975.

30. Napper, D.H. and Gilbert, R.G., _Makromol. Chem._, Macromol. Symp., 1987, **10/11**, 503.

31. Lichti, G., Gilbert, R.G. and Napper, D.H., _J. Polym. Sci._, _Polym. Chem._, 1983, **21**, 269.

32. Feeney, P.J., Napper, D.H. and Gilbert, R.G., *Macromolecules*, 1984, 17, 2520.

33. Feeney, P.J., Napper, D.H. and Gilbert, R.G., *J. Colloid Interf. Sci.*, 1985, 107, 159.

34. Feeney, P.J., Napper, D.H. and Gilbert, R.G., *Macromolecules*, 20, 2922.

35. Durbin, D.P., El-Aasser, M.S., Poehlein, G.W. and Vanderhoff, J.W., *J. Appl. Polym. Sci.*, 1979, 24, 703.

36. Ugelstad, J., El-Aasser, M.S. and Vanderhoff, J.W., *J. Polym. Sci. Poly. Lett.*, 1973, 111, 503.

37. Choi, Y.T., El-Aasser, M.S., Sudol, E.D. and Vanderhoff, J.W., *J. Polym. Sci., Polym. Chem.*, 1985, 23, 2973.

38. Azad, A.R.M., Ugelstad, J., Fitch, R.M. and Hansen, F.K., in *Emulsion Polymerization*, Piirma I. and Gardon, J.L., Eds, ACS Symposium Series, Vol.24, Am, Che. Soc., Washington, DC, 1976, p. 1.

39. Delgado, J., El-Aasser, M.S., Silebi, C.A. and Vanderhoff, J.W., *J. Polym. Sci., Polym. Chem.*, 1986, 24, 861.

40. Chamberlin, B.J., Napper, D.H. and Gilbert, R.G., *J. Chem. Soc. Faraday Trans.*, I, 1982, 78, 591.

41. Delgado, J., El-Aasser, M.S., Silebi, C.A. and Vanderhoff, J.W., *J. Polym. Sci., Polym. Chem.*, 1989, 27, 193.

42. Delgado, J., El-Aasser, M.S., Silebi, C.A. and Vanderhoff, J.W., accepted for publication, *J. Polym. Sci., Polym. Chem.* 1989.

43. Delgado, J., El-Aasser, M.S., Silebi, C.A., Vanderhoff, J.W. and Guillot, J., *J. Polym. Sci., Polym. Phys.*, 1988, 26, 1495.

44. Pelsynski, H.A., *Batch and Semicontinuous Coplymerization of Vinyl Acetate and Methyl Acrylate*, M.S. Report, Lehigh University, 1987.

45. Rodriguez, V.S., El-Aasser, M.S., Asua, J.M. and Silebi, C.A., accepted for publication, *J. Polym. Sci., Polym. Chem.*, 1989.

46. Guo, J.S., El-Aasser, M.S. and Vanderhoff, J.W., *J. Polym. Sci., Polym. Chem. Ed.*, 1989, 27, 691.

47. Johnson, P.L. and Gulari, E., *J. Polym. Sci., Polym. Chem.*, 1984, 22, 3967.

48. Rabagliati, F.M., Falcon, A.C., Gonzalez, D.A. and Martin, C., *J. Dispersion Sci. Technol.*, 1986, 7, 245.

49. Atik S.S. and Thomas, J.K., *J. Am. Chem. Soc.*, 1981, 103, 4279.

114

50. Jayakrishnan, A. and Shah, D.O., J. Polym. Sci., Polym. Chem., 1984, 22, 31.

51. Kuo, P.L., Turro, N.J., Tseng, C.M., El-Aasser, M.S. and Vanderhoff, J.W., Macromolecules, 1987, 20, 1216.

52. Kine B.B. and Redlich, G.H., in Surfactants in Chemical/ Process Engineering, Wasan, D.T., Ginn, M.E. and Shah, D.O., Eds., Marcel Dekker, Inc., 1988, 28, p.263-314.

53. Bartholome, E., Gerrens, H., Herbeck R. and Weitz, N.M., Z. Elektrochem., 1956, 60, 334.

54. Sutterlin, N., in Polymer Collids II, Fitch, R.M., Ed., Plenum, New York, 1978, p. 583.

55. Vanderhoff, J.W., J. Polym. Sci., Polym. Symps., 1985, 72, 161.

56. Bonner, F.J., Prepr. Org. Coatings Plastics Chem., 1980, 42, 181.

57. Al-Shahib, W.A. and Dunn, A.S., J. Polym. Sci., Polym. Chem., 1978, 16, 677.

58. Dunn, A.S. and Al-Shahib, W.A., in Polymer Colloids II, Fitch, R.M., Ed., Plenum, New York, 1980, p. 619.

59. Piirma, I. and Wang P.C. in Emulsion Polymerization, Piirma, I. and Gardon, J.L., Eds., Am. Chem. Soc. Symp., Series 24, Washington D. C., 1976, p. 34.

60. Piirma, I. and Chen, S.R., J. Coll. Interface Sci., 1980, 74, 90.

61. Fitch, R.M., in Science and Technology of Polymer Colloids, Poehlein, G.W., Ottewill, R.H. and Goodwin, J.W., Eds., Applied Sci., Series E: No. 67, Nijhoff, The Hague, 1983, p. 100.

62. Dunn, A.S. and Chong, L.C.H., Br. Polym. J., 1970, 2, 49.

63. Derjaguin, B. and Landau, L.P., Acta Physico. Chim., 1941, 14, 633.

64. Verwey, E.J.W. and Overbeek, J.Th.G., Theory of the Stability of Lyophobic Colloids, Elsevier, Amsterdam 1948.

65. Yeliseyeva, V.I. and Zuikov, A.V., in Emulsion Polymerization, Piirma, I. and Gardon, J.L., Eds., ACS Symp. Ser., No. 24, Washington D.C., 1976, p.62.

66. Vijayendran, B.R., J. Appl. Polym. Sci., 1979, 23, 733.

67. Ottewill, R.H., in Emulsion Polymerization, Piirma, I., Ed., Academic Press, New York, 1982, p. 1.

68. Fuchs, N., Z. Physik, 1934, 89, 736.

69. Hogg, R., Healy, T.W., Fuerstenau, D.W., _Trans. Faraday Soc._, 1966, **62**, 1638.

70. Lack, C.D., El-Aasser, M.S., Silebi, C.A., Vanderhoff, J.W. and Fowkes, F.M., _Langmuir_, 1987, 3, 1155 (1987).

71. Lack, C.D., _Emulsion Formation and Stabilization with Mixed Emulsifier Liquid Crystals_, Ph.D. Dissertation, Lehigh University 1985.

72. Ugelstad, J., _Makromol. Chem._, 1978, **179**, 815.

73. Higuchi, I. and Misra, J., _J. Pharm. Sci._, 1962, **51**, 459.

RADICAL CAPTURE EFFICIENCIES IN EMULSION POLYMERIZATION KINETICS

IAN A MAXWELL, BRADLEY R MORRISON, ROBERT G GILBERT AND DONALD H NAPPER
Departments of Physical and Theoretical Chemistry,
Sydney University,
NSW 2006, Australia

ABSTRACT

Experimental data and models are presented for initiator efficiency in emulsion polymerization systems in the absence of particle formation. The data show that a number of models are inapplicable, *viz.*, those assuming that the rate–determining step for free radical entry into a particle is either diffusional capture, surfactant displacement, or colloidal entry. The data however support the model (first suggested by Priest) of aqueous phase propagation to a critical degree of polymerization, whereupon capture of the resulting oligomeric free radical by a particle is instantaneous. Mutual aqueous phase termination of smaller species also occurs; one must take account of the fact that the rate coefficient for this is some two orders of magnitude greater than that for low molecular weight species in polymeric systems. This model is in quantitative and qualitative accord with the experimental dependences of the entry rate coefficient on the concentrations of initiator, of surfactant, of aqueous phase monomer, and of latex particles, as well as on particle size, on ionic strength and on temperature.

INTRODUCTION

Although it is often assumed to the contrary, it is now well established that radical capture efficiencies in emulsion polymerization systems can be very low: 1% is not atypical [1,2]. The observation of low efficiencies comes from extensive studies [2] of the kinetics of seeded growth, in Interval II (absence of particle formation and constant monomer concentration), employing (a) the steady–state rate, (b) the rate of approach to steady state, and (c) the rate of relaxation from steady state (the latter with polymerization initiated by γ radiolysis). The combination of all these data, in favorable cases, overdetermines the system, and enables one unambiguously to establish the rate coefficient for entry. Given that it is a gross error to assume 100%

capture efficiency in emulsion polymerization systems, a reliable model is required for this efficiency: i.e., for the pseudo—first—order rate coefficient ρ for the entry of free radicals into a latex particle. Because extensive and reliable data for ρ are now available, one can now test models in a way that can actually refute some mechanistic postulates which have been made.

ENTRY MODELS

The basic events for the entry of a free radical into a latex particle must include some or all of the following: creation of a free radical from aqueous—phase initiator decomposition; aqueous phase propagation of this free radical; attachment of the resulting species to the particle surface; penetration of the free radical species from the surface into the interior of a latex particle (i.e., into a region of the particle where the monomer concentration is relatively high); desorption of the species from the surface before penetration into the interior can occur; and aqueous—phase termination of the various free radical species involved. Various entry models, as follows, are based upon suppositions as to which of these might be rate determining.

Diffusion control

If the sole rate—controlling step is diffusion of the free radical to the particle surface [3,4], the kinetics can be quantified in terms of the following events [2,4,5]:

$$\text{initiator decomposition: I–I} \rightarrow 2\text{R}^{\bullet} \text{ (rate coefficient } k_d) \tag{1}$$

$$\text{aqueous phase termination: } 2\text{R}^{\bullet} \rightarrow \text{termination (rate coefficient } k_t) \tag{2}$$

$$\text{entry: R}^{\bullet} + \text{particle} \rightarrow \text{entry (rate coefficient } k_e) \tag{3}$$

with the second—order entry rate coefficient $k_e(\text{diffusion}) = (k_B T/3\eta)(r_S/r_R)$, k_B being Boltzmann's constant, η the viscosity of the medium, r_S the swollen radius of the particle, and r_R the radius of the entering species. One then has [4,5]:

$$\rho = (k_e^2/2k_t)[\{N^2 + (2k_t k_d[I]/k_e^2)\}^{\frac{1}{2}} - N] \tag{4}$$

where N is the latex particle concentration.

Surfactant displacement

It has been suggested [6] that displacement of surfactant from the particle surface is rate—determining. In that case, the foregoing mechanism must be expanded to include surface adsorption and desorption of the free radical $\text{R}^{\bullet}$, and a rate coefficient for radical absorption onto the surface, k_{ab}, which equals that for surfactant desorption

[7]. This gives an expression which is the same as eq 4, except that k_e therein is replaced by k_{ab}/N.

Colloidal entry

It has been suggested [5] that the entering species is a large oligomer, whose entry rate coefficient is governed by colloidal (DLVO type) considerations. If this is the sole rate-determining step, then eq 4 again holds, with k_e therein obtained from DLVO theory. This depends on the surface charge density and size of the entering colloidal species.

Aqueous phase propagation control

We shall treat this mechanism in more detail than those preceding, since it will emerge as the only one which is qualitatively and quantitatively consistent with all available data. Priest [8] suggested that addition of monomer units to the primary free radical is necessary to effect entry (i.e., to gain surface activity); entry is negligible if the aqueous phase free radical is below a critical degree of polymerization z, and instantaneous for degree of polymerization z (and greater). The quantification of this, given below, is virtually identical to that for the HUFT [4,9] model for homogeneous nucleation. Thus eq 1 is now replaced by:

$$I-I \rightarrow 2I^{\cdot} \text{ (rate coefficient } k_d) \tag{5}$$
$$I^{\cdot} + M_{aq} \rightarrow IM_{aq}^{\cdot} \text{ (rate coefficient } k_p^{\cdot}) \tag{6}$$

Here M_{aq} is monomer in the aqueous phase. The free radical formed from eq 6 may need to undergo further propagation until a critical degree of polymerization z:

$$I(M_{aq})_i^{\cdot} + M_{aq} \rightarrow I(M_{aq})_{i+1}^{\cdot} \text{ (rate coefficient } k_p) \tag{7}$$

(in actuality, the rate coefficients for eq 7 will depend, for small i, on the degree of polymerization i; this dependence is replaced here by the step function represented by eqs 6 and 7). In addition, one may have termination between all aqueous phase species, represented as:

$$I(M_{aq})_i^{\cdot} + T^{\cdot} \rightarrow \text{ termination} \tag{8}$$

where $T^{\cdot}$ is any aqueous phase free radical. Lastly, when an aqueous phase free radical achieves a degree of polymerization z, it is assumed to enter a particle instantly (i.e., the processes of attaching to and penetrating a particle are assumed not to be rate determining). Thus:

$$\rho = k_p[M_{aq}][I(M_{aq})_{z-1}{}^\bullet]/N \tag{9}$$

Since there is experimental evidence [11] that $k_p^\bullet \gg k_p$, eq 6 is not rate–determining. Thus one has the following rate equations for this model:

$$d[IM_{aq}{}^\bullet]/dt = 2k_d[I] - k_t[M_{aq}][T^\bullet] - k_p[IM_{aq}][M_{aq}] \tag{10}$$

$$d[I(M_{aq})_i{}^\bullet]/dt = k_p[M_{aq}]([I(M_{aq})_{i-1}{}^\bullet] - [I(M_{aq})_i{}^\bullet]) - k_t[I(M_{aq})_i{}^\bullet][T^\bullet] \tag{11}$$

$$[T^\bullet] = \sum_i [I(M_{aq})_i{}^\bullet] \tag{12}$$

The contribution of $I^\bullet$ to $T^\bullet$ can be neglected because eq 6 is so rapid. Eqs 10–12 can be solved using the steady–state approximation. This solution can be numerical, by iteration from a guessed initial value of (say) $[T^\bullet]$. Alternatively, one may use the result [4] that $[T^\bullet] = ((2k_d[I]-k_p[M_{aq}][I(M_{aq})_{z-1}{}^\bullet])/k_t)^{\frac{1}{2}}$ and . assuming (as is reasonable for all except a highly water–soluble monomer) the approximation that [4] $2k_d[I] \gg k_p[M_{aq}][I(M_{aq})_{z-1}{}^\bullet]$; in this case one obtains the following analytic expression:

$$\rho = (2k_d[I]/N)\left[(k_t[T^\bullet]/k_p[M_{aq}])+1\right]^{1-z} \tag{13}$$

where $[T^\bullet] \simeq (2k_d[I]/k_t)^{\frac{1}{2}}$. Note that while the numerical evaluation of eqs 10–12 is trivially accomplished, it is also found that eq 13 is quite accurate, even for a monomer such as vinyl acetate with a high k_p and fairly high M_{aq}.

TESTS OF MODELS

The various models discussed above will now be compared with extensive data on the seeded emulsion polymerization of styrene with persulfate initiation. Note that extracting the value of ρ from the data also yields the rate coefficient for radical exit (desorption); the value for this quantity so obtained was found to be in good accord with that obtained from γ radiolysis relaxation experiments [2], suggesting the correctness of the data interpretation.

Diffusion control

Evaluation of ρ from eqs 1–4 gives values that are orders of magnitude greater than those observed experimentally [5], even using values for the various rate parameters at the extremes of their likely range. This mechanism can therefore be discounted on quantitative grounds.

Surfactant displacement

This mechanism implies that ρ should depend on the surface coverage of the latex by surfactant. However, experiments [7] wherein the surface coverage was varied from 25 to 100% showed no significant change in ρ. This qualitatively refutes the supposed mechanism.

Colloidal entry

This mechanism [5] can be discounted for several separate reasons [7]. (1) Firstly, calculation of k_e using an exact [11] solution of the Poisson–Boltzmann equation for the coagulation of (say) a 5 nm oligomeric free radical with a latex particle gives a value of k_e that is diffusion controlled (ca. $10^9 - 10^{10}$ dm^3 mol^{-1} s^{-1}). As with the diffusional entry mechanism, this gives a grossly incorrect ρ, thereby refuting the mechanism on quantitative grounds. Note that the very large value for the hetero-coagulation rate coefficient (even though the surface charge density of the two species is quite high) arises because of partial attractions that the postively charged clouds of counter–ions (in the double layer of the larger particles) exert on the negative surface charges of the smaller particles. (2) The mechanism would predict that the value of ρ would change with surface charge; as stated, no change in ρ is observed with large changes in the surfactant coverage, thereby refuting the mechanism qualitatively. (3) Lastly, it is found that changing the ionic strength (while keeping all other factors constant) has no effect on ρ in a typical system. Since the coagulation rate should be altered under such circumstances, this again refutes the postulate qualitatively.

Aqueous phase propagation control

In testing this mechanism against experiment, it is essential to note that the oligomeric species involved are likely to be of low degree of polymerization: perhaps 2 – 4 for styrene. Now, it has been established experimentally [12] that the value of k_t for such species in the aqueous phase is high: e.g., 7×19^9 dm^3 mol^{-1} s^{-1} for acrylonitrile oligomers at 50°C. While equivalent data for the monomer in the present study (styrene) are not available, this value is adopted pro tem., since (being a diffusion controlled process) it should not be a strong function of monomer "chemistry". Now, this large value of k_t is some two orders of magnitude less than the "zero–conversion" value obtained from bulk and solution polymerization experiments. This apparent inconsistency can in fact be understood when it is realized that, in the latter system, species of degree of polymerization in the range 2 – 4 are in vanishingly small concentrations, and are kinetically insignificant (simply because they propagate very rapidly to much larger chains). The zero–conversion bulk/solution k_t values are in fact for chains sufficiently large that termination is controlled by segmental diffusion, rather than centre-of-mass diffusion. This degree of polymerization (say, $d_p \leqslant 50$) is detectable because the total population of chains with $d_p \leqslant 50$ is sufficiently high that there is a significant overall rate of termination in a bulk/solution system at zero conversion, whereas the rate of termination between (say) dimers is completely negligible in these systems. Such is not the case for the aqueous phase in question, since *only* very low d_p values are present.

The model is *qualitatively* consistent with all the general observations discussed above. For example, it assumes that the adsorption onto, and penetration into, the particle are *not* rate determining; this is consistent with the observation that changing the surface characteristics has no effect on ρ (within experimental uncertainty), at least for the styrene results discussed above.

The model was tested *quantitatively* by fitting eqs 9–12 against a range of experimental data. The parameters involved are k_d, k_t and z, the rest being established from other experiments [2]. The value of k_t was taken as described above. The value of k_d may depend on pH, and on surfactant and monomer concentrations. A value of $k_d = 1.6\times10^{-6}$ s^{-1} has been reported for persulfate in pure water at 50°C [13]. Because of the aforementioned uncertainties, k_d was also treated as a free parameter and fitted to the data, with the constraint that it should

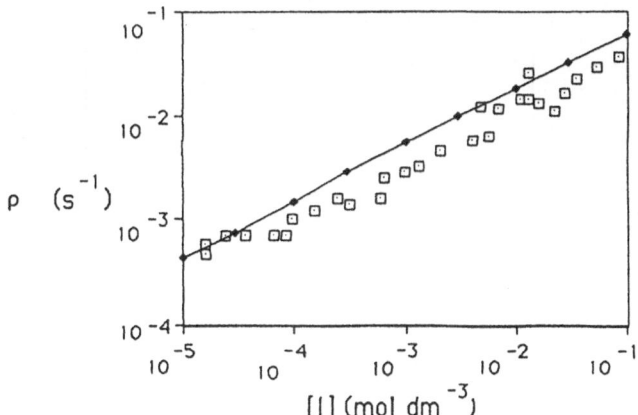

Figure 1. Variation of ρ with $K_2S_2O_8$ initiator concentration for seeded styrene emulsion polymerization at 50°C. Particle concentration 4.9×10^{16} dm^3; $r_S = 79$ nm; all data in Interval II, with $[M_{aq}] = 4.3\times10^{-3}$ mol dm^{-3}. Points: experiment [1,2]; line, fitted from eqs 9–12 with parameters as in text.

not deviate too far from the value just quoted. This gave $k_d = 0.8\times10^{-6}$ s^{-1} at 50°C. The observed activation energy [13] was employed to give k_d at other temperatures from this value. The optimal value of z was found to be z=2, although moderate fits could be found for z=3. A value of z=2 is physically reasonable: z=1 corresponds to styrene sulfonate, which is relatively soluble in water, whereas the species with z=2 (containing 16·carbons) would be strongly surface active.

We now show that eqs 9–12 fit a very wide range of data types. Figure 1 shows a typical fit to results for the variation of ρ with initiator concentration, all other quantities in the system being held fixed. Figure 2 shows similar data, except that now the swollen particle radius r_S is a factor of three larger than that in Figure 1, and there is also a large difference in particle concentration. In both cases an excellent fit is observed; note that these two data sets contain more independent information than the number of adjustable parameters in the model.

The observed temperature dependence of ρ [7] can also be fitted very well. The values of k_d for different temperatures were as discussed above. In addition, k_t, being diffusion controlled, was assumed temperature independent; k_p was given its observed temperature dependence [2], and the comparatively small temperature dependence of $[M_{aq}]$ was assumed to be negligible. One thus finds, from eqs 9–12, $\rho(s^{-1}) = 10^{1.5}exp(-25$ kJ mol^{-1}/RT), for the conditions of the experimental data [7]. The experimental result is $\rho(s^{-1}) = 10^{1.8\pm0.5}exp([-24\pm3]$ kJ mol^{-1}/RT). Again, the

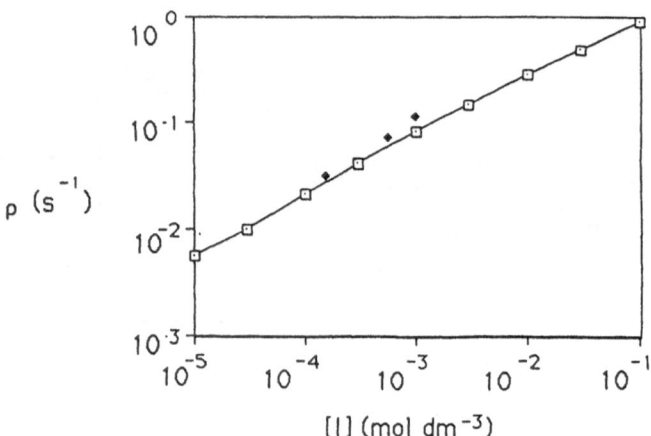

Figure 2. Variation of ρ with $K_2S_2O_8$ concentration for seeded styrene emulsion polymerization at 50°C. Particle concentration 3.2×10^{15} dm^3, $r_S = 240$ nm. Points: experimental [14]; line, fitted from eqs 9–12 with parameters as in text.

accord is good.

Figure 3 shows the computed and observed variation of ρ with particle number, this time for a seed with $r_S = 56$ nm. The agreement is not now as good. However, by increasing z from 2 to 3, the model can be made to give good accord with the data. This discrepancy can be assigned a number of causes. (a) It was assumed that the variation of k_p with d_p could be ignored for $d_p > 1$: $k_p^{\bullet}$ (dm^3 mol^{-1} s^{-1}) $\simeq 10^9$ ($d_p = 1$), $\simeq 250$ ($d_p > 1$), whereas in actuality $k_p(d_p)$ is more likely to decline from a very large to a small value over a wider range of d_p. (b) For sufficiently small particles, attachment and/or penetration must become rate determining; since the particles in this particular system are comparatively small, perhaps this event may start to be rate-determining for the conditions of Figure 3.

Finally, Figure 4 shows the variation of ρ with w_p, the weight fraction of polymer in the latex. This is a particularly stringent test of the model, for the following reason. The model implies that ρ should show a strong dependence on $[M_{aq}]$. This latter quantity can be varied by changing w_p, without changing any other quantities in the system; theory and experiment are then compared. In this comparison, it is necessary to determine $[M_{aq}]$ as a function of w_p experimentally, since partitioning is likely to be non-ideal. The measurements were carried out by GC analyses of monomer in both the whole system and a dialysis bag, initially filled

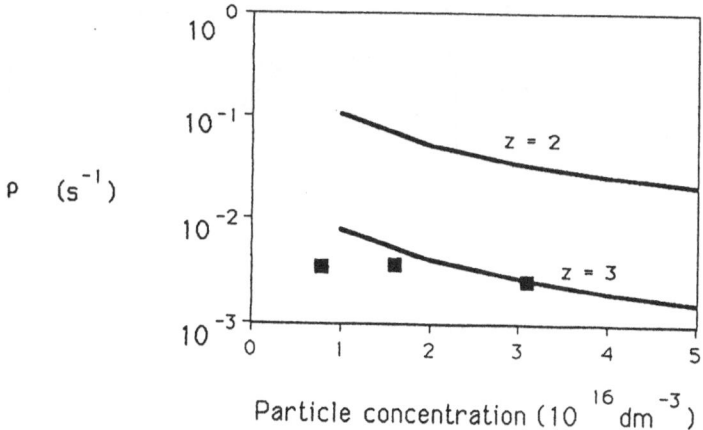

Figure 3. Variation of ρ with particle concentration for seeded styrene emulsion polymerization at 50 °C, with [I] = 1.3×10^{-2} mol dm^{-3}, $r_S = 56$ nm; Interval II data. Points: experimental [15]; line, fitted from eqs 9–12 with parameters as in text.

with pure water, and allowed to equilibrate with the latex. It was found for styrene that the particle:aqueous–phase monomer concentration obeyed the same (non–ideal) 0.6 exponent observed earlier with MMA [16], rather than the exponent of unity which would hold for an ideal system. Using the experimental [M_{aq}], observed [17] and calculated results for ρ shown in Figure 4 are in good agreement.

CONCLUSIONS

The results discussed here are able to refute a number of postulated mechanisms for free radical entry into latex particle: the rate–determining step is none of diffusion, colloidal coalescence, or surfactant displacement. A qualitatively and quantitatively wide range of data is able to be fitted by assuming that the rate–determining step is aqueous phase propagation to a critical degree of polymerization, as suggested *inter alia* by Priest, Fitch, Tsai, Ugelstad and Hansen. The quantification of the model must take into account the fact that aqueous phase termination of the small oligomers occcurs with a much larger rate coefficient than for "zero conversion" in solution or bulk polymerization.

It must be noted that the free radical attachment to, or penetration of, a particle, while apparently not rate determining in Intervals II or III, must become

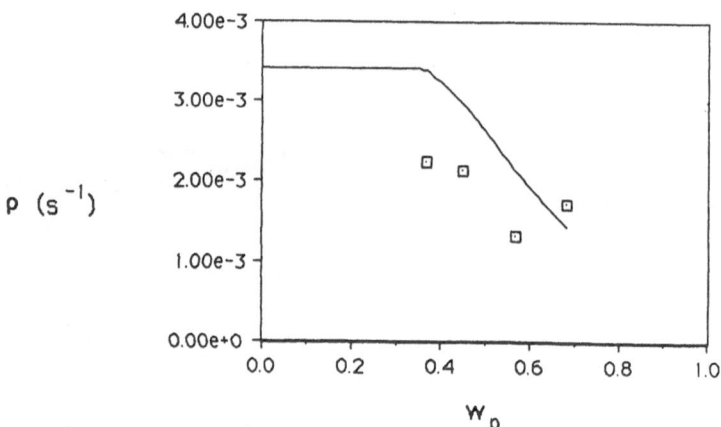

Figure 4. Variation of ρ with w_p for seeded styrene emulsion polymerization at 50°C. Particle concentration 5×10^{16} dm^3; r_S = 76 nm; [I] = 5×10^{-4} mol dm^{-3}. Points: experimental [17]; line, fitted from eqs 9–12 with parameters as in text, including directly–measured [M_{aq}].

significant in Interval I (i.e., during particle formation). This is an important area for future investigation.

The financial support of the Australian Research Council is gratefully acknowledged.

REFERENCES

1. Hawkett, B.S., Napper, D.H. and Gilbert, R.G., J. Chem. Soc., Faraday Trans. I, 1980, 76, 1323.

2. Gilbert, R.G. and Napper, D.H., J. Macromol. Sci.-Rev. Macromol. Chem. Phys., 1983, C23, 127.

3. Gardon, J.L., J. Polym. Sci. A1, 1968, 6, 2853.

4. Ugelstad, J. and Hansen, F.K. In Emulsion Polymerization, ed. I Piirma, Academic Press, New York, 1982, pp 55–93.

5. Penboss, I.A., Gilbert, R.G. and Napper, D.H., J. Chem. Soc., Faraday Trans. I, 1986, 82, 2247.

6. Yeliseeva, V.I. In Emulsion Polymerization, ed. I Piirma, Academic Press, New York, 1982, pp 247–288.

7. Adams, M.E., Trau, M., Gilbert, R.G., Napper, D.H. and Sangster, D.F., Aust. J. Chem., 1988, 41, 1799.

8. Priest, W.J., J. Phys. Chem., 1952, 56, 1077.

9. Fitch, R.M. and Tsai, C.H. In "Polymer Colloids", ed. R.M. Fitch, Plenum, New York, 1971.

10. Marathamutu, P., Makromol. Chem. Rapid Comm., 1980, 1, 23.

11. Barouch, E. and Matijevic, E., J. Chem. Soc., Faraday Trans. I, 1985, 81, 1797, 1819.

12. Dainton, F.S. and James, D.G.L., J. Polym. Sci., 1959, 39, 299; Dainton, F.S. and Eaton, R.S., J. Polym. Sci., 1959, 39, 313.

13. Kolthoff, I.M. and Miller, I.K., J. Am. Chem. Soc., 1951, 73, 5118.

14. Adams, M.E., PhD thesis, Sydney University, 1988; Adams, M.E., Russell, G.T., Napper, D.H., Gilbert, R.G. and Sangster, D.F., to be published.

15. Whang, B.C.Y., Napper, D.H., Ballard, M.J., Gilbert, R.G. and Lichti, G., J. Chem. Soc. Faraday Trans. I, 1982, 78, 1117.

16. Ballard, M.J., Napper, D.H. and Gilbert, R.G., J. Polym. Sci., Polym. Chem. Edn., 1984, 22, 3225.

17. Leslie, G.L., Gilbert, R.G. and Napper, D.H., unpublished data.

THE EFFECT OF EMULSIFIER-POLYMER COMPLEX FORMATION ON PARTICLE NUCLEATION IN EMULSION POLYMERIZATION

W.A.B. DONNERS
DSM Research, P.O. Box 18, 6160 MD Geleen, Netherlands
B. MIDGLEY
Dept. Pharm. Chem., Univ. Bradford, Bradford BD7 1DP, UK

ABSTRACT

Formation of water insoluble complexes between polyethyleneoxide (PEO) and polycarboxylic acids in aqueous solutions is a well known phenomenon. In certain emulsion polymerizations comprising carboxylic comonomers and ethoxylated emulsifiers circumstances are prevalent that favor a similar complexation between emulsifiers and carboxylic copolymers.
In a previous model study of PEO-methylmethacrylate/methacrylic acid systems (J. Polym. Sci. Polym. Lett. Ed., (1987) 25, 29) it has been shown that complexation effects affect particle formation. In the present paper more evidence from kinetic and conversion measurements is presented to support previous observations. The unexpected finding that the presence of PEO during the emulsion polymerization of MMA also affects particle size is shown to be caused by (partial) incorporation of PEO inside the PMMA later particles. Furthermore, evidence is presented that strongly suggests grafting of PMMA onto PEO to be absent in these systems.

INTRODUCTION

In the emulsion polymerization of acrylic monomers acrylic acid or methacrylic acid is often one of the comonomers, and ethoxylated emulsifiers are also used extensively. At the start of the polymerization thus conditions prevail (high temperature, low pH) that may favor formation of water in soluble complexes similar to those described to be formed between polyethyleneoxide (PEO) and polycarboxylic acids (1-3).
In a model study of the system PEO - methacrylic acid/methylmethacrylate (MAA/MMA) (4) it was shown that the presence of PEO during the emulsion polymerization of MAA/MMA mixtures with molar ratio varying from 0/100 to 100/0 led to larger particle sizes as compared with PEO-free recipes.

Apparently PEO causes flocculation of primary particles or successive entrapment of PEO and acid rich copolymer by existing particles via complexation. Surprisingly also in pure MMA recipes PEO increased the particle size of the latex.

It is the aim of this paper to present results, obtained by various techniques, that enable more definite conclusions to be drawn concerning the role of PEO in MAA/MMA polymerizations.

EXPERIMENTAL

The experimental details of the preparation of the latices and particle size measurements have been described elsewhere (4). Conversion was followed by gas chromatographic determination of residual monomer. Tg measurements of MAA/MMA copolymers were performed by DSC on dried latex samples.

RESULTS AND DISCUSSION

Particle size of MAA/MMA = 80/20 latices and the time final particle size is reached during polymerization increases with PEO concentration present in the recipe. If the time required to reach a plateau value for the particle size is taken as a measure for the moment that monomer conversion reaches its final value, then this reflects a retarding effect of PEO.

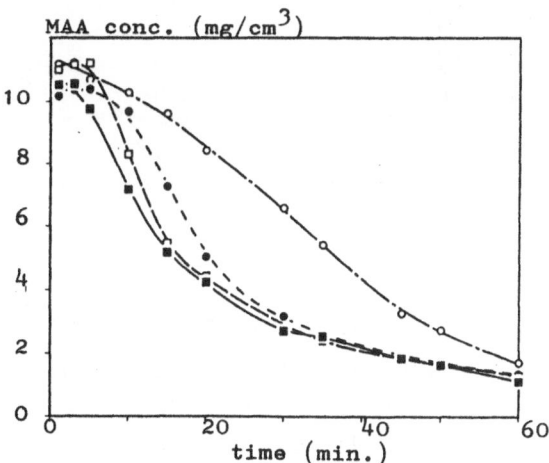

Figure 1. Residual MAA concentration vs. time during polymerization of MAA/MMA = 80/20 latices for various PEO concentrations. Filled squares: no PEO; open squares, filled circles, open circles = EO/monomer molar ratio is 32, 8 or 1 respectively. PEO molar mass = 20,000.

This retarding effect is also obvious from fig. 1 showing MAA conversion, and can be explained by coagulation of primary particles or entrapment of oligomers by existing particles via complexation.

Both processes lead to a lower number of active polymerizing centres and therefore a slower rate of conversion and a larger final particle size. MMA conversion in the systems of fig. 1, though somewhat faster, shows a similar trend. In preliminary experiments PEO concentration in the aqueous phase has been found to decrease at a similar rate as the free monomer concentration, showing that PEO is taken up by the latex particles.

Fig. 2 shows results for pure MMA recipes. It is clear that presence of PEO leads to a larger particle size as compared to the blank, much the same as has been found for high acid concentration recipes. This suggests that PEO retards the polymerization of pure MMA, but the kinetics of the polymerization of MMA in the presence of or absence of PEO did not show any difference.

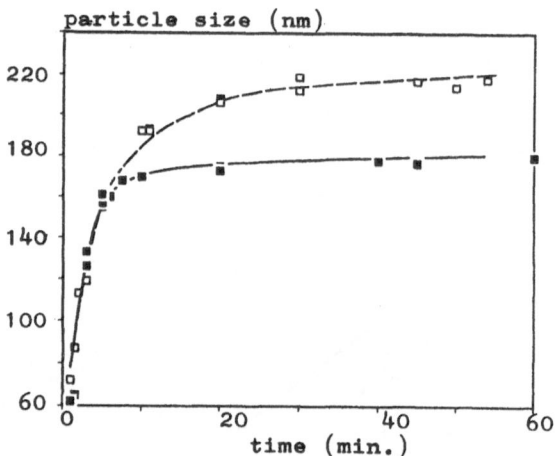

Figure 2. Particle size vs. time during the polymerization of MMA in the absence (filled squares) or presence (open squares) of PEO (EO/monomer molar ratio = 1). PEO molar mass = 20,000.

The difference in particle size of PMMA latices prepared in the presence and in the absence of PEO is too large to be explained by PEO adsorption. Tg measurements (shown in Table 1) on various samples after cleaning by serum replacement and drying suggest that PEO cannot be removed completely from the latex particles by serum replacement if its molar mass is sufficiently large. In those cases the PEO is taken up by the particles by some

mechanism. If grafting would be involved, one would expect a steric contribution to the stabilization mechanism of PMMA latices prepared in the presence of PEO even after removal by serum replacement of possibly adsorbed
PEO. However, critical coagulation concentrations determined for Na_2SO_4 of
such latices were similar to those of PMMA latices prepared in the absence
of PEO, which suggests that grafting is not involved in these systems.

TABLE 1

Tg of PMMA prepared by emulsion polymerization in the presence or absence of
PEO (after serum replacement and drying). Monomer/EO molar ratio is 1 in
samples B through E

Sample	PEO molar mass	PEO present during pol.	PEO added after pol.	Tg °C
A	-	no	no	121
B	600	yes	no	120
C	600	no	yes	122
D	20,000	yes	no	110
E	20,000	no	yes	124

The mechanism by which PEO is taken up by the PMMA particles is not clear
yet. Possibly PEO adsorbs to the growing PMMA-particles strongly enough to
be 'overgrown' by growing oligomers depositing from the aqueous phase on top
of the particles.
Particle sizes of 5/95 and 10/90 MAA/MMA latices develop in much the same
way as for PMMA latices. In these systems however there is a definite retardation by PEO. One might argue that 5 % of MAA should not be sufficient to
cause complex formation of the copolymer with PEO. The findings of Saito and
Fujiwara (5) and Thompson and McEwen (6) however support the hypothesis that
complex formation can occur. Furthermore we have found from kinetic measurements, that MMA polymerizes faster in these systems than MAA. For the 5/95
system for instance the MAA/MMA ratio shifts finally to 25/75. Such a latex,
prepared without PEO, appears to have a steric contribution to its stability, as can be expected if partially water soluble polymer is present at
the particle surfaces. XPS measurements from particle surfaces obtained
after freeze etching of a droplet of a cleaned 5/95 latex indeed show a
25/75 MAA/MMA monomer composition of the particle surfaces.
Preliminary results obtained with nonionic emulsifiers instead of PEO show
that the hydrophobic moiety generally reinforces the effects of complexation

in high acid concentration recipes. In low acid concentration recipes smaller particle sizes are obtained as compared with systems prepared with PEO, due to stabilization of (primary) particles by the emulsifier molecules.

In addition for higher molar mass emulsifier similar trends as for PEO's of comparable molar mass are observed.

CONCLUSIONS

PEO influences particle formation in MAA/MMA emulsion polymerization. In high acid concentration recipes this leads to larger particle sizes accompanied by retardation effects which are brought about by coagulation of particle and/or oligomers through complexation. Low acid concentration recipes show some retardation as well. PMMA polymerization is not retarded by PEO, but presence of PEO does lead to larger particles. PEO is taken up by the PMMA particles if its molar mass is high enough. In the latter effect grafting does not play a role.

REFERENCES

1. Tsuchida, E., Osada, Y. and Ohno, H., J. Macromol. Sci. Phys. B., (1980), 17, 683.

2. Bekturov, E. and Bimendina, L.A., Adv. Polym. Sci., (1981), 41, 99.

3. Kabanov, V.A. and Papisov, I.M., Polym. Sci. USSR, (1979), 21, 261.

4. Donners, W.A.B., J. Polym. Sci. Polym. Lett. Ed., (1987), 25, 29.

5. Saito, S. and Fujiwara, M., Colloid Polym. Sci., (1977), 255, 1122.

6. Thompson, L. and McEwen, A., J. Colloid Interface Sci., (1983), 93, 329.

EMULSION POLYMERIZATION OF BUTADIENE

PIERRE WEERTS, JOS VAN DER LOOS and ANTON GERMAN
Department of Polymer Chemistry,
Eindhoven University of Technology,
P.O. Box 513, 5600 MB Eindhoven, NL

ABSTRACT

The kinetics of the *ab initio* emulsion polymerization of butadiene
was investigated using several emulsifiers and dissociative
initiators. The steady polymerization rate in interval II was
found to be highly insensitive to variations of the initiator
concentration, suggesting a low initiator efficiency.
The development of particle number as a function of conversion
at several emulsifier concentrations, shows that limited
coagulation of latex particles is occurring. Increasing the
cation content in the recipe promotes coagulation and leads to
a higher value of the emulsifier dependency exponent towards
particle number. The average number of radicals per particle
was found to increase with particle size, while values
significantly less than 0.5 were calculated, indicating a
first-order radical loss process.

INTRODUCTION

Despite the great industrial importance of butadiene

(co)polymers prepared by emulsion polymerization (e.g. high

impact materials, synthetic rubbers, coatings, adhesives), very

little has been reported about the kinetics and mechanisms of

the emulsion (co)polymerization of butadiene [1,8]. The present

paper reports some kinetical aspects of the emulsion

polymerization of butadiene using dresinate 214 and sodium

dodecylsulfate as emulsifiers, and three dissociative

initiators, namely sodium or potassium persulfate (PPS),
4,4'-azo-bis-4-cyanopentanoic acid (ACPA) and 2,2'-azo-bis-
isobutyronitrile (AIBN).

The recipes and reaction conditions are given in Table 1. All
materials used are of high purity quality, except t-dodecyl
mercaptan which is a distillation fraction of C_{12}-isomers, and
dresinate 214 which is a disproportionated mixture of abietic
acid-type of derivatives with dehydro-, dihydro- and
tetrahydroabietic acid as main components and less than 0.2 %
abietic acid. From the gravimetrically determined conversion
data, and the average particle diameter measured with dynamic
light scattering (DLS, Malvern IIc) and transmission electron
microscopy (TEM, Philips 420 and Jeol 2000 FX), the particle
number and the average polymerization rate per particle were
calculated. Experimental details are given elsewhere [9].

TABLE 1
Standard polymerization recipes in parts by weight.

Ingredient	Recipe 1	Recipe 2
water	230	230
butadiene	100	100
dresinate 214	7.6	
sodium dodecylsulfate		7.6
K^+/Na^+ carbonate[a]	4.4	2.0
K^+/Na^+ persulfate[a]	0.8	0.7
t-dodecyl mercaptan	0.7	0.7
[initiator]	13 mmol.L^{-1}	
temperature	62°C	
pH	10.5-10.8	

[a] In recipe 1 potassium salts and in recipe 2 sodium salts were
used.

RESULTS AND DISCUSSION

Any variation of the initiator concentration [I] was
accompanied by an appropriate adjustment of the amount of
sodium or potassium carbonate to keep the initial ionic
strength constant. Coagulation phenomena can thus be ignored.

Figure 1 shows the effect of variation of the initiator
concentration for the initiators investigated on the rate of
polymerization in interval II (R_{pol}), using dresinate 214 as
emulsifier (recipe 1). It appears that the polymerization rate
is highly insensitive to changes in initiator concentration. We
found: $R_{pol} \propto [I]^{0.08}$. Comparable results have been obtained with
sodium dodecylsulfate (recipe 2) using sodium persulfate as the
initiator, viz. $R_{pol} \propto [I]^{0.04}$. Furthermore, the kinetics is
unaffected by the chemical nature and charge of the primary
initiator radicals, indicating that radical entry into latex
particles is determined by the hydrophobic moiety of surface-
active oligomeric species formed in the aqueous phase. Similar
results have been reported for styrene [10].

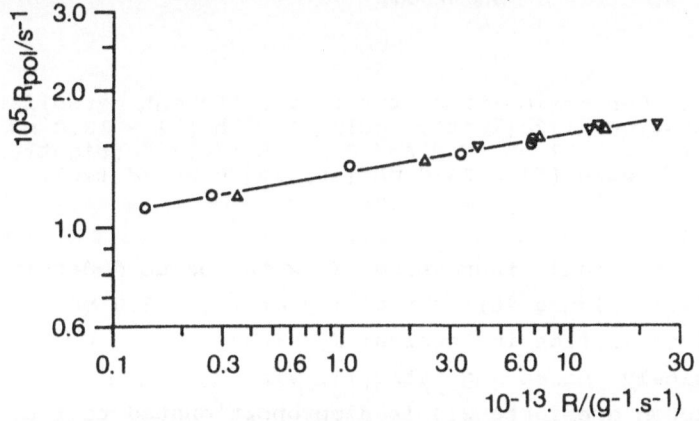

Figure 1. Variation of polymerization rate R_{pol} with radical
production rate per gram emulsion R for different initiators
using recipe 1 : PPS (O), ACPA ($\triangle$) and AIBN (∇).

Interval I in the conversion-time curves is unusually long,
typically some two hours or more (Figure 2), suggesting that
the initiator efficiency with regard to particle nucleation is
very low. Also the conversion-time trajectory in interval II is
remarkably linear while the initiator concentration is
decreasing significantly, again indicating a low efficiency.
Polymerizations with dresinate 214 (Figure 2a) behave normal in
that the duration of interval I increases with decreasing [I].

134

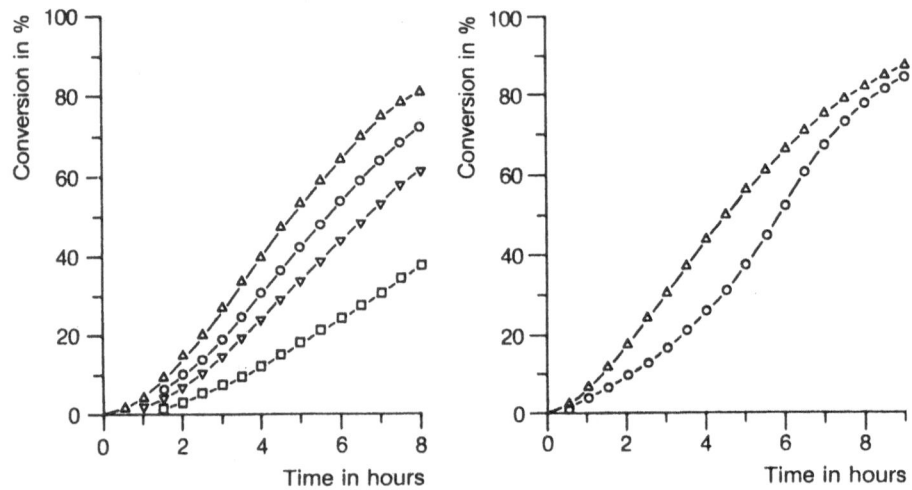

Figure 2. Conversion-time curves at different persulfate
concentrations : 2a(left), recipe 1 with [I] = 13.0 mM ($\triangle$),
2.6 mM ($\bigcirc$), 0.52 mM ($\triangledown$) and 0.26 mM ($\square$); 2b(right), recipe 2
with [I] = 26.0 mM ($\bigcirc$) and 0.26 mM ($\triangle$).

On the other hand, increasing [I] with sodium dodecylsulfate as
emulsifier (Figure 2b), especially at [I] $\geq$ 1.0 mM, slows down
the polymerization in interval I, although R_{pol} in interval II
keeps slowly increasing with [I], viz. $R_{pol} \propto [I]^{0.04}$.
Even though dresinate 214 is disproportionated to remove
retarding components, trace amounts of such substances are
still present and will affect the kinetics, especially at lower
initiator concentrations. The absence of retarders in sodium
dodecylsulfate makes the observed behaviour even more
surprising. Apparently aqueous phase events, such as hetero-
termination cannot be neglected, although the initiator
concentration (varied between 0.26 and 26.0 mM) is not
excessively high. This is consistent with the abovementioned
observation of low initiator efficiency.

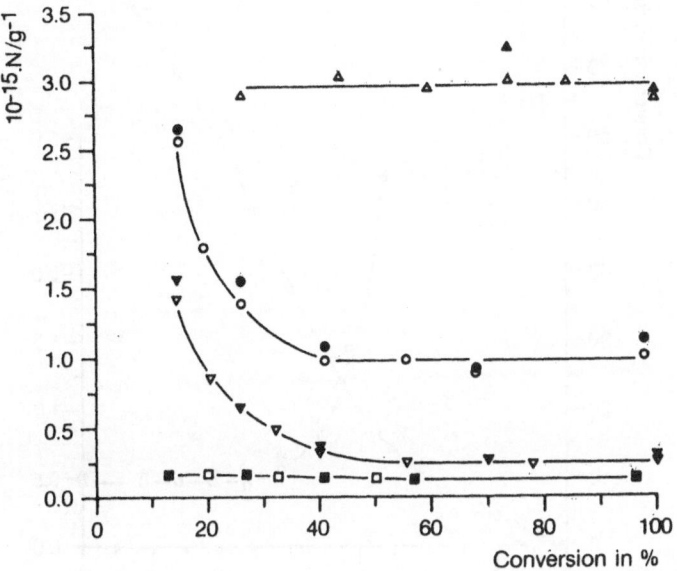

Figure 3. Variation of particle number N using DLS (empty symbols) and TEM (full symbols) for particle sizing, at different concentrations of dresinate 214 : 64.8 $g.L^{-1}$ ($\triangle$), 32.4 $g.L^{-1}$ ($\bigcirc$), 16.2 $g.L^{-1}$ ($\triangledown$) and 8.1 $g.L^{-1}$ ($\square$).

By variation of the concentration of dresinate 214 (recipe 1) an experimental value of 0.61 was found for the emulsifier exponent towards R_{pol}, which is in apparent agreement with simple micellar [11,12] and homogeneous nucleation theories [13] that ignore particle coagulation. However, for the dependency of the final particle number the exponent turned out to be much higher, viz. 1.6, caused by the limited coagulation of latex particles up to about 50 % conversion (Figure 3). This behaviour is in qualitative agreement with coagulative nucleation mechanisms as described by several other workers for various systems [e.g.14-16]. Similar results were obtained with sodium dodecylsulfate as emulsifier (Figure 4).

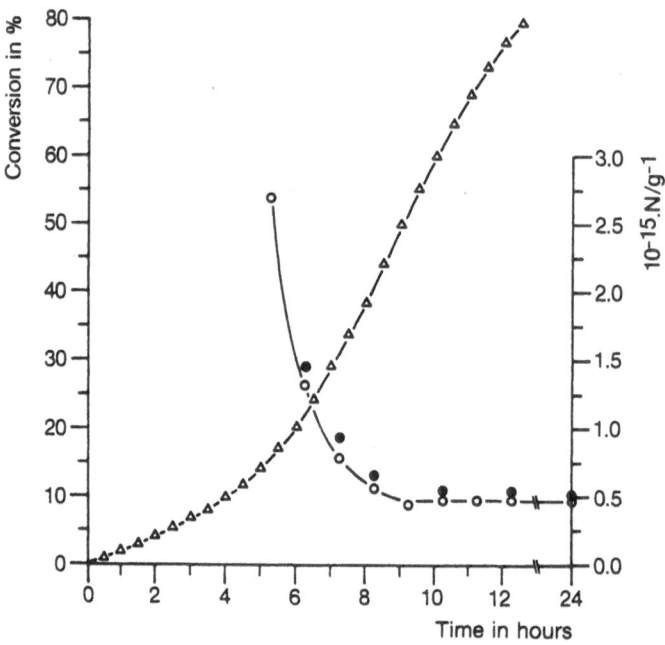

Figure 4. Conversion (△) and particle number vs. time, using
recipe 2 with 8.1 g.L⁻¹ sodium dodecylsulfate and 13.0 mM sodium
persulfate. Particle diameters measured with DLS are weight-
average (○) and with TEM volume-average (●).

Indeed, promoting coagulation by increasing the cation content
in the recipe will lead to a higher average particle diameter
(Figure 5). This is also reflected in the value for the
emulsifier dependency exponent towards particle number, which
becomes larger with increasing sodium-ion concentration (Table
2). Values for this exponent as high as 3 have been reported in
literature [17].

TABLE 2
Variation of the emulsifier dependency exponent towards
particle number as a function of the Na^+-ion concentration.

$[Na^+]$/mol.L⁻¹	0.1	0.15	0.2	0.3	0.4	0.5
exponent	0.5	0.8	1.4	2.1	2.2	2.4

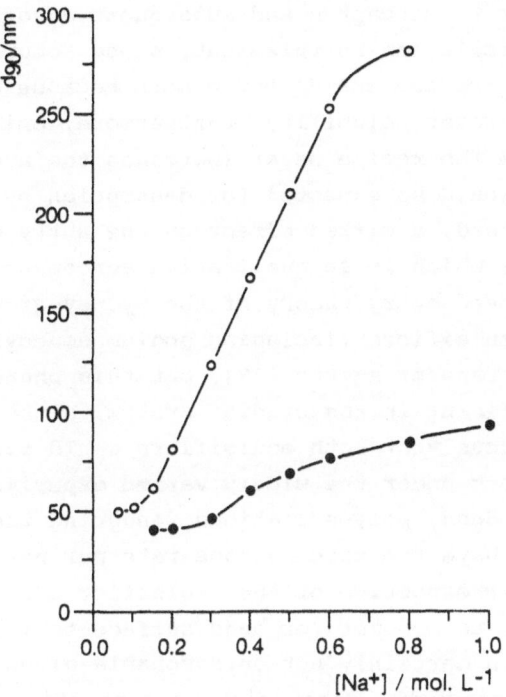

Figure 5. Particle diameter at 90 % conversion d_{90} vs. overall
sodium ion concentration at different sodium dodecylsulfate
concentrations : 32.8 g.L^{-1} (●) and 8.1 g.L^{-1} (O).

As shown in Figure 4 the particle number decreases as
polymerization rate increases, and as a consequence the average
number of radicals per particle (ñ) is a function of particle
size. This behaviour has been described previously for more
water-soluble monomers, such as methyl methacrylate [14], ethyl
acrylate [18] and vinyl acetate [19], and is rather typical of
low ñ-systems. Using the value for the propagation rate
coefficient k_p reported by Morton et al. [2] (100 dm^3.mol^{-1}sec^{-1}
at 62°C) gives values for ñ lower than 0.5, indicating radical
loss by chain transfer/desorption or first-order termination.
The latter is unlikely to occur because the glass transition
temperature of the polybutadiene (T_g = -86°C) is well below
reaction temperature.

Chain transfer to mercaptan and subsequent desorption of mercaptyl radicals can be ruled out, since Nomura et al. [20] showed that C_{12}-mercaptans do not desorb because of their extremely low water solubility. Furthermore, omitting the mercaptan from the recipe never increases the average rate per particle, as would be expected for desorption by mercaptyl radicals. Instead, a marked effect on the entry of radicals is observed [21], which is in qualitative agreement with the recently proposed entry theory of the Sydney-group [22]. Many common emulsifiers (including sodium dodecylsulfate) can act as chain transfer agents [23], but this phenomenon cannot be very significant in the studied system, since it would be highly fortituous that both emulsifiers would exert an identical effect under the widely varied experimental conditions. Indeed, polymerizations rendering the same final particle size have the same average rate per particle in interval II, irrespective of the emulsifier used.

Chain transfer to monomer (or dead surface-active oligomers for that matter) is certainly not unreasonable given the rather low value for k_p, since the desorption rate coefficient is inversely proportional to k_p [24,25]. Although further research is necessary to elucidate the exact kinetic scheme, the data presented here give a better understanding of the behaviour of this widely used monomer in emulsion polymerization.

CONCLUSIONS

The steady state reaction rate of the emulsion polymerization of butadiene is insensitive to changes in the initiator concentration in the studied region, because of the low initiator efficiency in particle nucleation and growth. Limited coagulation is occurring in the present system as a result of the high ionic strength of the aqueous phase. Promoting coagulation by increasing the cation content results in a higher value for the emulsifier dependency exponent towards particle number.

The polymerization rate per particle depends on particle size, while values for ñ are usually significantly less than 0.5. We think that radical desorption of monomeric (or oligomeric) species is responsible for this behaviour, and is consistent with a low value for the propagation rate coefficient.

ACKNOWLEDGEMENT

This work was supported by DSM Research, Geleen, The Netherlands and in part by the Netherlands Foundation "Emulsion Polymerization" (SEP).

REFERENCES

1. Morton, M., Salatiello, P.P. and Landfield, H., J. Polym. Sci., 1952, 8, 111-121.
2. Morton, M., Salatiello, P.P. and Landfield, H., J. Polym. Sci., 1952, 8, 215-224.
3. Morton, M., Salatiello, P.P. and Landfield, H., Ind. Eng. Chem., 1952, 44, 739-742.
4. Bhakuni, R.S., Ph. D. Thesis, University of Akron, 1964.
5. Wendler, K., Karim, N. and Fedtke, M., Plaste Kautsch., 1983, 30, 247-249.
6. Wendler, K., Pielert, L. and Fedtke, M., Plaste Kautsch., 1983, 30, 438-440.
7. Wendler, K., Würtenberg, R. and Fedtke, M., Plaste Kautsch., 1984, 31, 367-368.
8. Wendler, K., Elsner, H., Hergeth, W.D. and Fedtke, M., Plaste Kautsch., 1985, 32, 128-130.
9. Weerts, P.A., van der Loos, J.L.M. and German, A.L., Makromol. Chem., 1989, 190, 777-788.
10. Pennboss, I.A., Napper, D.H. and Gilbert, R.G., J. Chem. Soc., Faraday Trans. 1, 1983, 79, 1257-1271.
11. Smith, W.J. and Ewart, R.H., J. Chem. Phys., 1948, 16, 592-599.
12. Gardon, J.L., J. Polym. Sci., Part A-1, 1968, 6, 623-641.
13. Roe, C.P., Ind. Eng. Chem., 1968, 60, 20-33.
14. Fitch, R.M. and Tsai, C.H., Particle Formation in Polymer Colloids. In Polymer Colloids, ed. R.M. Fitch, Plenum, New York, 1971, pp. 73-102.
15. Hansen, F.K. and Ugelstad, J., J. Polym. Sci., Polym. Chem. Ed., 1978, 16, 1953-1979.
16. Feeney, P.J., Napper, D.H. and Gilbert, R.G., Macromolecules, 1984, 17, 2520-2529.
17. Fitch, R.M., Br. Polym. J., 1973, 5, 467-483.

18. Yeliseyeva, V.I., Polymerization of Polar Monomers. In
    Emulsion Polymerization, ed. I. Piirma, Academic Press,
    New York, 1982, pp. 247-288.
19. Dunn, A.S. and Chong, L.C.H., Br. Polym. J., 1970,
    2, 49-59.
20. Nomura, M., Minamino, Y., Fujita, K. and Harada, M.,
    J. Polym. Sci., Polym. Chem. Ed., 1982, 20, 1261-1270.
21. Weerts, P.A., van der Loos, J.L.M. and German, A.L.,
    Polymer Commun., 1988, 29, 278-279.
22. Maxwell, I.A., Morrison, B.R., Gilbert, R.G. and Napper,
    D.H., Radical Capture Efficiencies in Emulsion
    Polymerization Kinetics. In Integration of Fundamental
    Polymer Science and Technology, Part 4, ed. P.J. Lemstra
    and L.A. Kleintjens, Elsevier Appl. Sci., London, 1989.
23. Piirma, I., Kamath, V.R. and Morton, M., J. Polym. Sci.,
    Polym. Chem. Ed., 1975, 13, 2087-2102.
24. Hansen, F.K. and Ugelstad, J., Makromol. Chem., 1979,
    180, 2423-2434.
25. Nomura, M. and Harada, M., J. Appl. Polym. Sci., 1981,
    26, 17-26.

POLYMERIZATION AT THE SURFACE OF SUBMICRON TiO$_2$ PARTICLES IN EMULSION-LIKE SYSTEMS.

C.H.M. Caris, A.M. van Herk and A.L. German
Department of Polymer Chemistry
Eindhoven University of Technology
P.O. Box 513
5600 MB Eindhoven
The Netherlands

ABSTRACT

TiO$_2$ particles were modified with different titanates. The presence of these titanates at the particle surface was studied in several ways. "Emulsion" polymerizations were carried out in aqueous dispersions of these hydrophobic particles, stabilized with an anionic or a cationic surfactant. The effect of parameters like the kind of surface modification, surfactant concentration, initiator and ratio TiO$_2$ to monomer on polymerization kinetics and product composition was studied.

Depending on reaction conditions two competitive polymerizations can take place: one in which polymer is formed at the particle surface, and one in which free polymer particles are formed. By using TiO$_2$ that has been modified with a hydrophobic titanate, a physical bond between polymer and inorganic particle can be formed, whereas by using a copolymerizable titanate also a chemical bond can be obtained. These particles may offer interesting perspectives in those cases, where a good coupling between particles and matrix is important, for instance in latex paints and in polymer composites.

INTRODUCTION

By means of emulsion polymerization processes several types of particles with a core-shell morphology can be obtained. In this paper we describe a way to obtain particles with an inorganic core (TiO$_2$) an a polymer shell. The surface of the TiO$_2$ particles is hydrophilic [1] and contains several acidic OH-groups, which can react with titanate coupling agents. These organic compounds consist of a central titanium atom, one or two small hydrolysable groups (like isopropyl groups) and two or three long hydrophobic chains or functional groups. A reaction can take place in which an alcohol is formed, leaving the titanate coupled to the pigment surface [2]. The modified, now hydrophobic, particles are dispersed in an aqueous solution of a surfactant, like sodium dodecylsulphate. Part of the surfactant is adsorbed at the particle surface, forming a micelle like structure, in which an "emulsion polymerization" can be carried out. The effect of reaction parameters, like concentration of surfactant, TiO$_2$ or initiator, on polymerization kinetics is studied. The reaction product, both encapsulated TiO_2 and free polymer particles, is studied by means of dynamic light scattering, dark field microscopy and transmission electron microscopy.

The encapsulated inorganic particles may offer perspectives as pigments in latex paints, where they provide a better coupling between pigment and binder (preventing agglomeration of the pigment, thus improving film properties and gloss), but also as fillers for polymer composites, in flame retardants and as carriers for catalysts.

EXPERIMENTAL

The experiments were carried out with pure rutile (Kronos, RLK). This material was washed with distilled water, in order to remove some K_2SO_4, adsorbed at the surface, and dried under vacuum at 130 ^{O}C before use. The (z-average) particle diameter was determined with a Malvern Autosizer 2c (dynamic light scattering). The surface composition of TiO_2 and modified TiO_2 was studied by ESCA (Electron Scattering for Chemical Analysis), using an instrument of Physical Electronics Industries Inc.. Titanates KR TTS, KR 7, KR 212 and KR 26S of Kenrich Petrochemicals Inc. were used without further purification.

Modification of TiO_2 was carried out in dichloromethane (Merck p.a.) or diethylether: 30 g TiO_2 and 30 g glass pearls were added to the flask containing the titanate in the appropriate solvent in a concentration of 1.5 to 6.0 g/l (0.5 to 2.0 wt-% with regard to TiO_2) and the mixture was shaken for about two hours. Then the glass pearls were removed by filtration and the modified TiO_2 was isolated by centrifugation. The product was washed three times with solvent, and dried at room temperature.

The amount of titanate at the surface was determined by elemental analysis (TNO, Zeist, The Netherlands).

Stability against solvolysis was studied by means of elemental analysis, UV spectroscopy (using a Hewlett-Packard 8451A Diode Array Spectrophotometer) and conductometry (using a Radiometer CDM80 conductivity meter). The structure of titanates at the particle suface was determined by FTIR (diffuse reflection and transmission).

The adsorption of sodium dodecylsulfate (SDS) at the modified particles in dispersions was determined by conductometric titrations. SDS (Fluka Chemie A.G., puriss.) and hexadecyl trimethyl ammoniumbromide (CTAB) (Sigma, ca. 99%) were used without further purification.

Polymerizations were carried out with methyl methacrylate (Merck, p.a.), distilled at reduced pressure under nitrogen to remove the inhibitor. Most of the polymerizations were carried out with a radical initiator based on 4,4'-azo-bis-(4-cyanopentanoic acid) (ACPA) (Fluka A.G., purum). Because of its limited solubility in water we decided to use the sodium salt instead of the acid (prepared by reacting the acid with two equivalents of sodium methanolate in methanol). As a water soluble initiator also 2,2'-azo-bis-(2-amidino propane) hydrochloride (AIBA) (Polyscience) was used, without purification. Polymerizations were carried out in double walled thermostated (55 ^{O}C) reaction vessels, kept under a slight excess pressure of nitrogen. Dispersions of (modified) TiO_2 in an aqueous solution of SDS (concentration varying between 2 and 4 g/l) or CTAB were made with an Ystral type X1020 high shear stirrer, and added to the reaction vessel. Then monomer was added and the mixture was flushed with N_2 for about 45 to 60 minutes at 20 ^{O}C, in order to remove O_2. The mixture was stirred with a magnetic stirrer, kept at about 250 rpm for 30 minutes at a temperature of 55 ^{O}C before adding the initiator solution, thus settling the equilibrium. The reaction vessels were equipped with a 10 ml addition funnel, from which extra monomer, soap or initiator solutions were added dropwise to the reaction mixture.

Samples (ca. 0.2 ml) were taken every one or two minutes, during the entire course of the polymerization, through a septum in the reaction vessel, using a syringe. After each sampling the septum was capped to prevent any possible leakage. Samples were diluted with ca. 2 ml distilled water, containing some hydroquinone to stop the reaction. Conversion was determined by gas chromatography, with isopropanol (Merck, p.a., approx. 3 wt%) as an internal standard in the reaction mixture. Gaschromatograph: Hewlett-Packard 5890; 1/8 inch polyphenylether packed column;

column temperature 80 °C; injection port temperature 100 °C; FID temperature 150 °C; carrier gas N_2. All polymerizations were carried out in demineralized water. Polymerization products, consisting of both encapsulated particles and free polymer particles, were studied by Dark Field Microscopy and Transmission Electron Microscopy. Both products were seperated by centrifugation and washing with an aqueous solution of SDS (about 3 g/l) and distilled water. The free polymer particle diameter was studied by means of dynamic light scattering. The polymer content of the encapsulated TiO_2 particles was determined gravimetrically, by heating for about one hour at 400 °C.

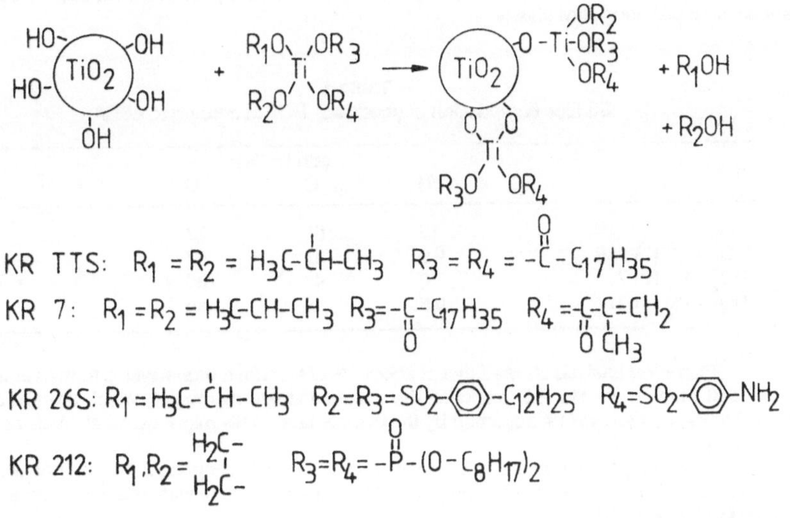

KR TTS: $R_1 = R_2 = H_3C\text{-}\overset{|}{C}H\text{-}CH_3$ $R_3 = R_4 = -\overset{O}{\overset{||}{C}}-C_{17}H_{35}$

KR 7: $R_1 = R_2 = H_3C\text{-}CH\text{-}CH_3$ $R_3 = -\overset{}{\underset{O}{\overset{||}{C}}}-C_{17}H_{35}$ $R_4 = -\underset{O\ CH_3}{\overset{||}{C}}\text{-}C\text{=}CH_2$

KR 26S: $R_1 = H_3C\text{-}\overset{|}{C}H\text{-}CH_3$ $R_2 = R_3 = SO_2\text{-}\bigcirc\text{-}C_{12}H_{25}$ $R_4 = SO_2\text{-}\bigcirc\text{-}NH_2$

KR 212: $R_1, R_2 = \underset{H_2C-}{\overset{H_2C-}{\underset{|}{}}}$ $R_3 = R_4 = -\overset{O}{\overset{||}{P}}\text{-}(O\text{-}C_8H_{17})_2$

Figure 1. Modification of TiO_2 with titanates.

RESULTS AND DISCUSSION

Modification of TiO_2 with titanates

The TiO_2 used consisted of pure crystalline rutile particles with a z-average diameter of 260 nm, a specific surface of 7.7 m^2/g and a total pore volume of 0.58 cm^3/g.

After washing with distilled water, to remove a small amount of adsorbed K_2SO_4, and drying, the surface composition according to ESCA was 18% Ti, 56% O and 22% C (in atomic %). The high carbon content was presumably caused by adsorption of CO, present in the instrument.

The TiO_2 was modified with titanates according to figure 1. The chemical structure and stability of the titanates was determined before [3,4]. The presence of titanates at the particle surface was determined by means of ESCA (table 1) and elemental analysis. The carbon content at the surface increases and the titanium and oxygen contents decrease as the surface is covered with titanates. In the case of KR 26S hardly any sulfur was detected, problably because the sulfur atoms are covered with long alkyl chains.

Table 1.
Surface composition of (modified) TiO_2 according to ESCA

	(atomic %)			
	Ti	O	C	P
TiO_2	18	56	22	-
TiO_2 + 1% KR 26S	9.9	33	47	-
TiO_2 + 1% KR 7	16	44	36	-
TiO_2 + 1% KR 212	9.5	46	44	1

Elemental analysis showed that at about 1% of titanate a monolayer is formed at the particle surface. If much more titanate is applied the excess titanate molecules, which are not chemically bound to the surface, will be adsorbed by the titanate layer. This might cause all kinds of (negative) effects [2].

When dispersed in an aqueous solution of SDS the hydrophobic TiO_2 will be stabilized by the surfactant. The amount of SDS adsorbed at the particle surface was determined by a conductometric titration with SDS.

According to BET-analysis the surface modification with titanates hardly influences the specific surface area of the TiO_2, as shown in table 2.

At 55 $^{\circ}$C 1 g TiO_2, modified with 1 wt% KR TTS, can adsorb $2.1*10^{-5}$ moles of SDS, when stabilized in an aqueous solution. The average adsorption area for SDS is about $52*10^{-16}$ cm^2 per molecule, so this amount of SDS can cover circa 6.5 m^2. Considering the experimental error in the specific surface area of modified TiO_2 and the accuracy of the adsorption area of SDS, this figure is in fairly good agreement with the surface area of the TiO_2 particles. Thus it can be concluded that the whole particle surface will be covered with surfactant molecules. A micelle like structure can be formed at the particle surface.

145

Table 2.
BET-analysis of (modified) TiO_2

	surface area (m^2/g)
pure TiO_2	7.7
TiO_2+1% KR 212	8.0
TiO_2+1% KR TTS	8.4
TiO_2+1% KR 26S	8.2

The structure of the titanates present at the particle surface was studied by means of diffuse reflection FTIR (figure 2). The presence of the SO_2-group in KR 26S was determined using transmission FTIR.

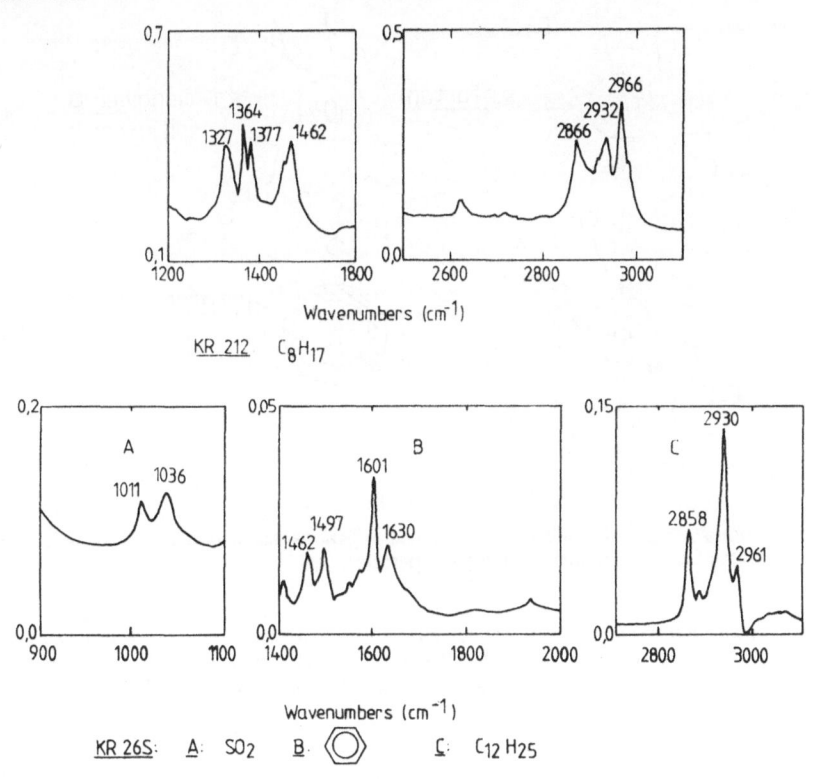

KR 212 C_8H_{17}

KR 26S: A: SO_2 B: ⬡   C: $C_{12}H_{25}$

Figure 2. FTIR signals of titanates at the TiO_2 surface.

Polymerizations at the surface of modified TiO_2

Polymerizations are carried out in dispersions of (modified) TiO_2, stabilized against coagulation by a surfactant. At the start of the polymerization monomer is present at four different places: a small part is dissolved in the water phase, the largest part is present in large monomer droplets, part is adsorbed in normal micelles, and part is adsorbed in "micelles" containing a TiO_2 particle in the core. Two types of polymerization can take place: a conventional emulsion polymerization in micelles, and a polymerization at the particle surface. Polymer formed by the latter polymerization can be physically bound to the TiO_2 (by entanglements with the titanate chains and by adsorption), or chemically by a copolymerization with a titanate containing monomeric moieties (for instance KR 7). These reactions are schematically shown in figure 3. In order to prevent polymerization in free micelles, the surfactant concentration in most experiments was kept at, or slightly above, cmc [5,6,7].

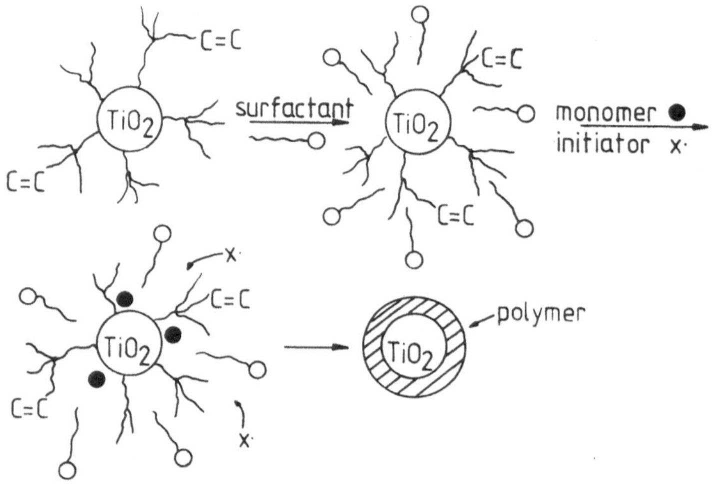

Figure 3. "Emulsion polymerizations" at the surface of TiO_2 particles; formation of a physical and/or chemical bond between polymer and inorganic particle.

It was determined that in a "normal" emulsion polymerization with ACPA as a water soluble radical initiator, pure TiO_2 particles hardly influence polymerization kinetics of methyl methacrylate. However, if modified, hydrophobic, TiO_2 is used, at a certain conversion polymerization rate suddenly decreases, and, after a few minutes, increases again. This effect has already been described for TiO_2, modified with KR TTS [3,4], but a similar effect is observed with TiO_2 modified with KR 26S, though at a lower conversion (figure 4).

This effect can be explained by assuming coagulation of TiO_2 particles. In the first part of the polymerization, reaction starts both at the particle surface and in free micelles. The growing particles need more surfactant to stay stabilized, so after a short period a deficiency of surfactant may arise, resulting in coagulation of TiO_2, and probably also some polymer particles. As the TiO_2 core cannot be penetrated by monomers, radicals, or oligomers, the TiO_2 particles with a thin polymer shell are likely to contain more than one radical at the same time, resulting in a high intrinsic reaction rate at the particle surface. So, when large coagulates of these latter particles are formed, poly-merization rate will decrease, because the radicals may get trapped inside. In the aqueous phase

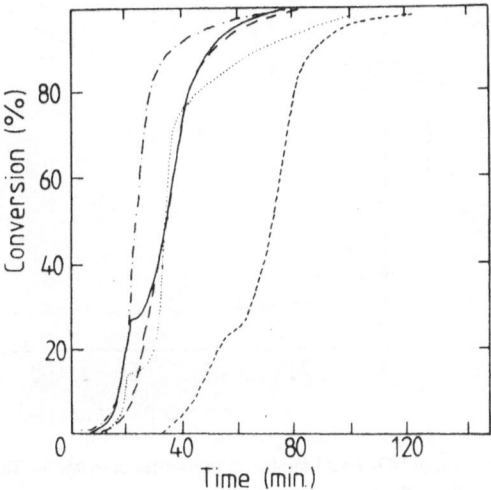

Figure 4. Effect of (modified) TiO$_2$ on polymerization kinetics. (modified) TiO$_2$: MMA = 1 : 3.

—·—·—·—.   MMA
— — — —   MMA + TiO$_2$
————————   MMA + TiO$_2$/KR TTS
·················   MMA + TiO$_2$/KR 26S
- - - - - - -   MMA + TiO$_2$/KR 7

new radicals are generated, and they will cause an increase in polymerization rate after some minutes. Thus, in the second part of the reaction, polymerization continues mainly in free polymer particles, in newly formed polymer particles, and at the surface of the agglomerates (though, as the total surface area of the agglomerates is less than the total surface area of the polymer particles, this polymerization does not play a very important role anymore). The agglomeration of TiO$_2$ particles, modified with KR TTS, was studied before by means of dark field microscopy. Also transmission electron micrographs of samples, taken at different conversions, indicate the formation of agglome-rates, and the formation of more free polymer particles in the second part of the polymerization. En-capsulated TiO$_2$ particles were seperated from the polymer particles, and their polymer content was measured, as a function of the conversion of monomer. Up to about 20% of conversion it increased rapidly, but after the decrease in polymerization rate it increased rather slowly, in agreement with the explanation above (figure 5). In the case of TiO$_2$, modified with KR 26S the "plateau" in the conver-sion-time curve occurs earlier during polymerization. This might be due to adsorption of more anionic surfactant, as the aromatic amine group can be positively charged, causing a deficiency of surfactant and thus agglomeration at a lower conversion.

As the surfactant concentration is increased, no temporary decrease in polymerization rate is observed, partly because contribution of polymerization in free micelles to the conversion-time curve plays a more important role, and partly because of a better stabilization of the inorganic par-ticles (figure 6A).

148

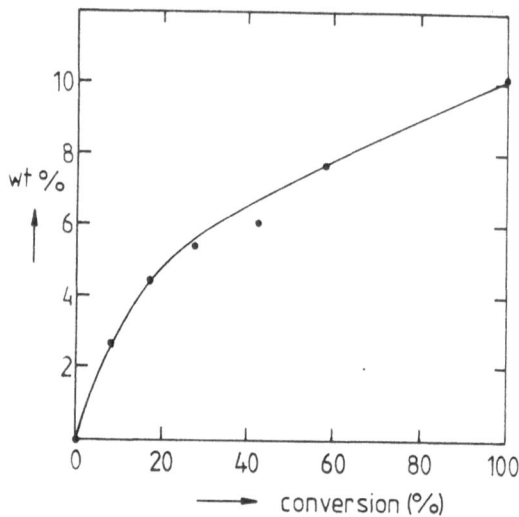

Figure 5. Polymer content of TiO$_2$ as a function of monomer conversion. TiO$_2$/1% KR TTS : MMA = 1 : 3; SDS concentration = 2.4 g/l

A decrease in initiator concentration has the largest effect on the first part of the polymerization, as shown in figure 6B. This would be expected, assuming that during that period polymerization on the particle surface is important, and that the inorganic particles can contain more than one radical at the same time. Contrarily to what one might expect, the decrease in polymerization rate occurs at a lower conversion. If coagulation is caused by a deficiency of stabilizing surfactant molecules, it would be expected that in case of a lower initiator concentration less free polymer particles are formed, resulting in a slowdown at a higher conversion. An explanation for this observed deviating behaviour might be that in this case the decrease in reaction rate is less compensated for by a polymerization in free polymer particles. Dark field microscopy has shown that coagulation occurs continuously during polymerization [3].

By using TiO$_2$, modified with 1 wt % of KR 7, a copolymerization was carried out. This reaction shows similar kinetics as the polymerizations on particles modified with different kinds of titanates, as shown in figure 1. The polymer content of the encapsulated particles strongly depends upon the titanate used (Table 3). The chemical bond between TiO$_2$ and polymer obtained by a copolymerization with KR 7, leads to a higher polymer content than can be obtained by formation of a physical bond, like in the case of KR TTS. The high polymer content obtained by KR 26S can be explained by electrostatic forces between positively charged amine groups at the particle surface and polymer chains, containing a negative charge.

In these cases the TiO$_2$ particles were stabilized with an anionic surfactant, resulting in a negative charge at the particle surface, and also the initiator carries a negative charge. If both electric charges are opposite one would expect an improvement of radical adsorption [7]. As the inorganic particles are likely to contain several radicals, this might result in more polymer being formed at the particle surface. Therefore polymerizations of MMA, with the same water soluble radical initiator, were carried out in an aqueous dispersion of TiO$_2$ (modified with KR TTS) stabilized by a

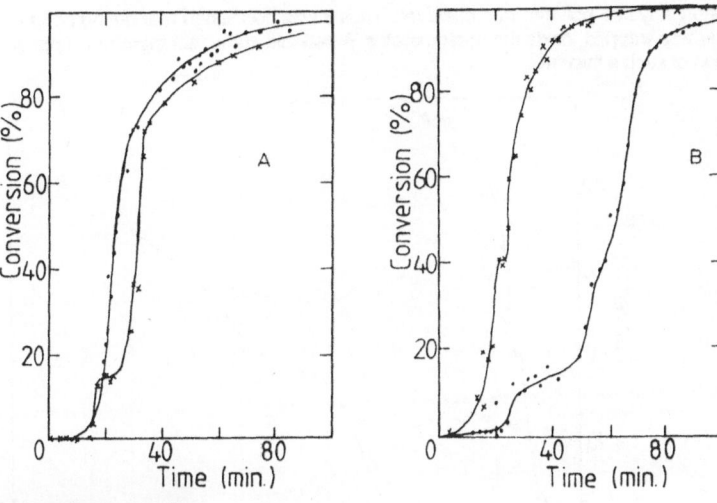

Figure 6. Polymerization kinetics.
A) Effect of SDS concentration. TiO_2/KR 26S : MMA = 1:3
      x SDS concentration = 2.49 g/l
      ● SDS concentration = 3.00 g/l
B) Effect of initiator concentration.
  TiO_2/KR TTS : MMA = 1:3
      x· ACPA concentration = 3.1 mmol/l
      ● ACPA concentration = 1.5 mmol/l

Table 3.
Polymer content of TiO_2 particles, modified
with different titanates.

titanate	[SDS] (g/l)	% polymer on TiO_2
KR TTS	2.37	16.05
KR 7	2.36	18.33
KR 26S	2.36	22.14

cationic surfactant (cetyl trimethyl ammonium bromide). At a conversion of about 35% (figure 7) severe coagulation occurred, resulting in a long "plateau". After about 30 minutes reaction rate started to increase, but notably slower than in the case of an anionic surfactant, as a result of a much higher observed viscosity of the reaction mixture (also the number of free polymer particles might be less, because of a lower surfactant concentration, which is limited by a lower water solubility of the CTAB) [8]. By centrifugation we tried to seperate free polymer from encapsulated TiO_2, and the polymer content of the TiO_2 particles was determined gravimetrically. The product appeared to be very inhomogeneous, but the average polymer content of the precipitated product was 75%, and hardly any free polymer was observed. Thus it was concluded that during coagulation also free polymer was trapped inside the agglomerates. A cationic surfactant therefore is not suitable for stabilization of such a system.

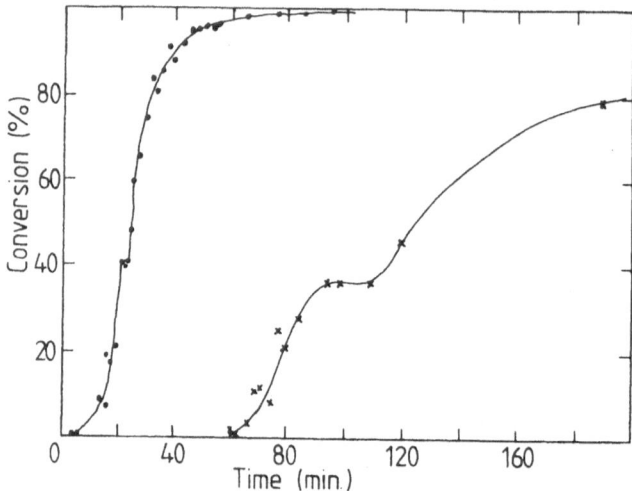

Figure 7. Effect of surfactant on polymerization kinetics.
TiO_2/KR TTS : MMA = 1 : 3
● SDS concentration = 8.2 mmol/l
✗ CTAB concentration = 1.1 mmol/l

Another possibility to obtain opposite charges is the use of a positively charged initiator, like AIBA. This system also showed some coagulation, but by adding the initiator solution slowly to the reaction vessel, after equilibrium had been set, immediate coagulation was prevented. As shown in figure 8 polymerization rate was extremely high. As the surfactant concentration is increased, the decrease in polymerization rate is observed at a higher conversion. This can be explainded by a better stabilization of the modified inorganic particles. A further increase of the surfactant concentration results in a smaller temporary decrease in reaction rate, partly because the contribution of polymerization in free polymer particles becomes more important. The surfactant concentration shows a large effect on polymerization rate during the first part of the polymerization. In the second part there is a smaller influence of surfactant concentration, in agreement with "normal" emulsion polymerization kinetics. A small increase in surfactant concentration above the cmc results in a large increase in the number of micelles and in their total surface area. Thus polymerization in "free" micelles becomes more important, and less polymer is formed at the particle surface. Though the same effect is observed for an anionic surfactant and a negatively charged initiator, in the case of two opposite charges the presence of free micelles appears to result in the formation of larger amounts of free polymer instead of encapsulated particles, as shown in table 4.

Formation of free polymer particles and agglomeration of encapsulated TiO_2 particles not only depends upon the surfactant or initiator concentration, but also upon the concentration of monomer. At low concentrations (starving conditions) only few free polymer particles are formed, but as the concentration increases, the ratio free to adsorbed polymer increases, though more

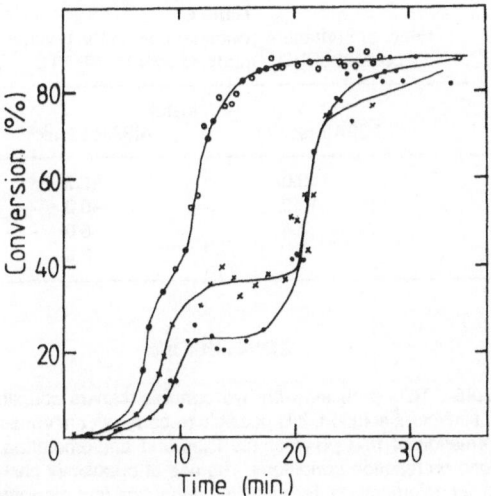

Figure 8. Initiation by AIBA.
TiO$_2$/KR TTS : MMA = 1 : 3
- ● SDS concentration = 2.43 g/l
- ✗ SDS concentration = 3.54 g/l
- ○ SDS concentration = 4.02 g/l

polymer is adsorbed at the particle surface as shown in figure 9. By carrying out the polymerization at a low monomer concentration agglomeration and the formation of large amounts of free polymer can be prevented (figure 10).

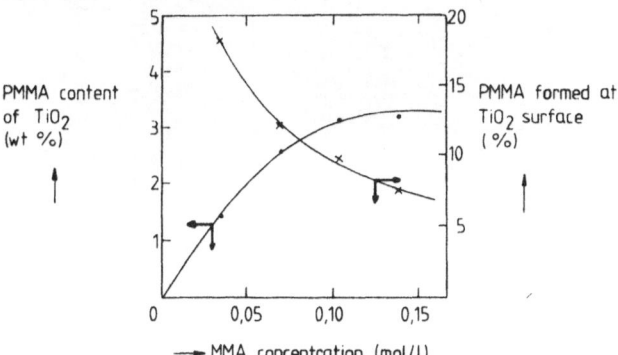

Figure 9. Effect of monomer concentration on polymerization product. SDS concentration = 2.4 g/l.

152

Table 4.
Effect of Surfactant concentration on the polymer
content of TiO$_2$, modified with 1% KR TTS.

[SDS] (g/l)	Initiator	
	ACPA(anionic)	AIBA(cationic)
2.41	16.0	82.3
2.81	19.0	48.0
3.50	7.4	5.0
4.08		2.6

CONCLUSIONS

By modifying hydrophilic TiO$_2$ particles with hydrophobic titanate coupling agents, and stabilizing them in an aqueous surfactant solution, it is possible to carry out an "emulsion polymerization" at the particle surface. Formation of free polymer particles and agglomeration of TiO$_2$ particles during polymerization depend on reaction conditions. The use of oppositely charged surfactants and radicals may easily invoke agglomeration. By adjusting surfactant and monomer concentration it is possible to prevent agglomeration and to encapsulate the inorganic particles with a thin polymer layer

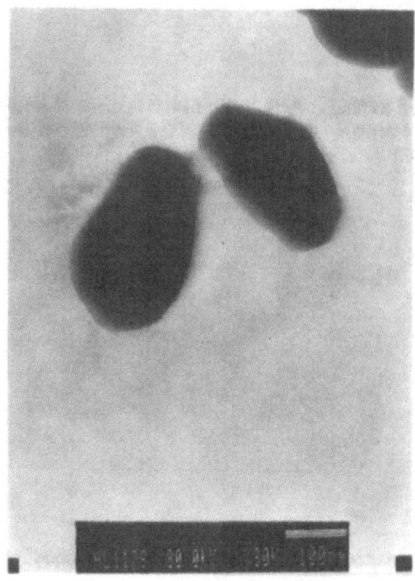

Figure 10. Transmission Electron Micrograph of two TiO$_2$ particles, coated with a thin PMMA layer. TiO$_2$/KR 26S : MMA = 10 : 1

153

ACKNOWLEDGEMENT

The authors would like to acknowledge financial support of the "Onderzoekstimuleringscommissie Verf (OSV)". Electron micrographs were made by mr. H.C.B. Ladan.

REFERENCES

1. Solomon, D.H. and Hawthorne, D.G., <u>Chemistry of Pigments and Fillers</u>, John Wiley & Sons, 1st Edition, New York, 1983

2. Monte, S.J. and Sugerman, G., <u>Ken-React Reference Manual: Titanate and Zirconate Coupling Agents</u>, Kenrich Petrochemicals Inc, Bayonne, USA, 1985

3. Carls C.H.M. Elven, L.P.M. van, Herk, A.M. van, German, A.L., Polymerization at the surface of inorganic submicron particles. <u>Presence and Future in Science and Technology of Coatings and their Components</u>, ed. E. Moeller, vol. III,, Ungheuer und Ulmer Kg GmbH, Ludwigsburg, 1988, pp. 341-354

4. Carls C.H.M., Elven, L.P.M. van, Herk, A.M. van, German, A.L., Polymerization of MMA at the surface of inorganic submicron particles, accepted for publication in <u>British Polymer Journal</u>

5. Hasegawa, M., Arai, K., Saito, S., The Rate of Heterogeneous Polymerization in Water for the Encapsulation of Inorganic Powder with Polymers., <u>J. Chem. Eng. Jap.</u>, 1988, **21**(1), 30-35

6. Hasegawa, M., Arai, K., Saito, S., Uniform Encapsulation of Fine Inorganic Powder with Soapless Emulsion Polymerization. <u>J. Polym. Sci.: Part A: Polym. Chem.</u>, 1987, **25**, 3117-3125

7. Hasegawa, M., Arai, K., Saito, S., Effect of Surfactant Adsorbed on Encapsulation of Fine Powder with Soapless Emulsion Polmerization., <u>J. Polym. Sci.: Part A: Polym. Chem.</u>, 1987, **25**, 3231-3239

8. Yamaguchi, T., Ono, T. Nozawa, M., Sekine, M. Iwai, T., Okada, T., Encapsulation Mechanism in the Polymerization of Methyl Methacrylate in the Presence of Inorganic Powder., <u>Kobushi Ronbunshu</u>, 1983, **40**(4), 259-266.

RUBBER-TOUGHENING OF POLY(METHYL METHACRYLATE) BY INCLUSION OF MULTIPLE-PHASE PARTICLES PREPARED BY EMULSION POLYMERISATION

P.A. LOVELL, J.McDONALD, D.E.J. SAUNDERS and R.J. YOUNG
Polymer Science and Technology Group, Manchester Materials Science Centre, UMIST, Grosvenor Street, Manchester, M1 7HS, United Kingdom

ABSTRACT

Poly(methyl methacrylate) (PMMA) has been toughened, without significant reduction in transparency, by blending with multiple-phase toughening particles which comprise radially-alternating rubbery and glassy layers. General requirements for design of the toughening particles are discussed, and the preparation of four-layer toughening particles is described. Blending of these particles with PMMA gave materials of greatly enhanced toughness, at the expense of only moderate reductions in modulus compared to the matrix PMMA. The dependence of deformation and fracture behaviour upon the weight fraction of toughening particles and strain rate has been evaluated and is discussed in terms of deformation processes. The results indicate that shear yielding, with cavitation of the toughening particles, is the dominant mode of deformation in the blends.

INTRODUCTION

Inclusion of a dispersed rubbery phase is well established as a method for toughening inherently brittle polymers [1,2]. Early attempts to rubber-toughen PMMA concentrated upon the use of suspension polymerisation to produce composite beads consisting of both PMMA and rubbery phases. Whilst these materials were commercialised, they were deficient in that the morphology, and hence toughness, of artefacts produced from them was very strongly dependent upon the moulding conditions employed. Substantial improvements have been realised by preparing separately the toughening particles and the matrix PMMA. The toughening particles are prepared by emulsion polymerisation and comprise two to four radially-alternating rubbery and glassy layers, the outer layer being glassy. They are crosslinked during their formation so that they retain their morphology and size upon subsequent blending with matrix PMMA and moulding of the blends. Hence, this route to rubber-toughened PMMA has the major advantage of allowing independent control of the properties of the matrix PMMA and of the composition, morphology and size of the dispersed rubbery phase. Although these improved materials were developed many years ago [3] and have been commercialised, there are few reports of investigations into their preparation and properties [4-8]. This paper reports the initial results from a major research programme aimed at identifying the mechanism(s) of toughening, and optimising toughening, in rubber-toughened

PMMA materials. The preparation and properties of materials comprising four-layer toughening particles are described.

MATERIALS AND METHODS

Sequential emulsion polymerisation was used to prepare four-layer toughening particles in which the rubbery layers consist of crosslinked poly[(n-butyl acrylate)-*co*-styrene] and in which the glassy layers are based upon poly[(methyl methacrylate)-*co*-(ethyl acrylate)]. Thus rubbery seed particles of 0.10 μm diameter were formed and then grown in three stages. In the first growth stage, crosslinked glassy polymer was formed around the seed particles, increasing the particle diameter to 0.20 μm. These particles were then grown to 0.28 μm diameter by formation of a second rubbery layer. Preparation of the toughening particles was completed by growth of an outer glassy layer of ca 15 nm thickness. The latex obtained was coagulated by controlled addition to aqueous magnesium sulphate solution to yield loose aggregates of the toughening particles which were isolated by filtration, washed and then dried at 70 $^{\circ}$C. These aggregates were blended with matrix PMMA on a twin-screw extruder, thus ensuring their complete disruption to yield a uniform dispersion of the toughening particles. A copolymer of methyl methacrylate with 10% n-butyl acrylate and with M_n = 47 kg mol^{-1} and M_w/M_n = 1.7 (GPC, polystyrene calibration) was used as the matrix PMMA. Four blends were prepared and have weight fractions (w_p) of toughening particles in the range 0.17 - 0.42. The morphology of the blends was confirmed by examining thin films (90 - 150 nm) in a Philips 301 Transmission Electron Microscope (TEM) operated at 100 kV.

The matrix PMMA and each of the blends were compression moulded into plaques of 3 mm thickness (for dynamic mechanical and tensile testing) and of 6 mm and 12 mm thickness (for fracture testing). Dynamic mechanical measurements were made in the temperature range -100 to 160 $^{\circ}$C using a Polymer Laboratories Dynamic Mechanical Thermal Analyser fitted with a double cantilever beam head, employing an operating frequency of 1 Hz and a heating rate of 5 $^{\circ}$C min^{-1}. Tensile stress-strain data were obtained at 20 $^{\circ}$C according to ASTM D638-84 employing a nominal strain rate of 2×10^{-3} s^{-1}. Critical stress intensity factors (K_{Ic}) and critical strain-energy release rates (G_{Ic}) were determined at a nominal strain rate of 3×10^{-3} s^{-1} using a 3-point bend test (ASTM E399-83) and at a nominal strain rate of 30 s^{-1} using an instrumented, falling-weight impact test (Charpy mode with impact velocity 1 m s^{-1}).

RESULTS AND DISCUSSION

A number of general requirements must be considered when designing the toughening particles. Many of the applications for rubber-toughened PMMA materials demand a high % transmission of visible light, i.e. comparable to that of the matrix PMMA. This is achieved by having a relatively small particle size and by selecting the compositions of the copolymers forming the rubbery and glassy phases so that their refractive indices match that of the matrix PMMA. The particles must have an outer glassy layer to prevent particle coalescence during their isolation and must be crosslinked so that they retain their integrity during melt processing. Grafting at the interfaces between the different layers is important and is provided by the crosslinking agent. A good interface between the toughening particles and the matrix PMMA results from mixing of the matrix PMMA with the outer glassy layer of the particles. The four-layer toughening particles reported here, are similar to those which have been used commercially.

Aliquots were removed during the formation of the toughening particles and were analysed gravimetrically to evaluate the overall % conversion as a function of reaction time. The aliquots were analysed also by gas-liquid chromatography to determine the % conversions of the individual monomers. This revealed that the glassy layers formed under monomer-starved conditions and are therefore of uniform composition. However, formation of the rubbery layers occurred under monomer-flooded conditions and was subject to copolymer composition drift. Hence there is a variation of refractive index within the rubbery layers. This is not serious and the blends show 90% transmission of visible light compared to 92% for the matrix PMMA.

Figure 1 shows a TEM micrograph of a blend of the toughening particles with matrix PMMA. The morphology of the toughening particles is clearly evident without the need for staining agents. The rubbery phases appear much darker than the glassy phases. The outer glassy layer of the particles is not visible because it is of almost identical composition to the matrix PMMA with which it mixes.

The presence in the blends of discrete rubbery phases is demonstrated by their dynamic mechanical spectra, which show two distinct glass transitions (see Figure 2). The mechanical damping due to the glass transition of the rubbery phases is weak, and shows a maximum in loss tangent (tan δ) over the temperature range -3 to -6 °C. The glass transition temperature of the glassy phases also shows no variation with w_p and occurs at 112±2 °C. Flexural storage moduli (E'), taken at 20 °C, decreased from 2.6 GPa for the matrix PMMA, to 1.2 GPa for the blend with

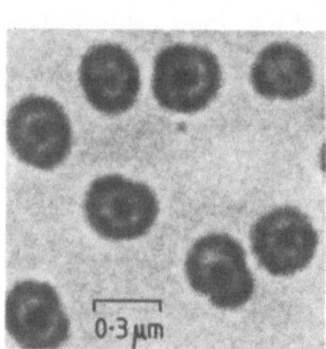

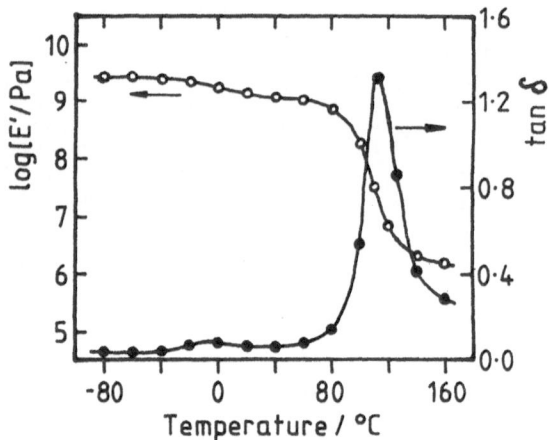

Figure 1. A TEM micrograph showing the internal morphology and size of the toughening particles

Figure 2. A dynamic mechanical spectrum representative of the toughened PMMA materials (w_p = 0.42)

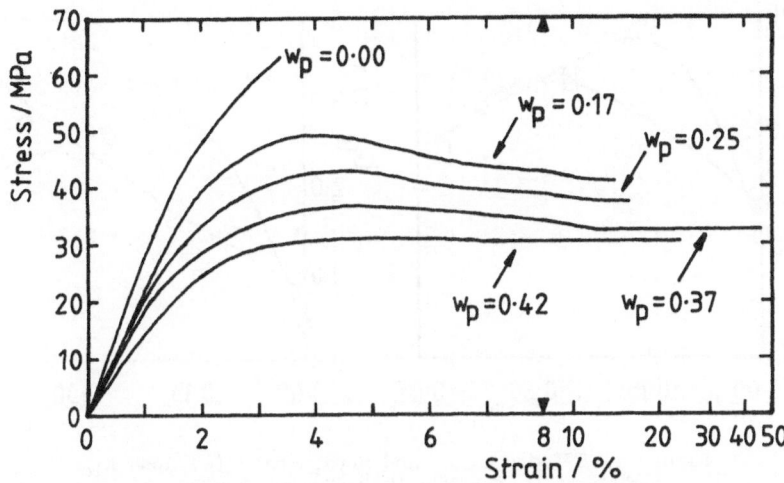

Figure 3. Mean tensile stress-strain curves showing the effects of w_p

w_p = 0.42. Tensile moduli, evaluated from the stress-strain curves presented in Figure 3, show a similar variation from 2.6 to 1.4 GPa over the same range of w_p. The stress-strain curves reveal also that the toughening particles induce macroscopic yielding of the matrix PMMA, giving rise to much greater fracture strains (ϵ_u). Thus the overall energy required to cause fracture is greatly increased. Yield stresses decrease from 49 to 31 MPa as w_p varies from 0.17 to 0.42, whereas ϵ_u passes through a maximum. The reduction in ϵ_u at high w_p is ascribed to the higher probability for defects arising from direct contacts between toughening particles. Unlike the matrix PMMA, at fracture the blends exhibit intense stress-whitening. This is first observed during the initial stages of yielding (at 3-4% strain), coinciding with the appearance of shear bands, and intensifies gradually as the strain increases. Volume strain measurements indicate that the stress-whitening is not due to crazing, confirming the observations of other workers [4,5]. Instead it is proposed that cavitation of the toughening particles is responsible. Evidence for this has been obtained by examining in the TEM toughening particles extracted from fracture surfaces. The absence of crazing, and the presence of shear bands, suggests that shear yielding is the dominant mode of deformation of the matrix PMMA in the blends.

The measurements of K_{Ic} and G_{Ic} were performed under conditions of plane strain and satisfied the criteria for application of linear elastic fracture mechanics. For the 3-point bend test, G_{Ic} values were calculated from K_{Ic} values using tensile moduli from the stress-strain measurements and Poisson's ratios from the volume strain measurements. Plots of K_{Ic} and G_{Ic} against w_p are shown in Figure 4 for each strain rate ($\dot{\epsilon}$) employed. Thus as w_p increases, K_{Ic} passes through a maximum the location of which moves to higher w_p as $\dot{\epsilon}$ increases. In contrast, G_{Ic} passes through a maximum at the lower $\dot{\epsilon}$ only; at the higher $\dot{\epsilon}$, G_{Ic} increases continuously with w_p over the range investigated. These results show that each of the blends has substantially greater resistance to fracture than the matrix

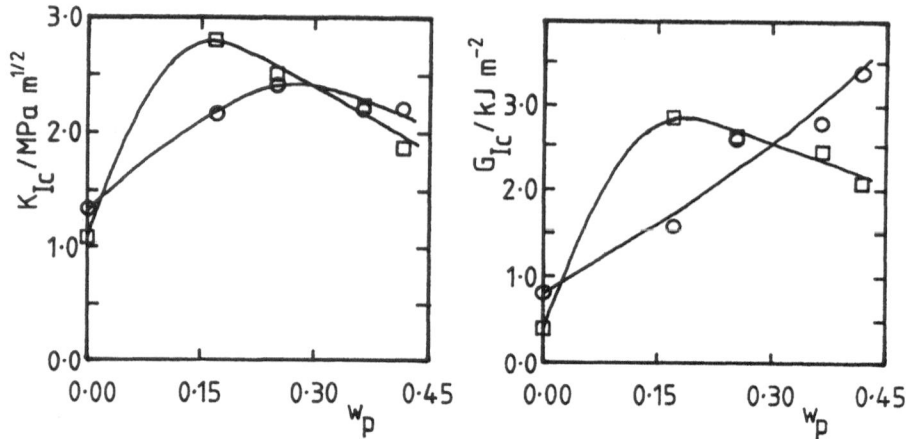

Figure 4. Effects of w_p and strain-rate ($\dot{\varepsilon}$) upon K_{Ic} and G_{Ic}: $\square$, $\dot{\varepsilon} = 3 \times 10^{-3}$ s^{-1}; o, $\dot{\varepsilon} = 30$ s^{-1}.

PMMA. The dependence of K_{Ic} and G_{Ic} upon $\dot{\varepsilon}$ highlights the need to consider carefully the end use before selecting a value of w_p for optimum performance.

ACKNOWLEDGEMENT

The authors express their thanks to the Science and Engineering Research Council and ICI plc for funding this research programme. The assistance of Pat Hunt, Bill Jung and Roy Moore of ICI plc is gratefully acknowledged.

REFERENCES

1. C.B. Bucknall, Toughened Plastics, Applied Science, London, 1977

2. A.J. Kinloch and R.J. Young, Fracture Behaviour of Polymers, Applied Science, London, 1983

3. BP 1,340,025 (1973); BP 1,414,187 (1975); GB 2,039,496A (1979)

4. C.J. Hooley, D.R. Moore, M. Whale and M.J. Williams, Plast. Rubb. Process. Appl., 1981, 1, 345

5. C.B. Bucknall, I.K. Partridge and M.V. Ward, J. Mat. Sci., 1984, 19, 2064

6. O. Frank and J. Lehman, Colloid. Polym. Sci., 1986, 264, 473

7. J. Milios, G.C. Papanicolaou and R.J. Young, J. Mat. Sci., 1986, 21, 4281

8. N. Shah, J. Mat. Sci., 1988, 23, 3623

SURFACTANT AND PROTEIN ADSORPTION AT SURFACE MODIFIED POLYSTYRENE LATICES

Wilfried M. Brouwer
Akzo Research Laboratories Arnhem
Department of Organic and Polymer Chemistry CRH
Velperweg 76
Postbus 9300
6800 SB Arnhem, The Netherlands

ABSTRACT

Various polystyrene latices, varying in the particle surface characteristics have been prepared batch wise. Three series of polystyrene latices were prepared: one according to the emulsifier free emulsion polymerization method yielding sulphate groups at the surface, another series using various acrylamide type comonomers yielding hydrophilic surfaces and a third using strong cationic comonomers to yield a positively charged particle surface. The surface charge densities were varied in the first and third series and the surface characteristics of the particles were studied by conductometric titration, colloidal stability measurements and contact angle measurements. Results of adsorption experiments using surfactants and a protein are discussed in terms of the surface characteristics.

INTRODUCTION

In many latex applications like coatings, adhesives, diagnostics the surface characteristics play a dominant role.
In this study the preparation of anionic and cationic polystyrene based latices of varying surface charge is described and the adsorption of surfactants and protein on these particles discussed in terms of the measured surface characteristics, such as surface charge density and surface polarity.

MATERIALS AND METHODS

Latex preparation

The preparation of polystyrene latices with sulphate surface groups
was batch wise according to the emulsifier-free method, employing
potassium persulphate as the initiator and sodium bicarbonate as the
buffer as has been described in more detail elsewhere [1].

The preparation of cationic polystyrene latices proceeded as follows. In
a thermostated stirred glass reactor water was heated to 70°C while
purging with nitrogen. The amount of water amounted (650-x)ml. Then 80 g
styrene and N-trimethyl N-ethylmethacrylate ammoniumiodide (TMA), the
latter dissolved in x ml water, were added and heated until 70°C under
stirring. Then the nitrogen purging was ceased and instead a nitrogen
blanket was applied. Subsequently 0.8 g 2,2 azobis (2-amidinopropane
hydrochloride) (AAP, Polysciences) dissolved in 50 ml water was added as
the initiator. The latex was allowed to react 20 h. Thereafter the latex
was routinely filtered over glass wool. The amount of x depended on the
amount of TMA added. For AES 1, TMA = 0 g, x = 0 ml; for AES 2 and 3 TMA
was 1 and 2 g, respectively, x = 50 ml; for AES 4 and 5 TMA amounted 5
and 10 g, respectively, x = 100 ml, TMA was 20 g and x = 150 ml for AES
6.

Negatively charged, hydrophilic polystyrene latex (AAJ 4) was prepared
batch wise in a thermostated, stirred reactor under a nitrogen blanket
and in the absence of emulsifiers. Potassium persulphate (0.8 g/l) was
used as the initiator and sodium bicarbonate (0.4 g/l) was added for pH
control. Styrene and acrylamide concentrations were 0.75 and 0.187 M,
respectively. Reaction time was 16 h at 70°C.

Analysis methods

Microfiltration [4] was used to purify the latices employing a 20 fold
volume of deionized, distilled water. Latex solids content during purification was 10 wt%. Nucleopore filters were used in stirred cells
(Amicon).

The size of the latex particles was determined by scanning electron
microscopy. The particle size could be measured within a 15 nm size
range.

The amount of covalently bound ionogenic surface groups was determined by
conductometric titration of the purified latices, which were transferred
in the H^+ form or in the OH^- form prior to titration in the case of
anionic and cationic surface groups, respectively. Titrants were sodium-
hydroxide and hydrochloric acid, respectively.

The contact angle of an octane drop was measured at the latex "film"/-
water interface to compute the polar component of the surface tension
using the methods described elsewhere [1,5]. The surface polarity as
defined by Wu [6] was derived: $X = \gamma/\gamma$, where γ is the surface tension
of the particles and γ the polar component thereof.

The critical flocculation concentration (cfc), was determined by
measuring the time dependence of the absorption of diluted latex samples
immediately after the addition of various amounts of sodium chloride [7].
Purified latex was brought into contact with sodium dodecyl sulphate
(SDS), Antarox CO-630 (ethoxylated nonyl phenolether with 10 ethylene-
oxide units in the polyethyleneoxide chain), hexadecyl trimethylammonium
bromide (CTAB) and bovine serum albumin (BSA). After equilibrium was

reached the remaining adsorbents in the continuous phase were determined by titration (SDS or CTAB), absorbance (Antarox) or size exclusion chromatography (BSA).

RESULTS AND DISCUSSION

Table I summarizes the particle surface characteristics and the adsorption data for all latices investigated.
Within each series of latices particle size did not vary much so that the effect of size on adsorption and colloidal stability characteristics can be neglected. Both anionic and cationic polystyrene latex particles could be prepared over a wide range of surface charge densities: between 0 and 2.5 µeq/m. The polystyrene-sulphate latices did not contain carboxylic weak acid groups, the cationic polystyrene latices did contain weak basic groups residing from the initiator as shown in Table I between the brackets.

Table I

Latex characteristics

Latex	D_{sem} (nm)	σ ueq m^{-2}	cfc mol dm^{-3}	contact angle	X^p	Γ_{sds}	$\Gamma_{Antarox}$ µmol m^{-2}	Γ_{CTAB}	Γ_{BSA} mg m^{-2}
AAJ 19 Dow LS	275	<0.2	0.15	120	0.166	5.64	3.38	-	-
1134B	305	0.24	0.38	107	0.232	4.11	2.76	-	-
XG 10	285	0.30	-		0.248	-	2.71	-	-
XG 9	286	0.31	0.40		0.250	4.01	-	-	1.27
VQ 94	299	0.40	0.40	104	0.264	2.60	1.91	-	1.11
VS 26	293	0.89	0.70	94	0.320	1.72	1.67	-	-
VQ 58	304	1.20	0.64			-	-	-	1.18
VQ 95	264	1.50	0.92	84	0.373	1.05	0.88	-	-
XG 5	309	1.60	-			-	-	-	1.25
XG 11	312	1.78	0.98	82	0.383	-	-	-	-
AAJ4	409	-	>1.0	27.5	0.66	2.90	1.17	-	0.0
AES1	567	0.00(0.6)	0.1	-	-	-	1.83	1.61	1.39
AES2	129	0.35(0.2)	0.3	-	-	-	0.78	2.63	1.00
AES3	119	0.38(0.3)	0.5	-	-	-	0.82	2.52	0.88
AES4	120	0.61(0.6)	0.8	-	-	-	0.87	2.98	0.62
AES5	170	1.39(1.0)	3.8	-	-	-	0.91	3.30	0.52
AES6	166	2.51(1.8)	4.3	-	-	-	0.93	3.66	0.42

1. see Ref. 2 for BSA adsorption. pH = 7.2, Sörenson buffer was used.
2. AAJ 19, 1134B, XG and VQ codes pertain to emulsifier free polystyrene latices with SO$_4^-$ charged groups.
3. AAJ 4 pertain to acrylamide styrene copolymer latices.
4. AES pertains to cationic latices.
5. σ is surface charge density

The colloidal stability of the particles increases with increasing surface charge density as indicated by the cfc values. The poly(acryl-amide-co-styrene) latices appear to be very stable, possibly due to extending polymer chains from the surface into the aqueous phase. Along with the surface charge density, surface polarity increases as derived from the contact angle measurements on the sulphate group bearing latices.

The saturation adsorption of SDS and Antarox (■ and □ , respectively, Fig. 1) on the anionic polystyrene latices can be described using the surface polarity as the only variable (Fig. 1), in accordance with the expression derived by Vijayendran [9] for surfactant adsorption on different latex types. This also holds true for Antarox adsorption on AAJ-4 latex, the latter carrying some weak acid groups as did the butylacrylate/vinylacetate latices, Vijayendran studied (Fig. 1). However, the adsorption of SDS on the styrene-acrylamide type latices is much higher than anticipated from the value X^P, perhaps due to for SDS accessible hydrophobic spaces between the extending hydrophilic chains at the particle surface.

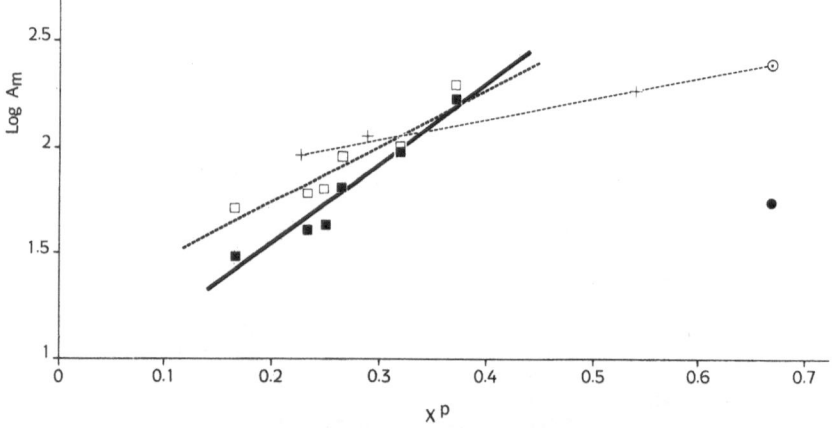

Figure 1. Saturation adsorption of SDS and Antarox CO-630 on various latices. A_m is area occupied by surfactant molecule ($Å^2$)
   ■ SDS on polystyrene (sulphate surface group) latex
   □ Antarox on polystyrene (sulphate surface group) latex
   + Antarox on butylacrylate/vinylacetate latex (ref. 9)
   o Antarox on poly(styrene-co-acrylamide) latex (AAJ 4)
   ● SDS on poly(styrene-co-acrylamide) latex (AAJ 4)

For the cationic latices the adsorption of CTAB increases with increasing surface charge, despite the repulsive forces between surfactant and latex surface. This result may be attributed to long chains extending into the continuous phase where CTAB can interact with. The Antarox CO-630 adsorption is more or less constant (except for AES 1) and rather low, which is indicative of a rather polar surface. The adsorption of bovine serum albumin seems to be governed rather by the possibility of hydrophobic interactions than by the electrostatic interactions. The saturation adsorption of BSA on anionic polystyrene latices hardly

depends on the sulphate charge density since always hydrophobic spots are
available to interact with hydrophobic parts of BSA.
Styrene/acrylamide latices have their hydrophobic surface entirely
shielded for the bulky molecules and no adsorption is observed. On the
oppositely cationic charged latices (BSA is negatively charged at pH =
7.2) adsorption of BSA decreases with increasing surface charge density.
Here hydrophobic spots are presumably shielded by polymer chains
extending into the aqueous phase.

CONCLUSIONS

The adsorption of surfactants on model polystyrene latices varying in
surface characteristics cannot be described with a single adsorption
model. The adsorption on sulphate groups bearing latices, prepared by the
emulsifier free method, is governed by the interfacial tension between
particle surface and the continous phase. The adsorption on copolymer
latices, core-shell in nature, is different for the charged and non-ionic
surfactants studied here. In general the adsorption of non-ionic Antarox
shows an inverse relationship with the surface polarity.
BSA adsorption is not governed by electrostatic interactions, nor by
overall surface polarity. The availability of hydrophobic spots at the
surface seems to be the key factor for adsorption of the large BSA
molecule.

REFERENCES

1. Brouwer W.M. and Zsom R.L.J., Colloids and Surfaces 1987 24, 195
2. Zsom R.L.J., J. Colloid Interface Sci., 111 (1986), 434
3. Brouwer W.M., Colloids and Surfaces, accepted for publication
4. Ahmed S.M., El-Aasser M.S., Pauli G.H., Poehlein G.W. and Vanderhoff
   J.W., J. Colloid Interface Sci., 73 (1980) 388
5. Hamilton W.C., J. Colloid Interface Sci., 40 (1972) 219, 47 (1974) 672
6. Wu S., J. Macromol. Sci. Rev. Macromol. Chem., C10 (1974) 1
7. Hiemenz P.C., Principles of Colloid and Surface Chemistry, Marcel
   Dekker, New York, Basel, 1977
8. Vijayendran in Polymer Colloids II, Fitch R.M. (ed), Plenum New York
   1978, p. 209
9. Vijayendran B.R., Bone T., Gajria C.; J. Appl. Polym. Sci 26 (1981)
   1351, 1981 26, 1351

164

CHARACTERIZATION OF IONIC OLIGOMERS FORMED DURING THE EMULSION POLYMERIZATION OF BUTADIENE BY MEANS OF ISOTACHOPHORESIS

J.L.AMMERDORFFER, C.G.PIJLS, A.A.LEMMENS AND A.L.GERMAN
Eindhoven University of Technology, Laboratory of Polymer
Chemistry, P.O.Box 513, 5600 MB Eindhoven, The Netherlands

ABSTRACT

The end products of several emulsion polymerizations of
butadiene with potassium persulfate as initiator were analyzed
by means of isotachophoresis. The results were compared with
the analyses of model compounds. The isotachopherograms of
the products of surfactant free emulsion polymerizations
as well as of polymerization with dresinate 214 as surfactant
show oligomers with one or two sulfate end groups and up to 5
monomeric units.

INTRODUCTION

It is generally accepted that in emulsion polymerization with
potassium persulfate as initiator, at least some oligomeric
radicals are formed in the water phase. Termination of these
radicals in the water phase gives oligomers with an amphipatic
structure. Qualitative and quantitative characterization of
these oligomers should give insight in the propagation and
termination reactions in the water phase as well as
information about the contribution of the oligomers (radicalic
and terminated species) to particle formation. Up to now there
are only few references relating to the actual proof of the
occurrence of these oligomers [1,2,3], and they are only
indicating their presence without any qualitative or
quantitative information.
Isotachophoresis (ITP) is an excellent method of analyzing
ionic components, qualitatively as well as quantitatively. Up

to now several workers have used ITP in applications related
to emulsion polymerization. Okubo and Mori [4] measured the
decomposition of potassium persulfate with ITP. Ogino,
Kakihara and Abe [5] used ITP to determine the CMC's of sodium
alkyl sulfates. Bolesse and Leising [6] showed the
possibilities of separating ionic monomers such as acrylic and
ithaconic acid.

ISOTACHOPHORESIS [7,8]

In ITP discrimination between components is based on
differences in electrophoretic mobility which is a function of
charge and a parameter involving geometric and interactive
components. Within a homologous series with the same ionic
group the latter parameter is proportional to molecular
weight, so ITP should be a reliable method to separate water
soluble oligomers according to number of monomeric units and
number of ionic end groups [9,10].
The description and theory of an isotachophoretic experiment
is given elsewhere [7,9]. In understanding the
isotachopherograms the following typical characteristics are
essential. After separation is completed in capillary
isotachophoresis all anionic components of the sample are
moving behind the leading anion to the anode, while remaining
in their own zone (Fig. 1), all cations are replaced by the
counter ions of the leading electrolyte. The successive zones
are then passing the detector in an order imposed by their

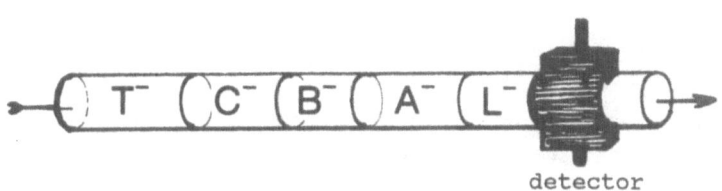

detector

Figure 1.Schematic view of ITP detection

electrophoretic mobility. With a conductivity cell as detector
the successive zones become visible as steps in the output
signal (Fig. 2). Each stepheight corresponds to a component of
the sample and thus provides us with qualitative information.
Mostly the stepheight is divided by that of the internal
standard to give a relative stepheight RSH. The length of each
zone, and with that of the steps, is a measure of the quantity
of each of the components.

EXPERIMENTAL

Polymerization conditions

Emulsion polymerizations of butadiene were carried out in a
stainless steel 1-liter reaction vessel with turbine-stirrer
at 62^0C. The monomer was freshly distilled. The other
components were analytical or polymerization grade and used as
purchased (Table 1).

TABLE 1
Emulsion polymerization recipe (in mol/l)

Batch	A1	P1	P5
Butadiene	.7	1.0	7.7
Dresinate 214	–	–	.11
Potassium persulfate	.010	.010	.013
Potassium carbonate	.001	.001	.14

Analysis

The products of the emulsion polymerizations were analyzed
by using the ITP equipment developed by Everaerts et al. [7].
In order to remove the latex particles, the samples were
ultrafiltrated over an anisotropic, hydrophilic YMT-membrane

on a polycarbonate carrier. For the ITP conditions
see (Table 2).

TABLE 2
ITP conditions

Leading electrolyte   :   Cl⁻   .005 M
pH   :   3.5
Terminator   :   acetate
Counter ion   :   β-alanine
Internal standard   :   maleic acid
Current   :   40 μA

In order to test the system several (mixtures of) model
components, resembling the species to be expected, were
analyzed.

RESULTS

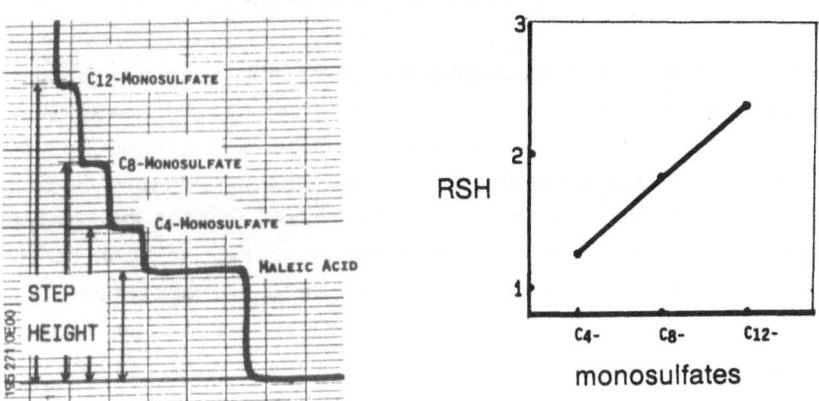

Figure 2. Separation of model compounds and the relation
between relative stepheight RSH and number of carbon atoms

Figure 2 shows the isotachopherogram of the model compounds
butyl, octyl and dodecyl sulfate. As expected there is a
linear relation between the relative stepheight and the
molecular weight.

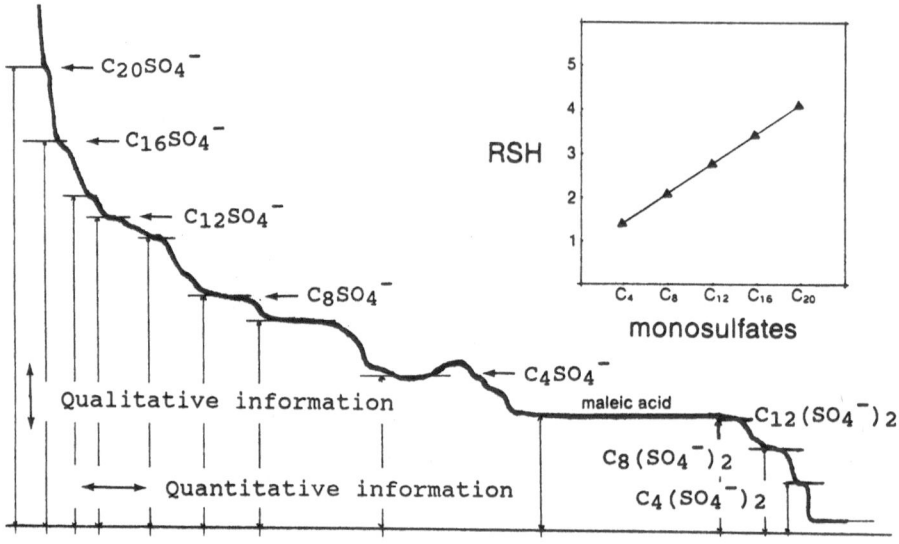

Figure 3. Isotachopherogram of oligomers from experiment P1,
and the relative stepheight versus number of carbon atoms of
the indicated components

Figure 3 shows an isotachopherogram of the reaction product of
experiment P1. The steps indicating monosulfates containing up
to 5 monomeric units and disulfates with 1-3 monomeric units,
can easily be recognised. The RSH of the components indicated
with an arrow again shows a linear relation with the molecular
weights, i.e. the number of monomeric units. The lowest RSH's
are in agreement with the RSH's of the model sulfates
(Fig. 2), so we consider it justified to assign the steps
indicated with an arrow (Fig. 3) to a series of <u>linear</u>
oligomers with one sulfate end group and 1 to 5 monomeric
units. Also the RSH's of the disulfate oligomers have a linear
relation with molecular weight and show good agreement with
the model disulfates. Some other steps, most of them showing
up near the octyl and dodecyl sulfates, remain to be
explained. It is known that branching and degree of

169

unsaturation has an influence on the electrophoretic mobility, so we assume the latter steps to be (mixtures of) oligomers with branched alkyl groups and/or differences in unsaturation due to various modes of termination.

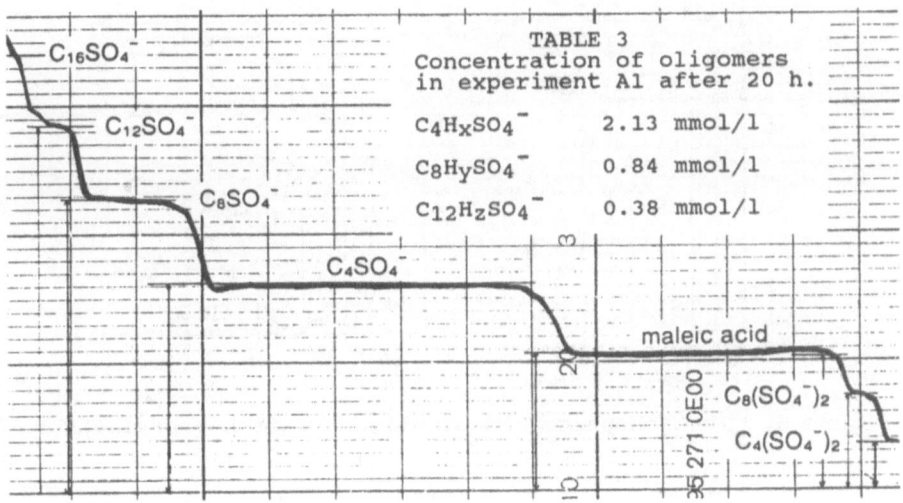

TABLE 3
Concentration of oligomers
in experiment A1 after 20 h.

$C_4H_xSO_4^-$	2.13 mmol/l
$C_8H_ySO_4^-$	0.84 mmol/l
$C_{12}H_zSO_4^-$	0.38 mmol/l

Figure 4. Isotachopherogram of oligomers from experiment A1

With experiment A1 reaction conditions were chosen in such a way as to minimize particle formation while still allowing oligomer production. As an consequence it was permitted to inject the reaction product directly into the ITP apparatus. Figure 4 gives the isotachopherogram of the reaction product after 20 hours. Since the actual reaction product was analyzed it becomes possible to obtain an accurate estimation of the quantity of monosulfates in the reaction product (Table 3). Until now it has not yet been possible to quantify the amounts of disulfates, but figure 4 strongly suggests that they build up to a significantly lesser amount than the monosulfates.

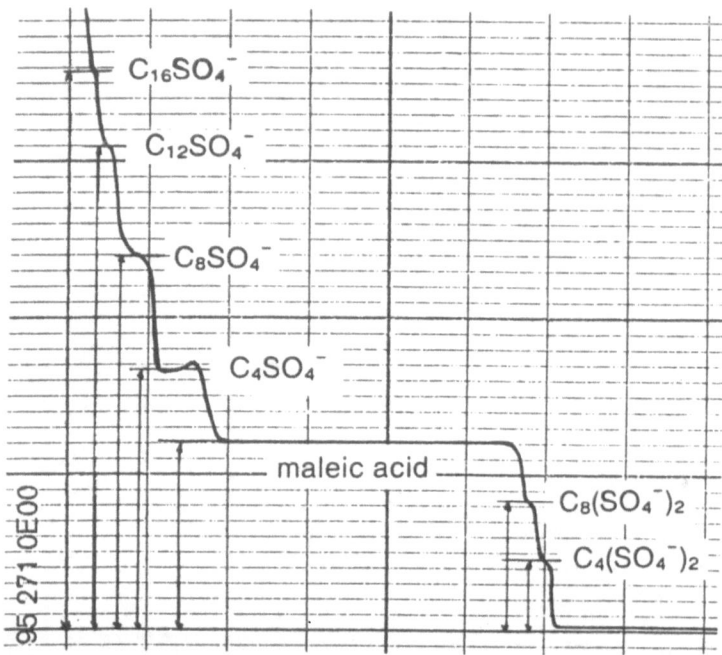

Figure 5.   Isotachopherogram of oligomers from experiment P5

Figure 5 shows the isotachopherogram of the end product of experiment P5. One of the advantages of ITP is, that by choosing the terminating electrolyte, one can mask the very slow ions in the sample. So in this case, by using acetate as terminator, the dresinate ions are way behind the other components and thus cannot interfere (for instance as mixed micels) with the sulfates.

Conclusions

The results of this investigation show that ITP is capable of qualitatively and quantitatively detecting oligomeric substances formed during the emulsion polymerization of butadiene. The quantitative results lead to the conclusion that, in butadiene emulsion polymerization, termination of the oligomeric radicals in the water phase by disproportionation and/or transfer is preferred to termination by combination. Furthermore, the amount of oligomeric species with a distinct interfacial activity, i.e. dodecyl and higher sulfates, is sufficient (in experiment A1 about .5 mmol/l) to expect some cosurfactant effect. At least in kinetic studies where the amount of surfactant is only slightly above the CMC, one should be aware of a possible contribution of amphipatic oligomers formed on nucleation and particle stabilization.

References

1. R.M.Fitch and C.H.Tsai, Polymer Colloids, Plenum Press, New York, 1971, pp. 73-102.
2. A.R.Goodall, et al, J.Pol.Sci.,Pol.Chem.Ed., 1977, **15**, 2193-2218.
3 C.Chen, I.Piirma, J.Pol.Sci., Pol.Chem.Ed., 1980, **18**, 1979-1993.
4 M.Okubo, T.Mori, Colloid Polym. Sci., 1988, **266**, 333-336.
5 K.Ogino, T.Kakihara, M.Abe, Colloid Polym. Sci., 1987, **265**, 604-612.
6 M.Bolesse, F.Leising, Makromol.Chem.Suppl., 1985, **10/11**, 305-309.
7 F.Everaerts, J.Beckers, T.Verheggen, Isotachophoresis, J.Chrom.Libr., **Vol 6**, Elsevier, Amsterdam, 1976.
8 A.A.G.Lemmens, Selectivity in Analytical Isotachophoresis, Ph.D.Thesis, Eindhoven University of Technology, 1989.
9 J.L.Ammerdorffer,A.A.G.Lemmens,to be published.
10 C.Pijls, M.S.Thesis (in dutch), Eindhoven University of Technology, 1989.

FLOW BEHAVIOUR AND VISCO-ELASTIC PROPERTIES OF EMULSIONS, STAINS AND PAINTS

C.A.J.J. van Rossum
W. van Gerresheim
DSM Resins BV,
Emulsion Research, Ceintuurbaan 5, Zwolle, The Netherlands

ABSTRACT

The rheological behaviour of acrylic emulsions and paints and stains prepared therefrom has been studied under conditions of oscillatory and steady shear flow. It was found that the rheological properties of the parent emulsions in general were not preserved in the products prepared therefrom. Only the influence of particle size remained nearly unchanged in the paints and stains. The changes in rheology on going from emulsions to emulsion products are caused by other additives, of which the thickener is the most important one.

INTRODUCTION

The rheological properties of products based on polymer emulsions can play a major role in their application properties. In turn the rheological properties are determined by physical and chemical interactions and processes in these systems and therefore by their compositions.

However the rheological profile of "ideal emulsion paints" is very complex, and a compromise in low shear viscosity and high shear viscosity is required. On the one hand a certain level of low shear viscosity is required to prevent sagging, but if it is not low enough flow and levelling will be insufficient and brushmarks remain in the dried coating. On the other hand a certain level of high shear viscosity is required to produce film build. High shear viscosities being too high will lead to wrist/arm fatigue of the painter.

Thus viscosity as a function of a broad range of shear rates from 10^{-2} s^{-1} (sagging/levelling) up to 10^4 s^{-1} (brushing, spraying) is an important rheological property in the application of paints and stains.

The degree of elasticity provides information about
(network) structures, that are mainly present at low shear.
These structures determine for instance the presence of low
shear Newtonian flow regions and the presence and magnitude
of yield stresses. An appropiate way to determine the
elasticity is the determination of the dynamic modulus Gd or
the loss angle delta as functions of oscillatory shear (1,2).

The major component of a paint or stain is the emulsion
itself. Consequently its rheological properties can largely
determine the rheological properties of the end product.
In this paper the effects of particle size, pH and amount and
type of acid monomer on the rheological behaviour of model
emulsions and its preservation in paints and stains prepared
therefrom are reported. Both viscosity as a function of
steady shear and visco- elasticity are dealt with.

TABLE I
Particle sizes (nm) of (meth)acrylic acid emulsions
as function of pH and amount of acid

Amount of methacrylic acid (%)

pH	0			2					5				
2-3	108	169	267	74	105	167	240	339	76	106	160	203	256
7	108	174	268	72	107	169	240	339	75	108	157	203	257

Amount of acrylic acid (%)

pH	0			2			5	
2-3	108	169	267	102	173	269	99	154
7	108	174	268	103	168	270	106	164

EXPERIMENTAL

Preparation
Acrylic emulsions (butylacrylate and methylmethacrylate) were
prepared by a standard emulsion polymerization process , with
appropiate control of particle size. The pH of the emulsions
was adjusted to 7 with a 5% NaOH solution.
Dry solids contents were 47.9% for methacrylic acid
respectively 49.0% for acrylic acid containing emulsions.
Paints and stains were prepared from the acrylic acid
containing emulsions with a pH of 7. A minimum of paint and
stain additives have been used (f.i. the only pigment used in
the paints was TiO_2).

174

Characterization

Particle sizes were measured at 23°C with a Malvern autosizer (Table I).
Rheological measurements were carried out with a Bohlin VOR rheometer at 23 ± 0.1°C.
The complex dynamic modulus Gd was determined with oscillation measurements (0.1 - 10 Hz) in the linear visco-elastic area (maximum strain = 0.002).
Viscosity was measured as a function of increasing, followed by decreasing shear rates (10^{-2} - 10^4 s^{-1}).

RESULTS AND DISCUSSION

Oscillatory shear measurements - determination of the degree of elasticity.

Emulsions: For methacrylic acid (5%) containing emulsions a change of the slope of the Gd-frequency curve from 0 to 1 with increasing particle size was observed (fig. 1.)
Thus an increase in particle size is accompanied by a change from elastic to Newtonian behaviour. Similar results were obtained for other methacrylic acid emulsions (0, 2% and 0-5% at pH=7) and for acrylic acid containing emulsions (0, 2, 5%) at a pH of 2.5 and 7.

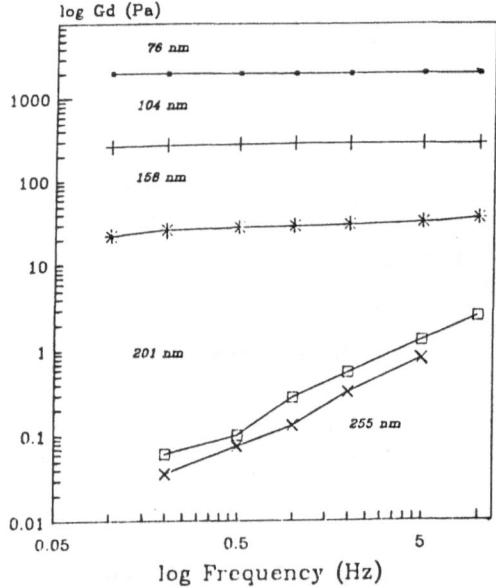

Figure 1. Dynamic modulus-frequency plots of emulsions
containing 5% methacrylic acid, pH=2-3.

Emulsions with particle sizes of 100 nm or less turned out to be elastic in nature, while emulsions with particle sizes larger than 200 nm were Newtonian in character. Consequently emulsions with particle sizes between 100 and 200 nm are visco-elastic in nature. It is believed that emulsions with particle sizes of 200 nm and more would only be visco-elastic at closed packed limit (3). It is evident that our emulsions with a dry solids content of 47.9% and 49% do not meet this criterium.

In all cases the slope of the Gd curves did not significantly change with the amount of acrylic acid. However in all cases an increase in pH resulted in a small increase of the slopes of the Gd curves, while the order in the curves had also changed. This increase is probably caused by the addition of NaOH to raise the pH, which leads to gradual destruction of the lattice formed by the emulsion particles (4). The change of the order of the Gd curves is most likely due to the fact that the amount of NaOH added was not equal in all cases. In order to adjust the pH to an equal value of 7, more NaOH was needed for the emulsions with higher acid monomer contents.

Paints and stains: For paints and stains the slopes of the Gd curves had increased significantly, when compared with the parent emulsions. This predicted a low shear Newtonian flow region for paints and stains compared to the parent emulsions (see under steady shear measurements). Again the order of the Gd curves had changed. The dominant effect of particle size in the emulsions (especially at pH = 2-3) was exchanged for the amount of acrylic acid as the most important parameter in the determination of the level of the Gd- curves.
The change from elastic properties in the emulsions to Newtonian properties in the paints and stains probably finds its origin in the different amount of emulsion. The weight share of emulsion in the paints and stains is 64% and 85% respectively. So to speak the emulsions have been "diluted" with paint and stain additives and the lattice formed by the emulsion particles has degraded even more than in the case of increasing the pH.
The change in the order of the Gd curves cannot be explained readilly. The differences in the emulsions (particle size, amount of acrylic acid) probably lead to different interactions of the emulsions particles with paint and stain additives, especially PU-thickener and pigments.

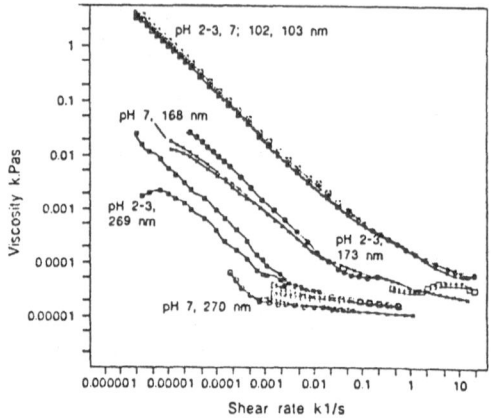

Figure 2. Flow curves of emulsions containing 2% acrylic
         acid, before and after adjustment of the pH to 7.

Steady shear measurements

   Emulsions: The flow curves of acrylic acid (2%)
containing emulsions are displayed in fig. 2. The
coincidence of the up and down curves indicated the absence
of thixotropic behaviour.This was found for all emulsions,
paints and stains described in this paper.

   It is evident that an increase in particle size leads to
a lowering of viscosity, especially at lower shear rates.
Thus shear thinning behaviour is less for emulsions with
larger particles. Similar trends in flow behaviour as a
function of particle size were observed for emulsions without
and with 5% acrylic acid and also for methacrylic acid
containing emulsions.

   A rise in pH did not effect the flow curves of emulsions
containing 2% acrylic acid. This phenomenon was also noticed
for emulsions without acrylic acid monomer and for all
methacrylic acid emulsions. Yet the behaviour of the 5%
acrylic acid emulsions, especially those with larger
particles was completely different. In this case
neutralization increased the viscosity for all shear rates
(figs. 3 and 4). At pH values of 2-3 the flow curves of
acrylic acid emulsions with equal particle sizes coincided.
So did, irrespective of the pH, all the flow curves of 0, 2
and 5% methacrylic acid containing emulsions.
Thus viscosity is only influenced for emulsions with more
than 2% acrylic acid, and after neutralization. This increase
in viscosity probably finds its origin in an increase in
particle size. After neutralization only the particle size of
the emulsions with 5% acrylic acid had increased (about 6%,
Table I).

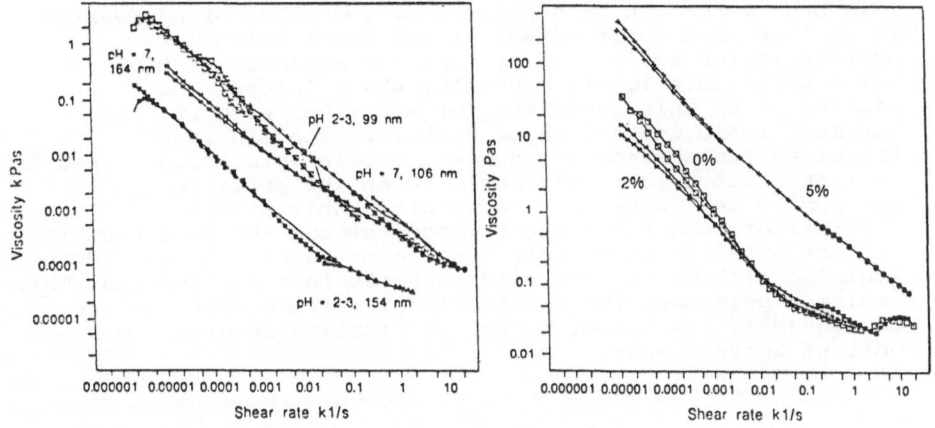

Figure 3. Flow curves of emulsions containing 5% acrylic
acid, before and after adjustment of the pH
to 7 (left)

Figure 4. Influence of amount of acrylic acid on the flow
behaviour of emulsions with particle sizes of ca
150 nm (pH=7) (right).

Paints and Stains: In general the trend of lower
viscosities for emulsions with larger particles was
reproduced in the paints and stains. However, for paints
based on emulsions with 5% acrylic acid the flow curves were
almost independent of the particle size.
Yet in both cases the influence of the amount of acrylic acid
on the flow behaviour was different from the parent emulsions
(f.i. fig. 5). The flow curves of the stains based on the
emulsion without and with 5% acrylic acid are positioned at
higher viscosities than the flow curve of the stain based on
the emulsion with 2% acrylic acid. On the other hand the
parent emulsions showed flow curves of 0 and 2% acrylic acid
emulsions that were equal and positioned at lower viscosities
than the flow curve of the 5% AA containing emulsion.(fig. 4)
The maximum in the flow curve was only noticed for stains
based on emulsions without acid monomer (150 and 270 nm) and
both for the upgoing as well as for the downgoing flow
curves.

This phenomenon can be explained by formation of aggregates at well defined shear rates. At low shear rates deflocculation is induced by Brownian motions. At high shear rates deflocculation is caused by shear forces. The limitation of this behaviour for emulsions without acid monomer, indicates a lack of stabilization. In these emulsions the absence of charged acrylic acid groups, results in less electronic repulsions between the particles. Apparently the remaining electronic stabilization (emulsifier) can not prevent flocculation. The fact that it was not possible to prepare a stain or paint from the 100 nm emulsion without acrylic acid supports this explanation. This emulsion possesses the largest total surface area and consequently the lowest number of stabilizing groups per unit of surface area.

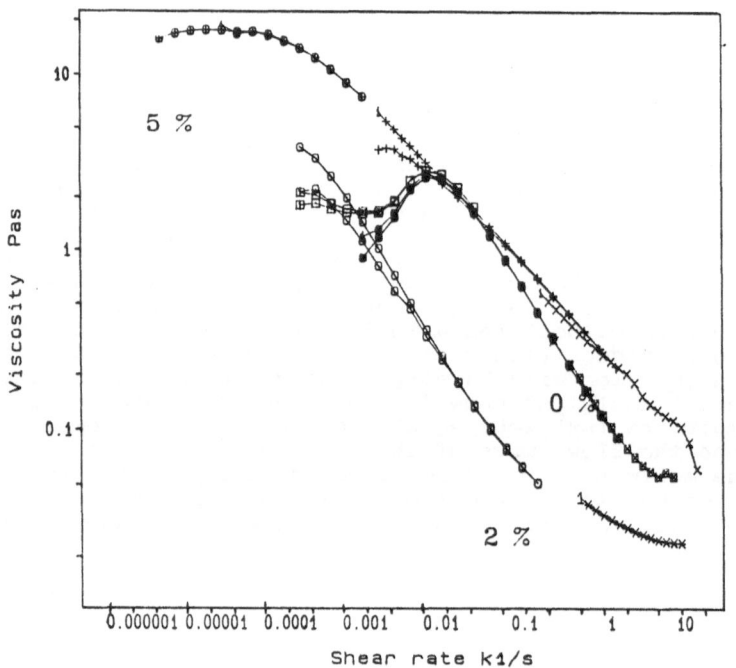

Figure 5. Influence of the amount of acrylic acid on the flow behaviour of stains, based on emulsions with particle sizes of ca 150 nm.

From figure 6 it becomes evident that the flow curves of stain and parent emulsion are very similar. Except at low shear rates where the viscosity of the stain is much lower. In contrast the viscosity of the paint is higher at all shear rates concerned. Similar pictures were obtained for the other acrylic acid containing emulsions, paints and stains. In contrast to the emulsions the paints and especially the stains possess a Newtonian flow region at low shear rates. This Newtonian flow region has already been predicted from oscillatory shear measurements as a decrease in elasticity and is probably induced by the employment of PU thickeners in the paints and stains. It is known that their non-specific hydrophobic interactions between latex and pigment particles result in associative networks (5, 6). The higher amount of PU-thickener in the stains accounts for their more pronounced Newtonian flow at low shear rates. The good resemblance of the flow behaviour of stains and parent emulsions at higher shear rates indicates that these networks are relatively weak. Finally the larger weight share of emulsion in the stains compared to the paints must account for the the better resemblance between stain and emulsion compared to paint and emulsion.

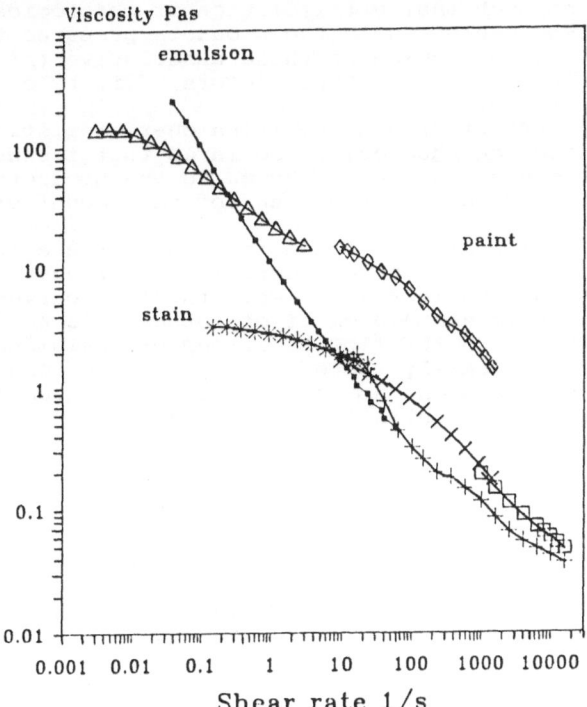

Figure 6. Comparison of the flow curves of paint and stain with the flow curve of the parent emulsion with 2% acrylic acid and 103 nm particles.

CONCLUSIONS

It is evident that the rheological behaviour of emulsion products cannot simply be extrapolated from the rheological behaviour of the parent emulsion. The preservation of the rheological behaviour of emulsions in products prepared therefrom is best at large deformations (high shear rates) and for higher shares of the emulsion in the end product.The largest differences in rheological behaviour for emulsions, paints and stains are found at low shear rates (small deformations). This indicates that the formation, weakening or disruption of networks is the main tool in the modification of the rheological profile of an emulsion (product). Oscillatory shear measurements provide a quick, sensitive and safe way to establish the changes in these network structures, while in contrast steady shear flow measurements always possess the danger of deformation of these delicate structures, even at low shear rates. Besides it has been shown that, in the low shear rate region, the results of oscillatory shear measurements and steady shear measurements correlate very well.
Some additives, especially thickeners, modify the rheological behaviour so much that extrapolation of rheological behaviour of the parent emulsions to the products prepared therefrom is hampered. Yet the action of these associative thickeners itself is influenced by other factors, f.i. by co-solvents and plasticizers.
However the effect of some emulsion characteristics, f.i. particle size, on rheology is so large that trends found in the parent emulsions are preserved in the products prepared therefrom. This was not the case for the amount of acrylic acid.
For the emulsions it was found that an increase in particle size decreases the elastic properties, decreases the shearthinning behaviour and lowers the flow curves of emulsions. Up to amounts of 5% of methacrylic acid and irrespective of pH the flow behaviour of emulsions is not affected. Additionally the effect of acrylic acid on the flow behaviour of emulsions is only noticed for amounts larger than 2% and after neutralization.

ACKNOWLEDGEMENTS

The authors are grateful for the contribution of Mr. J. Palmen (DSM Research BV, Geleen) in the rheological measurements on methacrylic acid containing emulsions and for his helpful discussion throughout the study.

181

REFERENCES

1. Bohlin, L., Proc. XIIIth Internat. Conf. in Org. Coat. Sci. & Techn., Athens, 1987, 35

2. Ferry, J.D., Visco-elastic Properties of Polymers, John Wiley & Sons, 2nd edition, New York (1972)

3. Buscall, R., Goodwin, J.W., Hawkins, M.W., Ottewill, R.H., J. Chem. Soc., Faraday Transl. 1, **78**, No. 10, 2873 (1982)

4. Blom, C., Mellema, J., Lopulissa, J.S., Reuvers, A.J., Colloid and Polymer Sc, **262**, 397 (1984)

5. Coatings Techn., **50**, 640 (1978)

6. Rokowski, J.M., Schaller, E.J., Aviles, R.G., Proc. 14 of XIVth Intern. Conf. in Org. Coat. Sci. and Techn., Athens, 1988, 37

7a. Jenkins, R., Silebi, C.A., El. Aaser, M.S., Vanderhoff, J.W., Graduate Research Progress Reports (Emulsion Polymers Institute, Le High), **30**, 141 (1988)
  b. idem, Progress Reports **31**, 174 (1988)
  c. Gebr. Borchers AG. Brochure No **MB-508**, 4

CONTINUOUS EMULSION POLYMERISATION IN A PULSED PACKED COLUMN

G.F.M. HOEDEMAKERS, D. THOENES
Department of Chemical Engineering
Eindhoven University of Technology
P.O.Box 513, 5600 MB Eindhoven, The Netherlands

ABSTRACT

The emulsion polymerisation of styrene was carried out in a pulsed packed column reactor, followed by a series of three continuous stirred tank reactors. The process was completely stable, and the products were identical to products made in a batch reactor under the same conditions, as regards particle number (per unit volume) and average molecular weight. Also the reaction rate was exactly the same.

INTRODUCTION

Emulsion polymerisation reactions are mostly carried out in batch reactors. They are easy to operate and there is a large amount of information about these reactor systems available in the literature. Recently there has been considerable interest in the use of continuous reactor systems for emulsion polymerisation. Economic incentives and better possibilities for controlling product quality are the main motives in the development of continuous emulsion polymerisation processes.

The reactor system that is mostly used in continuous emulsion polymerisation is a series of continuous stirred tank reactors (CSTR's). Although this reactor system is already used in the commercial production of rubber latices, the system has some disadvantages which prevent it to be used on a large scale for general applications.

Firstly there is the problem of particle formation. In a series of CSTR's the first CSTR can be regarded as the reactor in which the particle formation takes place: the seeding reactor. Because the rate of polymerisation is usually proportional to the number of polymer particles present, it is important to optimize the number of particles that is formed in the first reactor. However, a CSTR as the first reactor has shown to produce a much lower number of polymer particles than a batch reactor operated under the same conditions. A second problem of a CSTR used for a continuous emulsion polymerisation is that in a CSTR sustained oscillations in conversion and particle number are often observed. These oscillations mostly result in large fluctuations in product quality. It was found that both problems were caused by the large residence time distribution in a CSTR and could be avoided when a plug flow type reactor was used as the first reactor instead of a CSTR.

In this paper a new plug flow type reactor for an emulsion polymerisation is presented: a pulsed packed column. This reactor can be used as a seeding reactor for a series of CSTR's, but may also be used for emulsion polymerisations up to almost 100% conversion. The performances of the new reactor will be compared with those of the conventional reactor types: a CSTR and a batch reactor.

DEVELOPMENT OF A NEW PLUGFLOW TYPE REACTOR

One of the problems of a CSTR used as the seeding reactor is, that the number of polymer particles (per unit volume) produced in this reactor remains far below the number of particles produced in a batch reactor. Nomura et al. [1] have predicted the existence of a maximum in the number of polymer particles at a certain low value of the mean residence time of the CSTR. According to their theory, which was in good agreement with their experiments, this maximum number is only 57% of the number of polymer particles, per unit volume, formed in a batch reactor. The same value for this maximum is obtained by Poehlein [2] who based his calculation on a model of Gershberg and Longfield [3]. This relatively low number of polymer particles is mainly caused by the large residence time distribution in a CSTR.

This causes large particles to be mixed up with freshly added emulsifier. As a result a large amount of the emulsifier is consumed for covering the surface of the large particles.

Therefore only part of the emulsifier is available for the generation of new particles. Increasing the number of polymer particles can be realized by replacing the CSTR by a plug flow reactor, that is characterized by the absence of any residence time distribution. Normura et al. [1] have shown that the number of particles per unit volume can then be increased up to the batch level. They also showed, in accordance with results of Greene et al. [4], that this can avoid the problem of sustained oscillations that are often observed in a CSTR. However, the liquid flow in a plug flow reactor has to be turbulent, for three reasons:

1. avoiding coalescence and creaming up of the monomer droplets (de-emulsification);
2. prevention of reactor fouling and wall polymerisation;
3. sufficient radial mixing for removing the heat of reaction through the reactor wall.

The first and last reasons are especially important for scaling up of the reactor system. Because of the requirement of a turbulent flow in the reactor, it is not feasible to use an ordinary tubular reactor as the seeding reactor. In such a reactor the required turbulence can only be realized at very high liquid velocities. In emulsion polymerisations, where long residence times are necessary to obtain a high conversion, this would lead to the application of a number of extremely long tubes, in parallel, which is unpractical.

If we examine the plug flow type reactors that are used recently in continuous emulsion polymerisation researches, it appears that almost all systems are difficult to scale up. Greene et al. [4] have used a spiralized teflon tube, in which they created plug flow by alternately injecting nitrogen and emulsion in plugs. Gosh and Forsyth [5] and Lee and Forsyth [6] used spiralized stainless steel tubes with outer diameters of 1/2-inch and 1/4-inch respectively. Lin and Chiu [7] used a static mixer with 20 elements in a cilindrical pipe (Toray Hi-mixer) for avoiding de-emulsification.

A new reactor type, that can be scaled up rather easily and in which the problems of de-emulsification and heat transfer could be minimized, is a pulsed packed column (PPC). Until now a pulsed packed column was mainly used in extraction processes (Simons [8]), where it was successful because of its ability to combine the properties of a turbulent flow and a reasonable plug flow In one example, the PPC was used as a chemical reactor (Simons [9]). Figure 1 shows a schematic drawing of a pulsed

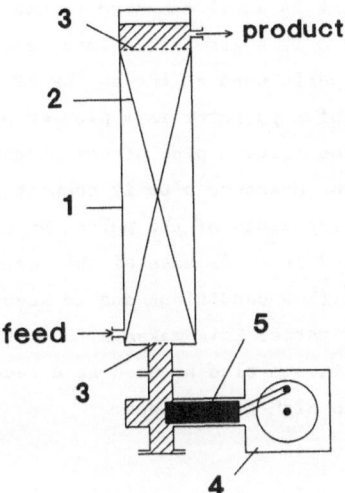

Figure 1. Pulsed packed column.
(1) column; (2) packing particles; (3) sieve plate; (4) pulsator; (5) plunger.

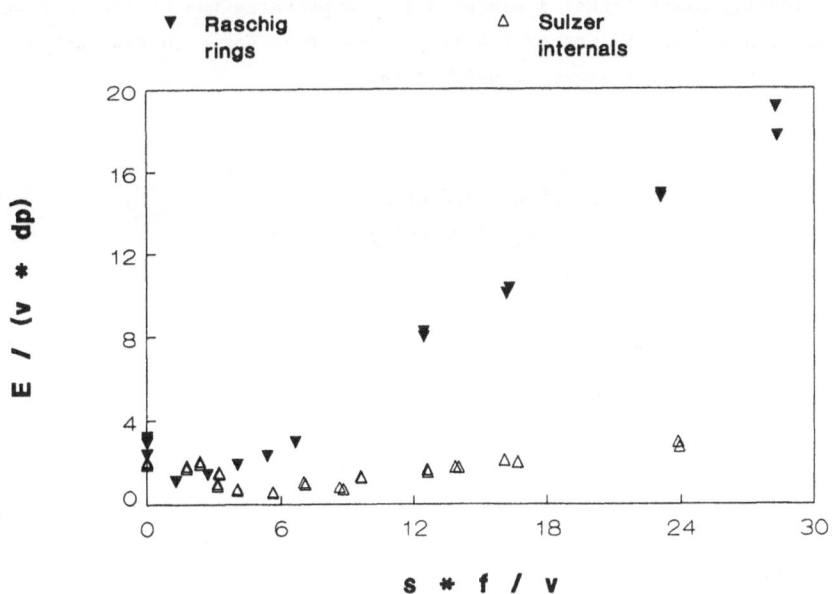

Figure 2. Axial dispersion coefficient (E) as a function of the pulsation velocity (s.f).

packed column. The column is filled with a packing material, e.g. Raschig rings, with a size that is small compared to the column diameter. At the top of the column there is a gas-liquid interface.

The pulsator is positioned at the bottom of the column. The most simple configuration of a pulsator is a plunger pump from which the valves are removed. The suction pipe of the plunger pump is blocked with a blind flange, and the pressure pipe is connected to the column. The strokelength and the frequency of the pulsation can be adjusted and are independent of the feed rate. Because of this construction it is possible to maintain turbulent flow conditions and to keep the monomer emulsified even at very low feed rates. This makes a PPC not only suitable as a seeding reactor, but it can also be used as a reactor for emulsion polymerisations, up to high conversions.

RESIDENCE TIME DISTRIBUTION IN A PULSED PACKED COLUMN

The residence time distribution in a vessel can be expressed in various ways. If it is not too large, one can consider the vessel as a series of N_T ideally mixed tanks. A second way of expressing the residence time distribution is the use of Peclet-numbers (Pe). For a pulsed packed column the Peclet-number is defined as

$$Pe = \frac{L \cdot v}{E} \tag{1}$$

where: E = axial dispersion coefficient
v = interstitial liquid velocity
L = column length

If $N_T \geq 10$, the Peclet-number is related to N_T by

$$N_T = \frac{Pe}{2} \tag{2}$$

As can be seen from equation (1) there are two contributions that determine the residence time distribution in a PPC of known length: the interstitial liquid velocity (v) and the axial dispersion coefficient (E)

The axial mixing was measured experimentally, as a function of pulsation velocity (i.e. the product of the frequency f and the stroke length s) with two types of internals: 10 mm Raschig rings and Sulzer packing SMV8-DN50. The results are shown in fig. 2. It appears that axial mixing increases with increasing pulsation velocity (above a certain minimum), and that the Sulzer packing gives much better results than the Raschig rings.

EXPERIMENTAL

Materials

The monomer used was industrial grade styrene. The inhibitor t-parabutyl catechol was removed by extraction with 5 weight% sodium hydroxide solution. The monomer feed ratio was 0,5 g per g of water. Sodium laurylsulfate was used as emulsifier (0,05 mol/l water). Potassium persulfate was used as initiator (0,01 mol/l water). The solution was buffered with sodium hydroxide (to P_H - 10).

Apparatus

Figure 3 shows the equipment that was used for the emulsion polymerisation experiments. The water phase and the monomer phase were prepared separately. The water phase was prepared by batchwise dissolving the emulsifier, the initiator and the buffer in distilled water. Both phases were fed continuously to a premixer in which the emulsion was prepared.

The premixer had a height of 5 cm, a diameter of 10 cm and was equipped with a 6 cm eight-bladed turbine impeller. The impeller speed was 500 rpm. The temperature in the premixer was kept constant at 10°C.

From the premixer the emulsion was fed to the column. In the experiments two type of columns were used. One had an inner diameter of 50 mm and was packed with 10 mm glass Raschig rings; the fraction free volume in the column was about 78%. The other column had an inner diameter of 55 mm and was packed with stainless steel Sulzer SMV8-DN50 internals; the fraction free volume in this column was about 99%. Both columns were made of glass and consisted of 5 sections of 1 m each. Samples could be drawn from the top and the bottom of the column and between all sections. The pulsator had a frequency range from 0 to 3.5 s^{-1} and a volume displacement that could be adjusted between 0 and 20 ml, which corresponds to a stroke length between 0 and 13.0 mm for the Raschig rings and between 0 and 8.5 mm for the Sulzer internals.

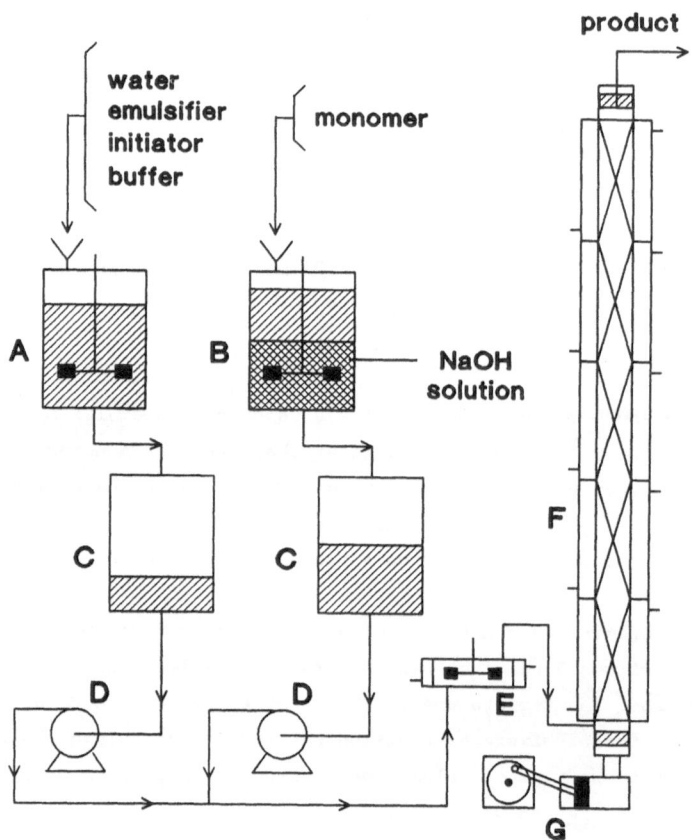

Figure 3. Equipment used for the emulsion polymerization experiments.
(A) mixing vessel for preparation of the aqueous phase; (B) mixing vessel for
sodium hydroxide washing of the monomer; (C) storage vessels; (D) metering
pumps; (E) premixer; (F) pulsed packed column; (G) pulsator.

Analysis

Monomer conversion was determined gravimetrically. The volume average
diameters of the polymer particles were measured by dynamic light-
scattering with a Malvern autosizer IIc. The number of polymer particles
was calculated from the monomer conversion and the volume average
diameters of the particles. Residence time distributions in the pulsed
packed columns were determined by conductivity measurements with sodium
chloride in distilled water used as a tracer.

Results and discussion

In fig. 4 the experimental results are shown for polymerisation
experiments in a batch reactor and a PPC. In order to find the influence
of residence time, samples were taken along the length of the column. It
appears that the experiment with the higher flow rate gives the same
conversion/time-relationship as the batch experiments. The relatively
lower conversions at the lower flow rate (at the same residence times)
are attributed to the effect of axial mixing.

In fig. 5 the results are shown for two values of the pulsation
velocity, compared with batch experiments. From these two graphs follows
that the ppc behaves as a plug flow reactor, if the flow rate is above a
certain minimum, and if the pulsation velocity is below a certain maximum.

In the figures 6 and 7 the performances of the PPC are compared with
the performances of CSTR's and a batch reactor. The black squares are
results from experiments carried out at different flow rates
corresponding to meean residence times as indicated on the abscissa. The
Peclet-numbers of these experiments ranged from 10 to 50. The figures
show clearly that at a pulsation velocity of 45.5 mm/s the PPC behaves
exactly like a batch reactor up to residence times of 60 minutes (Pe =
20). At larger residence times a slight decrease can be observed in
reaction rate and number of polymer particles formed, which is attributed
to axial mixing. However, even at these large residence times the
performances of the pulsed packed column are much better than those of a
CSTR. A second effect that can be seen in both figures is the fact that
the number of polymer particles formed in a CSTR has a maximum at a very
low residence time and that this maximum is about 55% of the number of
particles obtained in a batch reactor as predicted by Nomura et al. [1].
Increasing the number of CSTR's in series shows to have only marginal
effect on the reaction rate and number of particles formed.

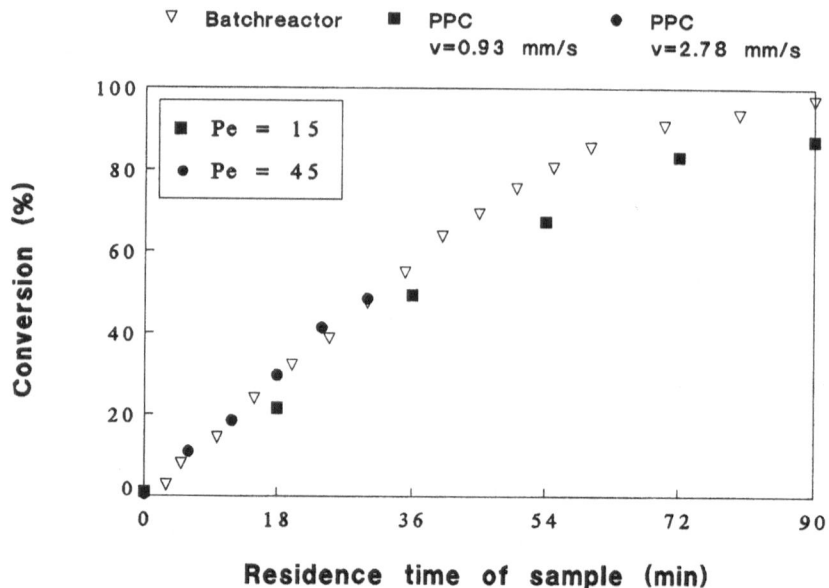

Figure 4. Influence of flow rate on styrene emulsion polymerization. Raschig rings packing; pulsation velocity s.f = 45.5 mm/s.

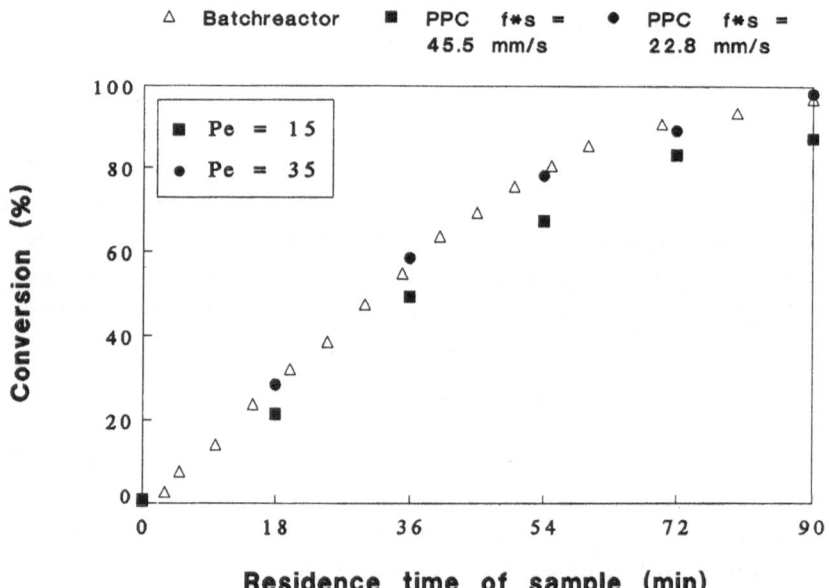

Figure 5. Influence of pulsation velocity on styrene emulsion polymerization. Raschig rings packing; pulsation velocity s.f = 45.5 mm/s.

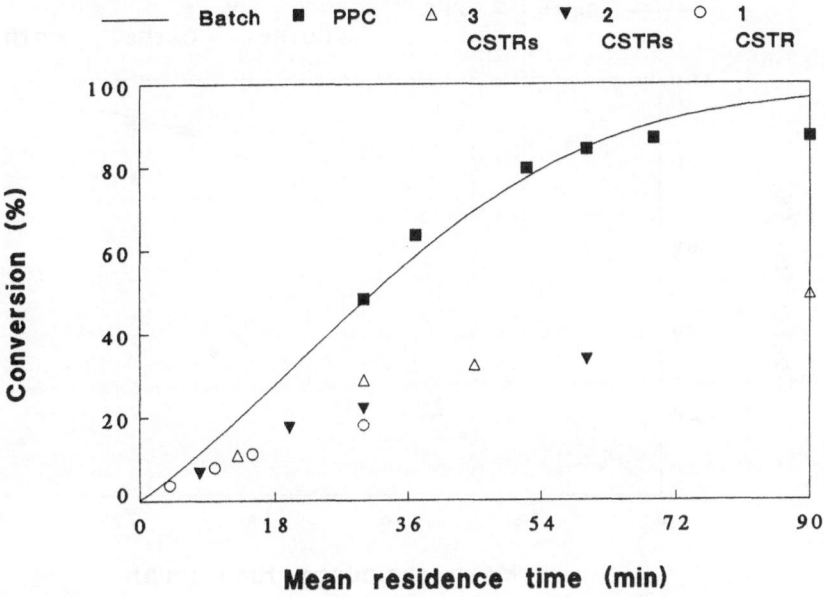

Figure 6. Comparison of reactor types: conversion as a function of mean residence time. Raschig rings; pulsation velocity s.f = 45.5 mm/s.

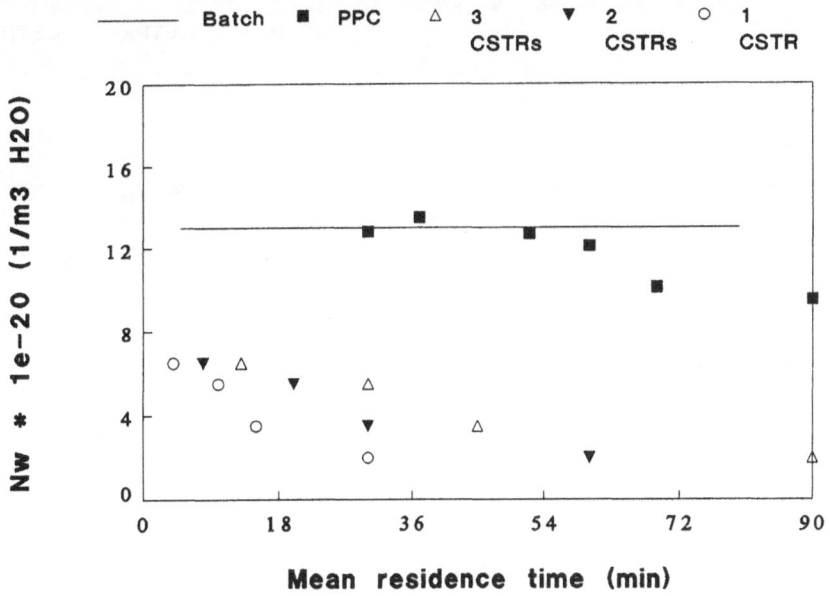

Figure 7. Comparison of reactor types: number of particles (Nw) as a function of mean residence time. Raschig rings; pulsation velocity s.f = 45.5 mm/s.

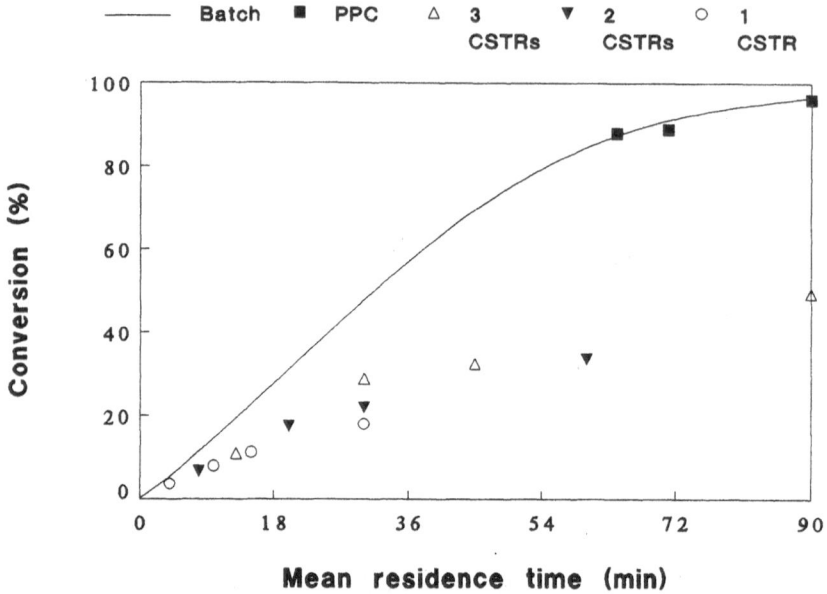

Figure 8. Comparison of reactor types: conversion as a function of mean residence time. Sulzer internals; pulsation velocity s.f = 29.8 mm/s.

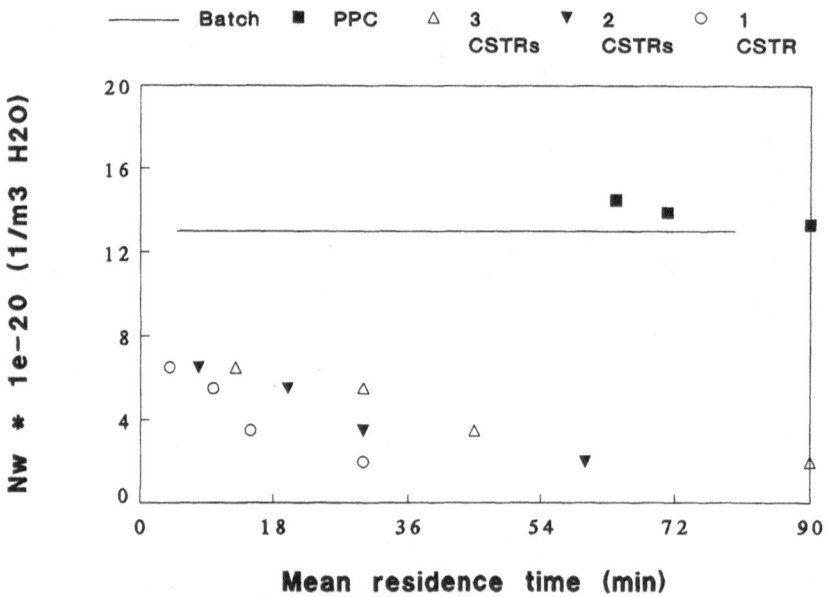

Figure 9. Comparison of reactor types: number of particles (Nw) as a function of mean residence time. Sulzer internals; pulsation velocity s.f = 29.8 mm/s.

The figures 8 and 9 show results of experiments with the Sulzer internals, with a pulsation velocity of 29.8 mm/s. The Peclet-numbers of the experiments ranged from 90 to 130. The lower axial mixing in this type of column makes it possible to apply much longer residence times (lower flow rates) and still approach plug flow.

CONCLUSIONS

The main conclusion from these experiments is, that a pulsed packed column is a very suitable reactor system for a continuous emulsion polymerisation. In our experiments, the reactor was completely stable during runs that lasted up to ten hours. The performance of this reactor type is equal to that of a batch reactor. At very long residence times, where the reaction rate and the number of particles formed in the column decline slightly, the performance of the pulsed packed column is still much better than that of one or several CSTR's in series.

The residence time distribution, which is the main cause of a declining performance of the reactor, can be minimized by increasing the flow rate or decreasing the pulsation velocity. However, a certain minimum pulsation velocity is necessary to avoid de-emulsification and fouling, and to guarantee a sufficient heat transfer rate to the reactor wall. Regularly ordered packing types such as Sulzer internals give better results than Raschig rings.

REFERENCES

1. M. Nomura, H. Kojima, M. Harada, W. Eguchi, S. Nagata, J.Appl.Polym. Sci., 15, 675-691, (1971)
2. G.W. Poehlein, "Emulsion Polymerisation in Continuous Reactors" Chapter in Emulsion Polymerisation, I. Piirma, Ed. (1982), New York, Academic Press
3. D.B. Gershberg, J.E. Longfield, Simp.Polym.Kinetics and Catalyst Systems. Preprints 10, 45th AIChE Meeting, New York, (1961)
4. R.K. Greene, R.A. Gonzalez, G.W. Poehlein, ACS Symp.Ser., 24 (Emulsion Polymerisation), 341-358, (1976)
5. M. Gosh, T.H. Forsyth, ACS Symp.Ser., 24 (Emulsion Polymerisation), 367-378, (1976)
6. C.K. Lee, T.H. Forsyth, ACS Symp.Ser., 165 (Emulsion Polym.Emulsion Polym.), 567-575, (1981)
7. C.C. Lin, W.Y. Chiu, J.Appl.Polym.Sci., 27, 1977-1993, (1982)
8. A.J.F. Simons Ph.D. Thesis, Steady-state and transient behaviour of systems in pulsed packed columns for liquid-liquid extraction, Geleen, The Netherlands, (1987)
9. A.J.F. Simons, Chem.Ind.nr. 19, Oct. 7, 748-757 (1978)

Part 4

LIQUID CRYSTALLINE POLYMERS

BIPHASIC BEHAVIOR OF MELTS OF LIQUID CRYSTALLINE COPOLYESTERS CONTAINING BOTH MESOGENIC AND NON-MESOGENIC UNITS

SUBRATA BHATACHARYA AND ROBERT W. LENZ
Polymer Science and Engineering Department,
University of Massachusetts, Amherst, MA 01003

ABSTRACT

A series of thermotropic, liquid crystalline copolyesters was prepared which contained both mesogenic and non-mesogenic units. The melt state anisotropy of the copolymers varied greatly with composition, and within an intermediate compositional range, the copolymers formed biphasic melts, which contained both nematic and isotropic phases. The amount of the nematic phase and the continuity of that phase decreased with increasing non-mesogenic unit content until no anisotropic phase could be observed by polarized light microscopy. The term "degree of liquid crystallinity" is proposed to describe and estimate the fraction of the anisotropic phase present in the biphasic compositional range. For one biphasic copolyester sample, the two phases were separated by solvent fractionation and characterized for their unit composition and liquid crystalline properties.

INTRODUCTION

Thermotropic liquid crystalline, LC, homopolymers and copolymers, which are composed entirely of rigid, linear, aromatic ester units, generally have very high melting temperatures, T_m, and very poor solubilities, even in the most aggressive solvents [1]. For this reason, either flexible spacers or non-linear aromatic ester units are often inserted into these polymers to lower their melting temperatures in order to improve their melt processing characteristics without eliminating their thermotropic behavior [2]-[5]. However, for both types of modifications the aspect ratio of the mesogenic sequences of these homopolymers or copolymers is generally substantially decreased so that the thermal stability of the LC state can be significantly lowered. When flexible spacers are used for this purpose, an isotropization temperature, T_i, is often observed, but with non-linear, non-mesogenic rigid spacers, the T_i may still be above the thermal decomposition temperature of the polymer. Nevertheless, if a sufficiently large fraction of rigid, non-mesogenic units is included in such a copolymer, a spontaneous phase separation can occur in the melt at a temperature well below T_i, and, as a result, the LC phase can coexist with an isotropic phase, possibly in equilibrium.

Several reports have appeared recently describing the biphasic behavior of thermotropic copolyesters containing both mesogenic and non-mesogenic units. In several of these reports, considerable attention was given to the phase separation behavior which occurs, in particular, in the copolyester containing 60 mole % of mesogenic p-oxybenzoate, OB, units and 40 mole % of non-mesogenic ethylene terephthalate, ET, units [6]-[12]. Wunderlich and coworkers [6] observed two glass temperatures, T_g, for this copolymer at approximately 60° and 180°C, and assigned these to separate phases, which were PET-rich and POB-rich, respectively. Baird and coworkers [7] observed directly the formation of two phases, above a T_m of approximately 200°C, by electron microscopy and by polarized light microscopy. One of the phases had an observable nematic texture while the other appeared isotropic. They concluded that the nematic phase was rich in POB units. Very similar results and interpretations were also reported by Sawyer [8]. Lewis and coworkers [9] detected two distinct relaxation peaks for this copolyester by dynamic mechanical thermal analysis, DMTA, and also assigned these to PET-rich and -poor phases. The fractional amounts of these two phases have been determined both by Nicely, based on broad line NMR measurements [10], and by Economy and coworkers, based on solvent fractionation and gravimetric determinations [11]. Both groups of investigators were able to estimate the repeating unit compositions of the two phases, and each concluded that the isotropic PET-rich phase contained between 55 and 65% ET units while the nematic POB-rich phase contained between 72 and 80% OB units, depending on the method of analysis. Recently, in a joint investigation with Porter and Sun, we have also been able to observe the two phases present in this copolymer by polarized light microscopy and by light transmission properties and to characterize their T_g behavior by DMTA [12].

Isotropic-nematic biphasic behavior has also been reported for similar thermotropic copolyesters with flexible spacers by Bilibin and coworkers [13] and by Lenz and coworkers [14][15]. The latter proposed the concept of "degree of liquid crystallinity" to describe the observed biphasic behavior and to estimate the fraction of the LC phase present as a function of the molar ratio of mesogenic, M, to non-mesogenic, N, units in such

copolymers [16]. Stupp and coworkers, on the other hand, have suggested
for their "chemically disordered" copolyesters with flexible spacers that
the N units form only defects in the LC phase, and they refer to their copolymers as having a "fringed micelle LC" morphology [17][18].

Several groups have described a similar biphasic behavior for all
aromatic M-N copolyesters, which contained only rigid, non-linear N units
[3][19][20]. Our earliest report on such a system was based primarily on
a polarized light microscopy study of several series of copolyesters of
chlorohydroquinone terephthalate, M, containing varying amounts of different non-linear biphenols, N, of the following general structure [5]:

$$\left[O\text{-}\underset{Cl}{\underset{M}{\bigcirc}}\text{-}O\overset{O}{\overset{\|}{C}}\text{-}\bigcirc\text{-}\overset{O}{\overset{\|}{C}}\right]_m \left[\sim\sim O\text{-}\bigcirc\text{-}R\text{-}\bigcirc\text{-}O\overset{O}{\overset{\|}{C}}\text{-}\bigcirc\text{-}\overset{O}{\overset{\|}{C}}\right]_{1-m}$$

in which R = O, CH_2, SO_2 and $C(CH_3)_2$ and m = 0.2-1.0. The visually observed intensity of the stir-opalescence of the copolyesters in these
series appeared to correlate closely with the mole fraction of N units,
above a certain threshold amount of these units, and the threshold amount
varied with the type of N unit in the copolyester.

RESULTS AND DISCUSSION

Table 1 lists the compositions, inherent viscosities, phase transition

TABLE 1
Properties of M/N copolyesters with R = $C(CH_3)_2$ in N

Polymer No.	M/N Mole Ratio	η_{inh} [a] dl/g	Stir [b] Opalescence	T_m °C	ΔH_m cal/g	T_c °C	ΔH_c cal/g
1	100:0	--	Very Strong	367	6.23	340	3.59
2	90:10	--	Very Strong	353	2.77	322	3.36
3	80:20	--	Very Strong	349	0.15	318	0.14
4	70:30	0.57	Strong	346	0.9	312	1.23
5	60:40	0.55	Weak	347	0.22	312	0.57
6	56:44	0.49	Weak	335	0.19	314	0.65
7	53:47	--	Very Weak	313 [c]	0.36	--	--
8	50:50	0.42	None	305 [c]	0.39	--	--
9	40:60	0.46	None	303 [c]	0.54	--	--
10	20:80	0.51	None	344 [c]	9.08	--	--
11	0:100	0.51	None	354 [c]	7.11	--	--

[a] Inherent viscosity measured on a 0.5% solution in p-chlorophenol at 50°C.
[b] Stir-opalescence observed at T_m + 20°C.
[c] Transition temperatures determined by DSC in the first heating cycle.

temperatures and the qualitative measure of liquid crystallinity by stir-opalescence for a series of copolyesters in which R is $C(CH_3)_2$ in N. All of the copolymers containing 30 mole % or more of bisphenol-A were soluble in p-chlorophenol, and the higher the amount of bisphenol-A in the copolymers, the greater was their solubility in p-chlorophenol. No quantitative study of the molecular weights of the samples was made, but it is believed that because the solution viscosities of all of the samples in the series were very close they were of approximately the same molecular weight and had similar molecular weight distributions.

For Polymers 7 to 11, no exotherms were observed in the cooling cycles of the DSC thermograms, and consequently, no endotherms could be found in the subsequent heating cycles. The melting temperature, T_m, was found to decrease with an increase in the amount of bisphenol-A up to Polymer 9, but thereafter T_m increased with further increase in bisphenol-A content. The enthalpy of melting was found to decrease steadily up to Polymer 6, then increased afterwards. Similarly, the temperature and enthalpy of crystallization were found to decrease with increasing amounts of bisphenol-A in the copolymers, but the unusually low enthalpy values for melting and crystallization of Polymer 3 cannot be explained. Table 2 lists the elemental analyses of all of the copolymer samples.

TABLE 2
Elemental analyses of polymers in Table 1

Polymer No.	M:N Mole Ratio	Elemental Analyses[a]	
		C	H
1	100:0	60.57 (61.22)	2.65 (2.57)
2	90:10	63.01 (63.23)	2.91 (2.89)
3	80:20	63.25 (65.12)	3.36 (3.18)
4	70:30	67.59 (66.9)	3.63 (3.46)
5	60:40	69.22 (68.6)	3.96 (3.73)
6	56:44	69.24 (69.25)	3.77 (3.83)
8	50:50	70.44 (70.2)	3.98 (4.02)
9	40:60	71:59 (71.72)	4.34 (4.22)
10	20:80	74.28 (74.53)	4.79 (4.66)
11	0:100	76.0 (77.01)	4.95 (5.06)

[a]The bracketed figures denote the theoretically calculated values.

The stir-opalescence, which is a qualitative measure of the melt anisotropy, was found to be very strong for Polymers 1, 2 and 3. After that it was found to decrease steadily from Polymers 4 to 7. Polymers 8 to 11 did not show any stir-opalescence.

Figure 1 shows the wide angle X-ray diffraction (WAXD) patterns of the

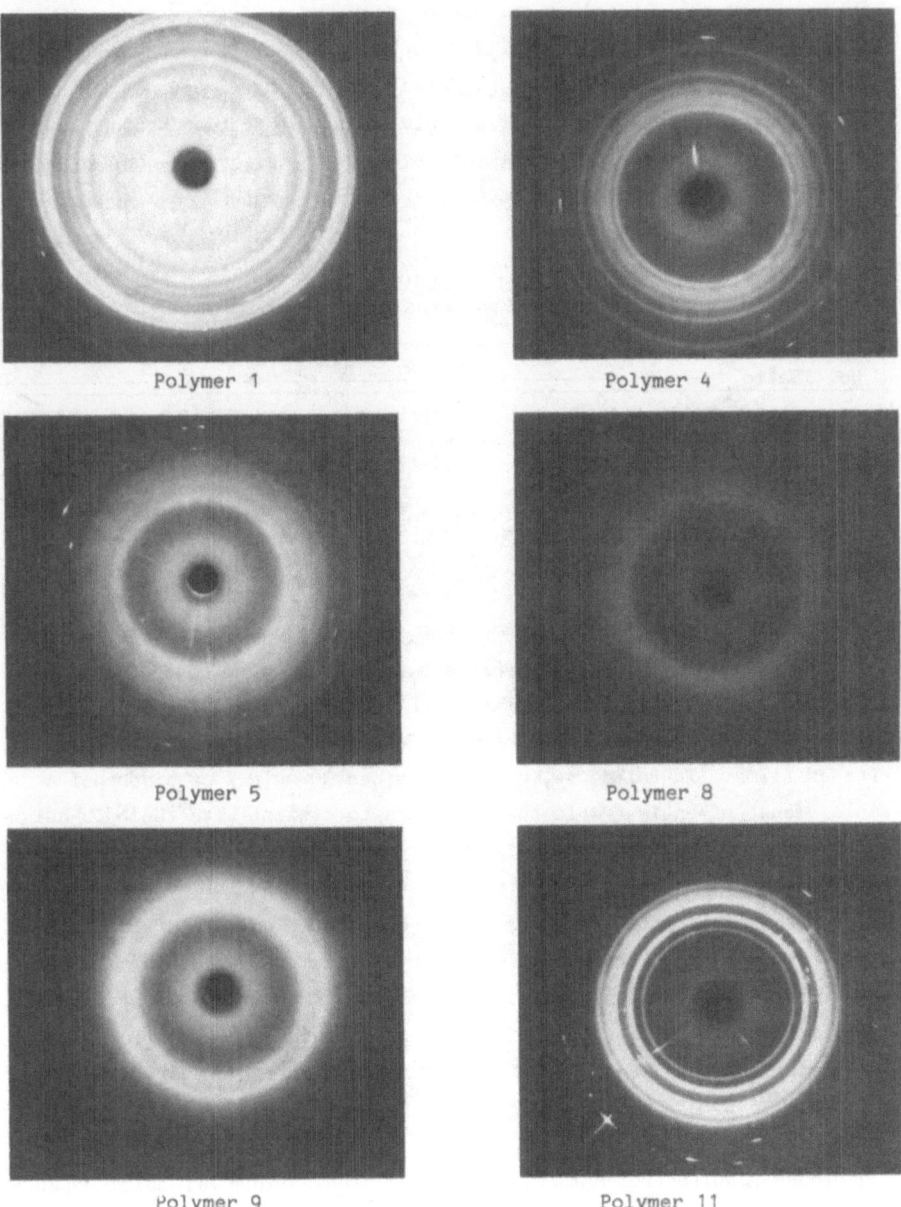

Figure 1. Wide angle X-ray diffraction patterns of representative polymers
in Table 1.

copolymer samples which contained different levels of bisphenol-A, and Table 3 lists the d-spacings calculated from the diffractograms. With increasing bisphenol-A the crystalline patterns of poly(chlorohydroquinone terephthalate) became weaker as expected, and Polymer 8 did not have a well defined crystalline phase; that is, only remnants of chlorohydroquinone terephthalate crystallinity could be observed. For samples containing more than 50 mole % of bisphenol-A, the crystalline structure of bisphenol-A terephthalate became prominent.

TABLE 3
Analysis of WAXD patterns of related polymers

Polymer No.	M:N Ratio	d Spacings (A)									
1	100:0	25.63	14.92	--	11.45	7.43	--	6.05	5.46	4.54	4.32
3	80:20	25.63	14.92	--	11.45	7.43	--	6.05	5.46	4.54	4.32
4	70:30	25.63	14.92	--	11.45	7.43	6.45	6.05	5.46	4.54	4.32
5	60:40	25.63	14.92	--	--	7.43	6.45	6.05	5.46	4.54	4.32
8	50:50	25.63	--	12.87	--	7.43	--	--	5.46	4.54	--
9	40:60	25.63	--	--	12.87	--	7.43	--	--	5.46	--	3.94
10	20:80	25.63	--	12.87	11.32	7.43	6.41	5.84	5.46	--	3.94
11	0:100	25.63	--	12.87	11.32	7.43	6.41	5.84	5.46	--	3.94

Figure 2 shows the photomicrographs of the polymers at the same relative temperature: T_m + 20°C. Polymers 1, 2 and 3 were found to show a homogeneous, volume filling, nematic Schlieren texture. The lines appearing in the photomicrograph of Polymer 1 are due to cracks in the melt-pressed films. In Polymer 4, which contained 30 mole % of bisphenol-A, a small amount of an isotropic phase appears to coexist with the bulk anisotropic phase. The amount of this anisotropic phase was found to decrease gradually with an increase in bisphenol-A content until Polymer 7, for which the nematic phase could no longer be identified. The illuminated streaks appearing in the photomicrograph of Polymer 7 were probably caused by a shearing of scattered anisotropic domains when pressing the film in the melt state. Polymer 8 showed only birefringent spots, while Polymer 9 was almost entirely isotropic with only a few birefringent spots barely visible. No birefringence was observed for Polymers 10 and 11 under the same thermal conditions.

These results from optical microscopy studies were strongly supported by the stir-opalescence observations. Thus, it appears that the type and extent of melt state anisotropy of the chlorohydroquinone-bisphenol-A terephthalate copolymers, under comparable thermal conditions, are direct functions of the bisphenol-A composition of the copolymers, and the higher

the amount of the non-mesogenic units, N, the lower the extent of melt
anisotropy until a composition is reached where further inclusion of non-
mesogenic units destroys the anisotropy altogether. That is, the various
copolymer compositions can be differentiated in terms of the extent of
their melt anisotropy, and we propose to refer to this variation with copolymer compositions as an indication of the "degree of liquid crystallinity" of the sample.

In small angle light scattering (SALS) studies of these copolymers,
circularly symmetrical scattering patterns were obtained above T_m for
Polymers 1 to 5 under H_v polarization, but the scattering intensity was observed to decrease gradually from Polymer 1 to Polymer 5. Polymer 8 showed
a rod-like scattering pattern with lobes at 45° to the polar directions.
Polymer 9 also showed a very faint rod-like scattering pattern of the same
type as that of 8. No scattering patterns were observed for Polymers 10
and 11 above T_m. The SALS studies were carried out with the same samples
used in microscopic studies so that the scattering patterns represent the
scattering behaviors of the copolymer samples at their respective temperatures of T_m + 20°C. The circularly symmetrical scattering patterns observed in Polymers 1 to 5 are believed to arise from the scattering from
the various LC domains as a function of their individual orientation
vectors, with the result that randomization of the total scattering from
the system occurred.

The gradual decrease in scattering intensity from Polymers 1 to 5 may
mean that either the concentration of the scattering domains decreased
gradually, or that the order parameter of the LC domains decreased gradually, or both as the amount of bisphenol-A increased in the copolymers.
At very high compositions of bisphenol-A in the copolymers, the scattering
from the few individual domains became prominent. The domains at that
composition mainly consisted of a relatively small number of chlorohydroquinone terephthalate-rich polymer molecules with a net orientation in one
direction, so a rod-like scattering pattern with an intensity gradually
decreasing from the center along the azimuthal direction was observed. Indeed, this type of scattering pattern usually occurs as a result of scattering from rod-like, one-dimensional scattering entities. Polymers 10 and
11, containing 30 and 100 mole % of bisphenol-A, did not exhibit anisotropy
in the melt state, and, consequently, no scattering patterns were observed
for those samples. The SALS observations very closely support the stir-
opalescence and microscopic results.

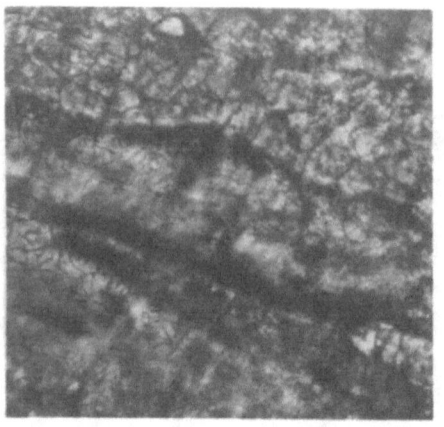

Polymer 1

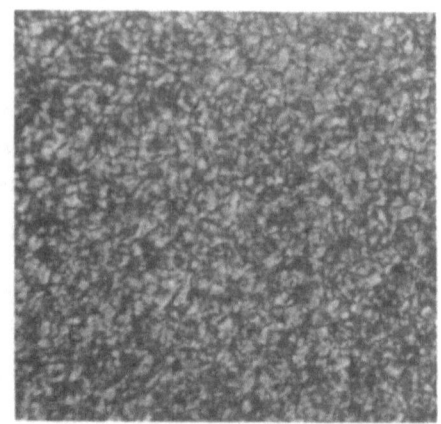

Polymer 3

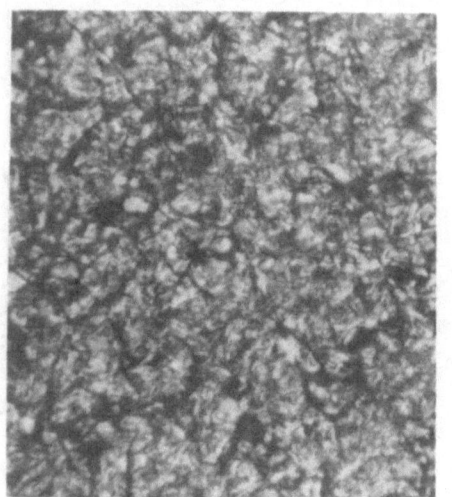

Polymer 4

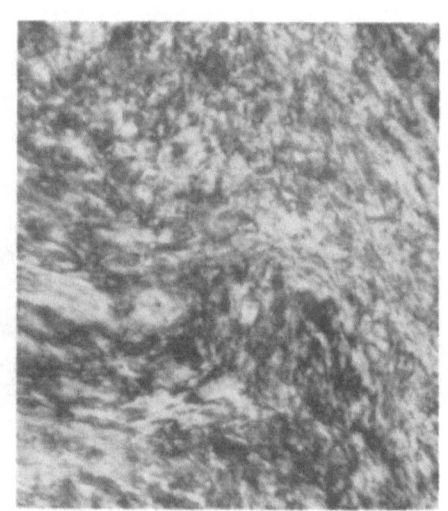

Polymer 5

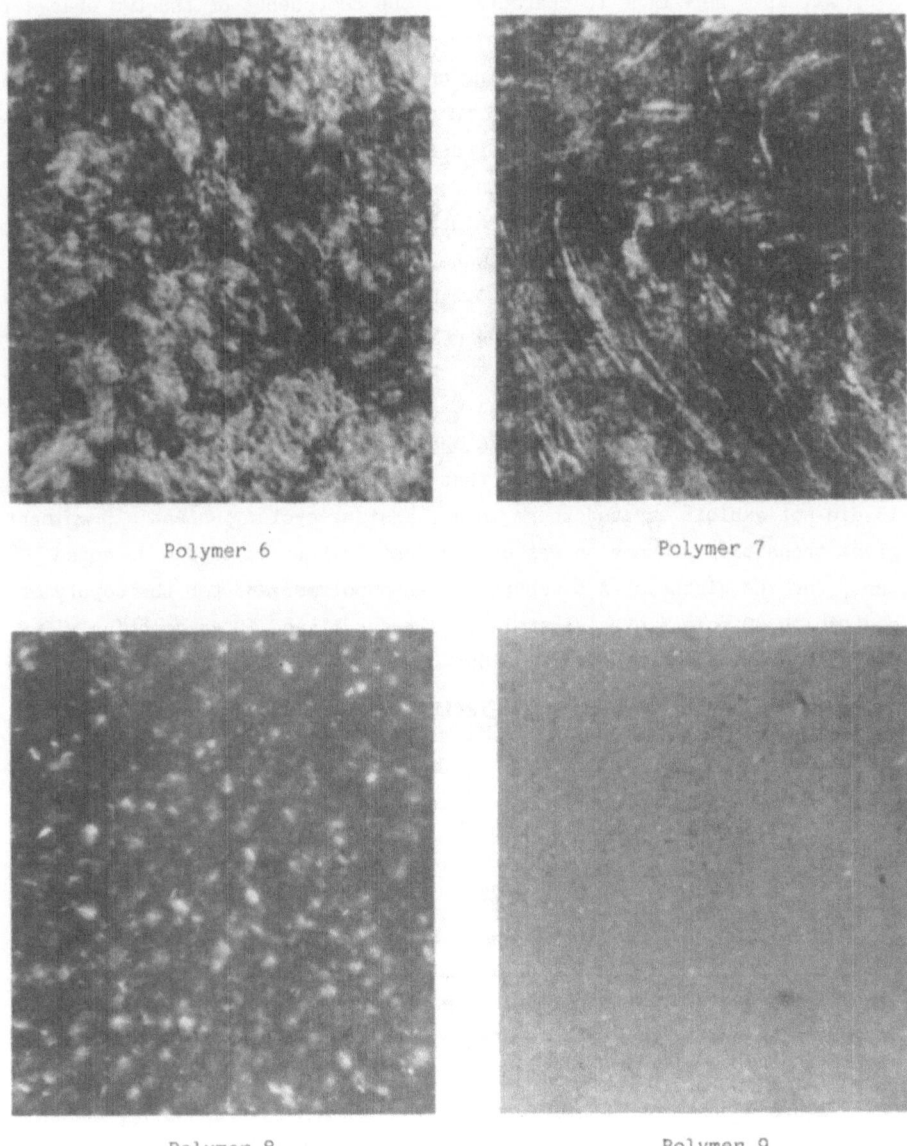

Polymer 6

Polymer 7

Polymer 8

Polymer 9

Figure 2.   Photomicrographs of representative polymers in Table 1 at the same relative temperature: T_m + 20°C (320X).

Attempts were made to characterize the components of the two phases in the biphasic systems. It was previously observed, as discussed above, that in an LC copolymer system exhibiting anisotropic-isotropic biphase, the co-monomers were unevenly distributed in the two phases. The physical separation of the two phases by solvent fraction of Polymer 6, which contained 56 mole % of chlorohydroquinone, is described in the Experimental section. The chloroform-insoluble and the chloroform-soluble fractions (Samples 12 and 13, respectively, of Table 4) were studied by DSC and optical microscopy. Figure 3 shows the DSC thermograms of Samples 6, 12 and 13 in the first heating cycles. It is observed that both Samples 6 and 12 showed well-formed binodal endotherms of the same shape, with maxima occurring at around 310°C. The endotherm in the case of Sample 12 was, however, larger in size than that for Sample 6. In addition, the enthalpy of melting of Sample 12 was 4.17 cal/g, whereas that of Sample 6 was 1.41 cal/g. Sample 13 did not exhibit any endotherm in any heating cycle; instead a prominent glass transition phenomenon was observed at 184°C. It should be noted that, for the bisphenol-A terephthalate homopolymer and for the copolymer containing 80 mole % of bisphenol-A (Polymers 11 and 10, respectively), a glass transition was observed at 185°C.

TABLE 4
Fractionation of Polymer 6 anbd preparation of equivalent copolymers

Polymer No.	M:N Mole Ratio	Stir Opalescence	T_g[a] °C	T_m[a] °C	ΔH_m cal/g	Texture
6	56:44	Weak	--	311	1.41	Biphasic
12	78:22	Very Strong	--	310	4.17	Nematic
13	25:75	None	184	--	--	Isotropic
14	78:22	Very Strong	--[b]	327	1.96	Nematic
15	25:75	None	--	--	--	Isotropic

[a]Transition temperatures determined from first heating cycles.
[b]A very weak endothermic slope change of the DSC scan, starting from about 160°C and continuing up to about 190°C, was observed in this sample.

Figure 4 shows the photomicrographs of Samples 6, 12 and 13, respectively, at 325°C. It is clearly seen that while Sample 6 showed a biphasic texture, Sample 12 showed a volume-filled nematic texture, and Sample 13 was isotropic.

From the DSC and optical microscopy results on Samples 6, 12 and 13, it is clear that the chloroform-soluble fraction (Sample 13), which formed an isotropic melt, is essentially identical in properties to the bisphenol-A

207

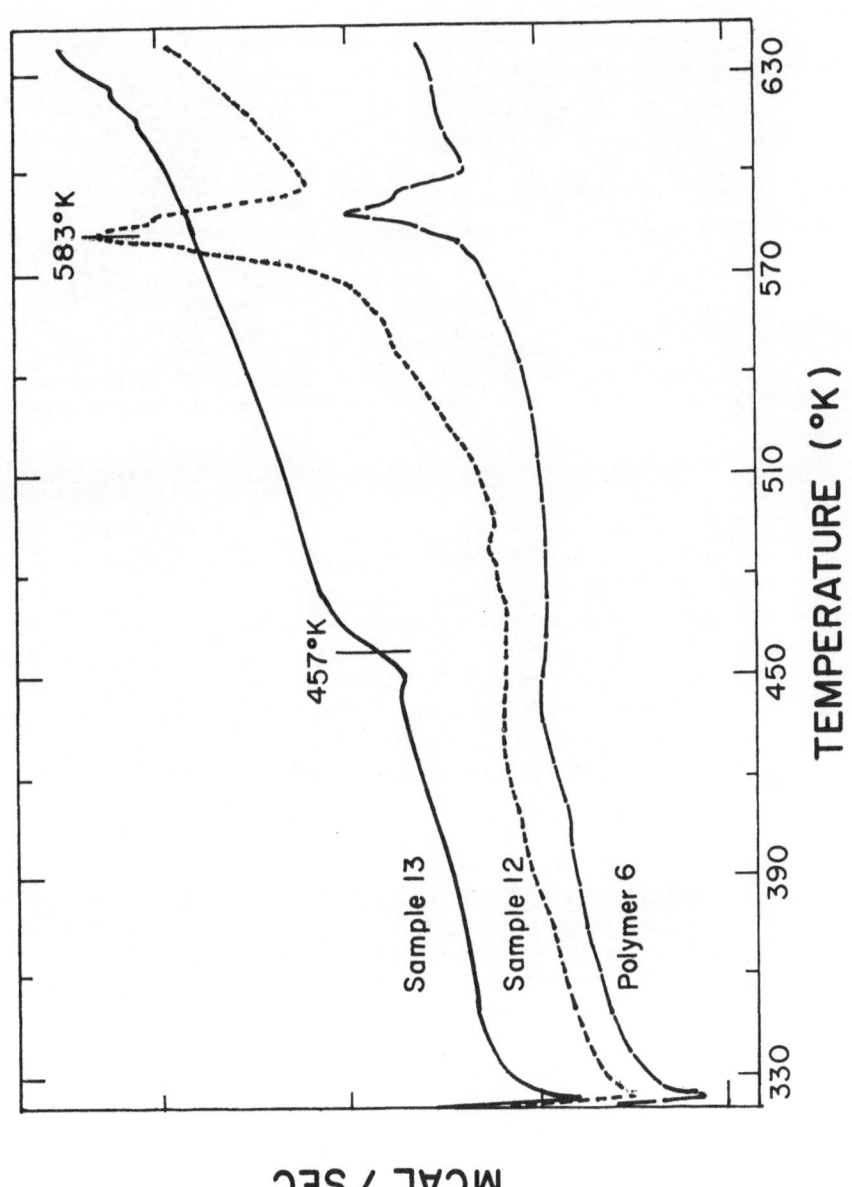

Figure 3. DSC thermograms of Polymer 6 and fractionated Samples 12 (insoluble and 13 (soluble) of Table 4; first heating cycle.

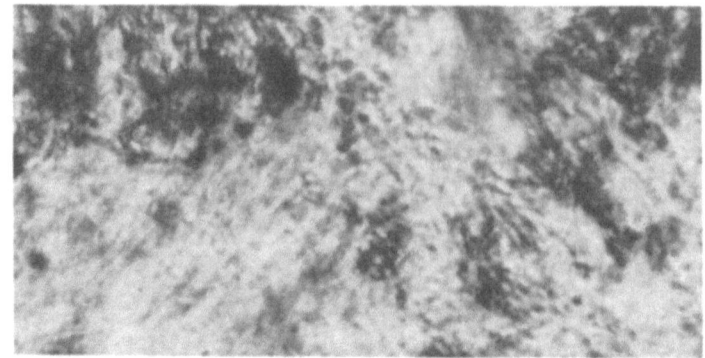

Polymer 6

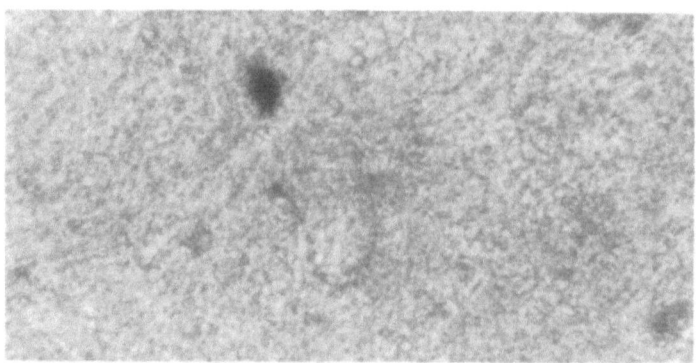

Sample 12

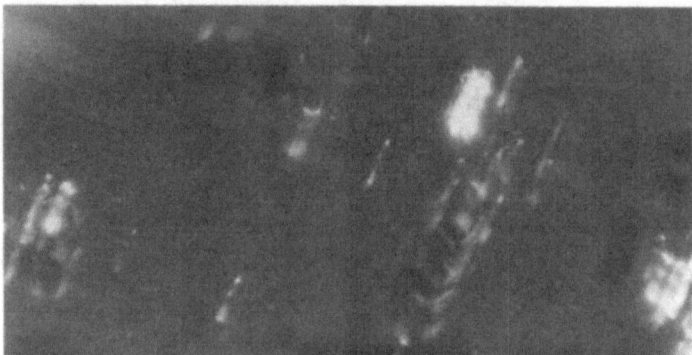

Sample 13

Figure 4. Photomicrographs of Polymer 6 and fractionated Samples 12
(insoluble) and 13 (soluble) of Table 4 at 325°C (320X).

terephthalate homopolymer. Conversely, the chloroform-insoluble fraction is essentially identical to the chlorohydroquinone terephthalate homopolymer and exhibited a volume-filled nematic texture on melting. The elemental analyses of Samples 6, 12 and 13 indeed confirmed this conclusion. Sample 12 was found to contain 78 mole % of chlorohydroquinone terephthalate, while Sample 13 contained 75 mole % of bisphenol-A terephthalate. Table 4 lists the characteristic parameters for Samples 6 and 12-15. A physical separation of the two phases was also carried out on Polymer 4, which contained 70 mole % chlorohydroquinone, and the chloroform-insoluble and -soluble fractions were found to contain 81 mole % and 11 mole % of chlorohydroquinone terephthalate, respectively.

The observation of the non-uniform distribution of the two comonomers in the two phases leads to the conclusion that individual copolymer molecules, although random in composition, did not exist in two phases simultaneously. That is, depending on their overall composition, the copolymer molecules exist in either one phase or another. This arrangement would differ from that of a fringed micelle morphology of the type that occurs for crystalline polymers. The uneven distribution of the two constituent units in the two phases is reasonable for nematic polymers in which long-range orientational order is imposed on the mobile polymer molecules in the nematic environment, which could then result in an exclusion of non-mesogenic rich copolymer molecules from the nematic domains.

Polymers 14 and 15 have the same composition as Samples 12 and 13, the chloroform-insoluble and -soluble fractions from Polymer 6, respectively. An endothermic peak at 327°C was observed in the first heating cycle of Polymer 14, and no endothermic peaks were observed for Polymer 15 under the same conditions. Sample 16 was a 1:1 by weight mixture of Polymers 14 and 15. Figure 5 shows the photomicrographs of Polymer 14 and Sample 16 at 335°C. Polymer 15 was found to be completely isotropic, while Polymer 14 showed a volume-filled nematic texture, and Sample 16 showed a biphasic texture similar to Polymer 6.

CONCLUSIONS

The similarity in the terms "degree of liquid crystallinity" and "degree of crystallinity" may seem to imply that these two terms originate from the same type of molecular concepts and from similar justifications. This implication, however, is not intended. While the definitions of the two terms

210

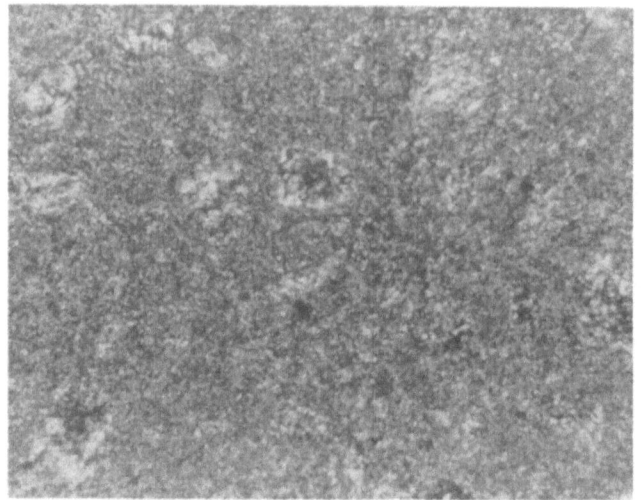

Polymer 14

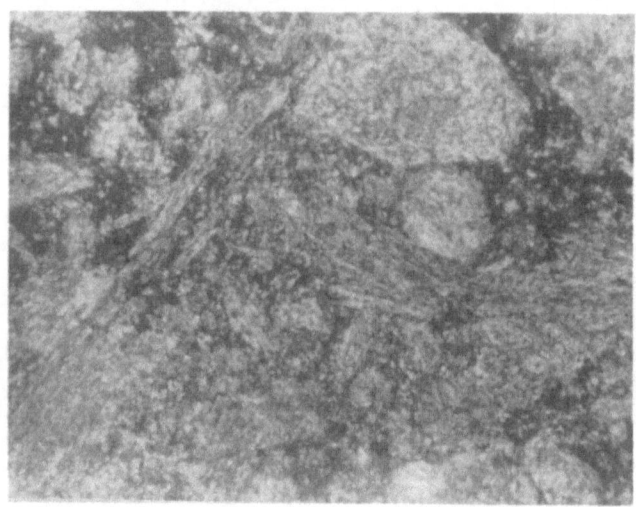

Sample 16

Figure 5.   Photomicrographs of Polymer 14 of Table 4 and Sample 16 at
335°C (320X).

originate from the same type of necessity (that is, of categorizing copolymer systems in terms of their extent of molecular aggregation in any physical state), and while it is possible to estimate both parameters quantitatively by measuring some physical properties of the copolymers under various conditions, the two terms differ considerably in their fundamental significance.

There can only be a three-dimensional molecular aggregation in the crystalline phase of a semi-crystalline polymer, but in the nematic state, the state of order will vary with any change in the extent of molecular orientation or order parameter. This variation will in turn affect some of the physical properties of the system. Hence, the relative amounts of the anisotropic and the isotropic phases present in any LC system will affect the physical properties in a similar manner as do the relative amounts of the crystalline and amorphous phases present in a semi-crystalline polymer. Likewise, the molecular weight and molecular weight distribution are known to affect the biphasic behavior of both crystalline and LC systems. Also, as with biphasic semi-crystalline copolymers, it is clear that in LC biphasic systems, in addition to molecular weight distribution, there will also be a compositional distribution of the mesogenic and the non-mesogenic units in the two phases, and this compositional heterogeneity can fully account for the presence of the two phases in the biphasic systems.

Recently, Stupp and coworkers have proposed the concept of "polyflexibility" to explain the effects of compositional heterogeneity. They have observed experimentally that a "monoflexible" but polydisperse LC copolymer has sharp transition temperatures, thereby supporting the fact that compositional heterogeneity in random copolymers plays at least as important a role as the molecular weight distribution in determining the phase behavior in such systems [18].

EXPERIMENTAL

Copolymer Preparation

The copolymers were synthesized by melt polymerization reactions from the diacetates of the two diols and terephthalic acid. Chlorohydroquinone diacetate and bisphenol-A diacetate were purified by recrystallization from toluene and ethanol, respectively. Terephthalic acid, obtained from Amoco Chemical Co., was used without further purification. After melt polymerization a final reaction was carried out under high vacuum at temperatures slightly less than the melting point of the polymers for an hour to increase the molecular weight of the polymers. Eleven samples containing various amounts of the two comonomers were synthesized including the chlorohydroquinone terephthalate and bisphenol-A terephthalate homopolymers. The polymers were serially numbered from 1 to 11 with decreasing amounts of chlorohydroquinone in their compositions as listed in Table 1.

Copolymer Fractionation

The separation of the two phases present in Polymer 6 containing 56 mole % of M units was carried out by refluxing a 0.5 g sample with a large volume of chloroform followed by filtration. This procedure was repeated several times until no polymer was found in the filtrate. The chloroform-soluble fraction was reprecipitated in methanol. The chloroform-insoluble and -soluble fractions were then washed thoroughly with acetone, dried for 48 hours in vacuum and serially numbered Samples 12 and 13, respectively. Two more copolymer samples were synthesized with the same chemical compositions as Samples 12 and 13; these were Polymers 14 and 15, respectively. Polymers 14 and 15 were mixed in 1:1 proportions by weight to give Sample 16. The mixing was carried out by grinding equal amounts of the two copolymers in a mortar for 15 minutes to a very fine powder and homogenizing the mixture by mechanical stirring.

Characterization

Inherent viscosities were measured with a Cannon-Ubbelhode type viscometer. Thermal properties were determined with a Perkin-Elmer DSC-2, using heating and cooling rates of 20°C/min. Unless otherwise specified, the transition temperatures and other thermodynamic parameters were determined from the second heating and cooling cycles. A qualitative comparison of stiropalescence of the copolymers was carried out on a Fisher-Jones melting point apparatus at T_m + 20°C for all samples. A Carl-Zeiss polarizing microscope was used, and film samples for microscopy were prepared in the Fisher-Jones apparatus by melting the polymers at T_m + 20°C, pressing into films, and quenching the films. The sample films were approximately 15-25 μm in thickness. Small angle light scattering studies were carried out on the same samples using an He-Ne gas laser (6358 A). Wide angle X-ray difraction patterns of powdered samples were obtained with a Staton II camera using point collimated nickel filtered CuK radiation and a typical exposure time of 3 hours.

ACKNOWLEDGEMENT

The support of this work by the Materials Research Laboratory of the

University of Massachusetts, which is funded by the National Science Foundation, is gratefully acknowledged.

REFERENCES

1. Goodman, I., McIntyre, J. E. and Stimpson, J. W., U.S. Patent 3,321,437, 1967; Goodman, I., McIntyre, J. E. and Aldred, D. H., U.S. Patent 3,368,998, 1967; Cottis, S. G., Economy, J. and Nowak, B. E., U.S. Patent 3,637,595, 1972.

2. Ober, C. K., Jin, J.-I. and Lenz, R. W. in Advances in Polymer Science, Vol. 59, M. Gordon and N. A. Plate, Eds., Springer-Verlag, Berlin, 1984, p. 103.

3. Griffin, B. P. and Cox, M. K., Brit. Polym. J., 1980, 12, 147.

4. Antoun, S., Lenz, R. W. and Jin, J.-I., J. Polym. Sci., Polym. Chem. Ed., 1981, 19, 1901.

5. Lenz, R. W. and Jin, J.-I., Macromolecules, 1981, 14, 1405.

6. Menczel, J. and Wunderlich, B., J. Polym. Sci., Polym. Phys. Ed., 1980, 18, 1433; Meesiri, W., Menczel, J., Guar, U. and Wunderlich, B., ibid., 1982, 20, 719.

7. Joseph, E., Wilkes, G. L. and Baird, D. G., Polymer, 1985, 26, 689.

8. Sawyer, L. C., J. Polym. Sci., Polym. Lett. Ed., 1984, 22, 347.

9. Benson, R. S. and Lewis, D. N., Polym. Commun., 1987, 28, 289.

10. McFarlane, F. E., Nicely, V. A. and Devis, T. G. in Contemporary Topics in Polymer Science, Vol. 2, E. M. Pearce and J. R. Schaefgen, Eds., Plenum Press, New York, 1977, p. 109 ff; Nicely, V. A., Dougherty, J. T. and Renfro, L. W., Macromolecules, 1987, 20, 573.

11. Quach, L., Hornbogen, E., Volksen, W. and Economy, J., J. Polym. Sci.: Part A, Polym. Chem., 1989, 27, 775.

12. Sun, T., Bhattacharya, S. K., Lenz, R. W. and Porter, R. S., submitted for publication.

13. Bilibin, A. Y., Pashkovsky, E. E., Tenkovtsev, A. V. and Skorokhodov, S. S., Makromol. Chem., Rapid Commun., 1985, 6, 545.

14. Chen, G. and Lenz, R. W., J. Polym. Sci., Polym. Chem. Ed., 1984, 22, 3189.

15. Zhang, W., Jin, J.-I. and Lenz, R. W., Makromol. Chem., 1988, 189, 2219.

16. Lenz, R. W., Farad. Disc. Chem. Soc., 1985, 79, 21.

17. Moore, J. S. and Stupp, S. I., Macromolecules, 1987, 20, 282.

18. Stupp, S. E., Moore, J. S. and Martin, P. G., Macromolecules, 1988, 21, 1217, 1222 and 1228.

19. Kleinschuster, J. J., Pletcher, T. C., Schaefgen, J. R. and Luise, R. R., Ger. Offen., 2,520,820, 1975.

20. Wunder, S. L., Ramachandran, S., Gochanour, C. R. and Weinberg, M., Macromolecules, 1986, 19, 1696.

POLYMERIZED DISCOTIC LIQUID CRYSTALS

J.F. VAN DER POL, E. NEELEMAN, R.J.M. NOLTE, J.W. ZWIKKER
AND W. DRENTH*
Department of Organic Chemistry, University at Utrecht,
Padualaan 8, 3584 CH Utrecht, The Netherlands

ABSTRACT

Octaalkoxyphthalocyanines with six to twelve carbon atoms in
each alkoxy chain are liquid crystalline. On heating, a
transition to the mesophase occurs in the region between 80 and
120 °C. In this phase, the disc-like molecules are arranged in
ordered stacks. These stacks show one-dimensional transport
properties. When the alkoxy chains are provided with
(meth)acryloyloxy groups at their terminal positions, the
molecules can be polymerized in their mesophase. As a
consequence of this polymerization the stack structure is
retained on cooling to ambient temperature.
Polymerization of an octaalkoxyphthalocyanine with an
acryloyloxy function in only one of the alkoxy groups yields an
acrylic side chain polymer with stacked phthalocyanine units.
A third polymer has been prepared, which has a central -Si-
O-Si-O- chain, each silicon atom being in the centre of an
octaalkoxyphthalocyanine.

INTRODUCTION

The property of liquid crystallinity is observed in a number of
organic compounds, the molecules of which are either rod- or
disc-like. In polymer chemistry most attention is paid to rod-
like molecules. Our group is engaged in a study of liquid
crystalline compounds consisting of disc-like molecules, viz.
alkoxy substituted phthalocyanines; see also Simon et al. [1].
The aim of our study is to find out whether these compounds are
liquid crystalline and if so whether they have columnar stacks
in their liquid crystalline phase and what the properties of
such stacks are.

RESULTS AND DISCUSSION

We prepared two series of octaalkoxy substituted phthalocyanines, one with a copper ion in the centre of the molecule (**1**) and one without a metal ion (**2**) [2,3]. In each series the number of carbon atoms in the side chains ranges from 6 to 12. The compounds were investigated by polarizing microscopy and differential scanning calorimetry. They are in a liquid crystalline state, the so-called mesophase, above a transition temperature which is in the region between 80 and 120 °C, depending on the number of carbon atoms in the chains and on the presence or absence of the copper ion. X-ray data reveal that in the mesophase these compounds have a so-called ordered hexagonal columnar structure, abbreviated as D_{ho}. 'Ordered' means that in a column all discs have equal interplanar distances. The columns are hexagonally arranged with respect to each other. (Figure 1).

Why do we prepare these compounds? The phthalocyanine rings have a conjugated π-system. Each column consists of a stack of these π-systems with an intracolumnar distance of 0.34 nm, surrounded by a hydrocarbon mantle. Therefore, the columns can be expected to show one-dimensional transport properties Transport of energy has indeed been concluded from photoluminescence experiments between 4.2 and 400 K. Photoluminescence is observed in the crystalline phase. On heating it almost vanishes at the transition to the mesophase. This behaviour has been explained by a relatively fast transport of excitons through the mesophase columns. The faster the transport, the higher the chance for an exciton to be trapped by a quenching site [4].

The electric conductivity of the samples has been measured by impedance spectroscopy. The conductivity is quite low; at 175 °C $\sigma = 6 \times 10^{-8}$ S/m [2]. This low conductivity is not surprising; it is due to the fact that each orbital is either almost completely filled or empty. The measured conductivity is probably related to the few electrons that at 175 °C have moved from the valence band to the conduction band. The conductivity, however, increases by several orders of magnitude when a sample is doped with iodine [3].

The conductivity has also been investigated by the Time Resolved Microwave Conductivity technique [5]. These measurements show an appreciable increase in conductivity when going from the crystalline phase to the mesophase.

When cooling down from the mesophase to ambient temperature the structure of the mesophase is lost at the transition temperature. Our aim was to retain the structure of the mesophase on cooling. We solved this problem by introducing a carbon-carbon double bond in each alkoxy chain [6]. For instance, phthalocyanine **4** was obtained by first synthesizing **3** and treating it with acryloyl chloride in dry chloroform at ambient temperature. Its phase behaviour is similar to that of the phthalocyanines with saturated chains. On heating, **4a** rearranges to the mesophase at 85 °C. In its mesophase it is completely miscible with compound **2** (n = 12) indicating that the acrylate functions do not appreciably effect the mesophase structure. Phthalocyanine **4a**, pure as well as in its mixtures with **2**, could be polymerized while in its mesophase with AIBN as initiator. With less than 0.5 mol % AIBN relative to the acrylate groups, a brittle polymer was obtained from pure **4a**. With 2 mol % of the initiator the product was elastic. Both polymers were totally insoluble in organic solvents and in concentrated sulfuric acid, indicating that polymerization had indeed taken place. Photo acoustic spectroscopy shows that some double bond absorption at 1620 cm^{-1} is still present. Thus, not all acrylate functions have reacted.

Small and wide angle X-ray diffraction pictures of the polymer at 25 °C are similar to those of the monomer in its mesophase. This similarity shows that the columnar order of the mesophase is indeed preserved on cooling the polymer to ambient temperature. Also other properties related to this order are preserved. For instance, the quantum efficiency of photoluminescence is low because of a fast transport of excitons to quenching sites. Remarkably, the electric conductivity is relatively high; at 175 °C $\sigma = 1 \times 10^{-5}$ S/m. This enhanced conductivity is probably not caused by defects in the phthalocyanine core. We checked this by refluxing the polymer in *tert*-butanol in the presence of potassium *tert*-butanolate. Starting material **3** could be recovered, indicating that the phthalocyanine core had not been attacked during the polymerization.

The foregoing phthalocyanines all had eight equal chain substituents. Another group of phthalocyanines, **5** and **6**, was prepared that contain apart from six alkoxy chains a methoxy group and one chain with an acryloyloxy or methacryloyloxy function at its terminal position [7]. On heating, these monomers show a transition from the crystalline phase to a D_{ho} mesophase, e.g. **5** at 52 °C. Polymerization could be achieved in oxygen-free solution in benzene at a temperature of 45 °C applying 12.5 % AIBN. In this way the polymer of **5** had an apparent molecular weight of 45,000 as determined by gel permeation chromatography, calibrated by polystyrene standards. The purified polymers are soluble in chloroform, toluene, and THF, but insoluble in polar solvents such as methanol and DMF. UV/Vis spectra of the solutions reveal that the pendent phthalocyanines are aggregated. X-ray diffraction shows that at ambient temperature the polymer of **5** is at least partially crystalline and that the crystalline structure consists of hexagonally arranged columns of phthalocyanine units. On heating up to 200 °C the columnar arrangement becomes more perfect. Between 0 and 200 °C neither a transition to a mesophase nor a glass transition has been observed by DSC, although some softening of the sample is observed at approximately 180 °C.

Stacking of the phthalocyanines in these polymers is in agreement with their transport properties. For instance, just as explained above, migration of excitons is fast and, accordingly, the quantum efficiency of luminescence is low.

Finally, condensation polymers, **7**, have been obtained from octaalkoxy substituted dihydroxysiliconphthalocyanines; see also Wegner *et al.* [8,9]. The average degree of polymerization was estimated from small angle X-ray scattering measurements to be 100. At ambient temperature the polymer has a D_{ho} phase. The polymerization reaction is accompanied by a total loss of luminescence intensity. According to our model, this loss implies a very efficient migration of excitons through the stacks.

ACKNOWLEDGEMENTS

The authors acknowledge important contributions by Professor G. Blasse and his coworkers, and Dr. J.C. van Miltenburg, University at Utrecht, Drs. J.M. Warman and M.P. de Haas, University at Delft, and Dr. N. Heijboer and his associates, Akzo, Arnhem. The SAXS experiments were skilfully performed by Dr. Th. Sauer, Max-Planck-Institut für Polymerforschung, Mainz.

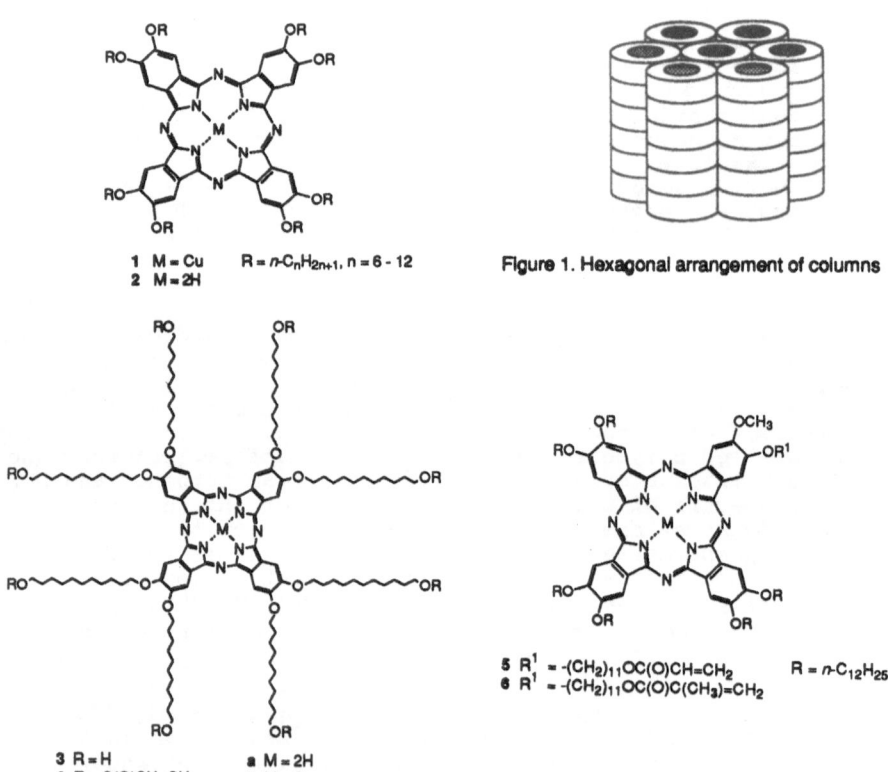

Figure 1. Hexagonal arrangement of columns

1 M = Cu   R = n-C_nH_{2n+1}, n = 6 - 12
2 M = 2H

3 R = H   a M = 2H
4 R = C(O)CH=CH$_2$ b M = Cu

5 R^1 = -(CH$_2$)$_{11}$OC(O)CH=CH$_2$ R = n-C$_{12}$H$_{25}$
6 R^1 = -(CH$_2$)$_{11}$OC(O)C(CH$_3$)=CH$_2$

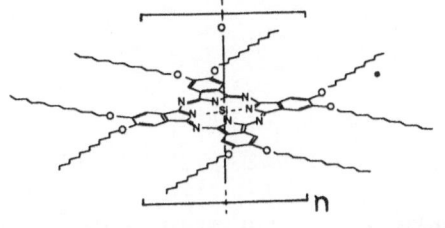

7   REFERENCES

1. Masurel, D., Sirlin, C. and Simon, J., Highly ordered
    columnar liquid crystal obtained from a new octasubstituted
    phthalocyanine mesogen. *New. J. Chem.*, 1987, **11**, 455-456

2. Van der Pol, J.F., Neeleman, E., Zwikker, J.W., Nolte,
    R.J.M. and Drenth, W., Evidence of an ordered columnar
    mesophase in peripherally octa-*n*-alkoxysubstituted
    phthalocyanines. *Recl. Trav. Chim. Pays-Bas*, 1988, **107**,
    615-620.

3. Van der Pol, J.F., Neeleman, E., Zwikker, J.W., Nolte,
    R.J.M., Drenth, W., Aerts, J., Visser, R. and Picken, S.J.,
    Homologues series of liquid crystalline metal free and
    copper octa-*n*-alkoxyphthalocyanines. Submitted to *Liq.
    Cryst.*

4. Blasse, G., Dirksen, G.J., Meijerink, A., Van der Pol, J.F.,
    Neeleman, E. and Drenth, W., Luminescence and energy
    migration in the solid state and in the ordered columnar
    mesophase of peripherally octa-*n*-dodecoxy-substituted
    phthalocyanine. *Chem. Phys. Lett.*, 1989, **154**, 420-424.

5. Warman, J.M., De Haas, M.P., Van der Pol, J.F. and Drenth,
    W., Electron tunnelling betweeen columnar aggregates of
    peripherally substituted phthalocyanines. Submitted to
    Nature.

6. Van der Pol, J.F., Neeleman, E., Van Miltenburg, J.C.,
    Zwikker, J.W., Nolte, R.J.M. and Drenth, W., A polymer with
    the mesomorphic order of liquid crystalline phthalocyanines.
    Submitted to *Macromolecules.*

7. Van der Pol, J.F., Neeleman, E., Nolte, R.J.M., Zwikker,
    J.W. and Drenth, W., Asymmetrically substituted liquid
    crystalline phthalocyanines and side-chain polymers derived
    from them. Submitted to *Makromol. Chem.*

8. Caseri, W., Sauer, Th. and Wegner, G., Soluble
    phthalocyanato-polysiloxanes: Rigid rod polymers of high
    molecular weight. *Makromol. Chem., Rapid Commun.*, 1988, **9**,
    651-657.

9. Sauer, Th. and Wegner, G., Control of the discotic to
    isotropic transition in alkoxy-substituted silicondihydroxo-
    phthalocyanines by axial substituents. *Mol. Cryst. Liq.
    Cryst.*, 1988, **162B**, 97-118.

SYNTHESIS AND THERMAL PROPERTIES OF POLYURETHANES WITH MESOGENIC UNITS AND SILICON CONTAINING FLEXIBLE SPACERS IN THE MAIN CHAIN

L. Willner, F. Braun, M. Heß and R. Kosfeld

Universität -GH- Duisburg, FB 6/Physikalische Chemie

Lotharstr.1, 4100 Duisburg 1, FRG

ABSTRACT

Polyurethanes with mesogenic units and flexible spacers in the main chain were prepared by addition reaction of aromatic diisocyanates and bis(ω-hydroxyalkyl)oligodimethylsiloxanes. The thermal properties were investigated by means of differential scanning calorimetry, thermogravimetry and polarizing microscopy. Liquid crystalline behaviour was observed for polyurethanes consisting of a 4,4"-p-terphenylylene moiety and of disiloxane containing spacer units. The LC polyurethanes possess mesophases in a temperature range not far from thermal decomposition. Some markedly different results were obtained by changing the alkyl/siloxane ratio. Thus, elongation of the alkyl part leads to an increase in thermal stability, while an extended siloxane segment causes the loss of the liquid crystalline properties.

INTRODUCTION

The thermotropic liquid crystalline behaviour of polyesters is well documented but only a few papers are concerned with the synthesis and thermal properties of LC polyurethanes [1]. This is partly due to the fact that urethanes possess comparatively high phase transition temperatures, caused by the high polarity of the carbamate linkage. An investigation of the mesophase behaviour close to the decomposition range is interrupted by thermal degradation of the material. Furthermore the existence of the liquid crystal range of urethanes [2] is reduced in comparison with corresponding esters [3].

In this paper we deal with polyurethanes of the type I (Scheme 1) with mesogenic units and silicon containing flexible spacers in the main chain. In order to lower the clearing and melting points we have introduced the highly flexible siloxane segments which are known for their strong reducing effects on phase transition temperatures [4, 5].

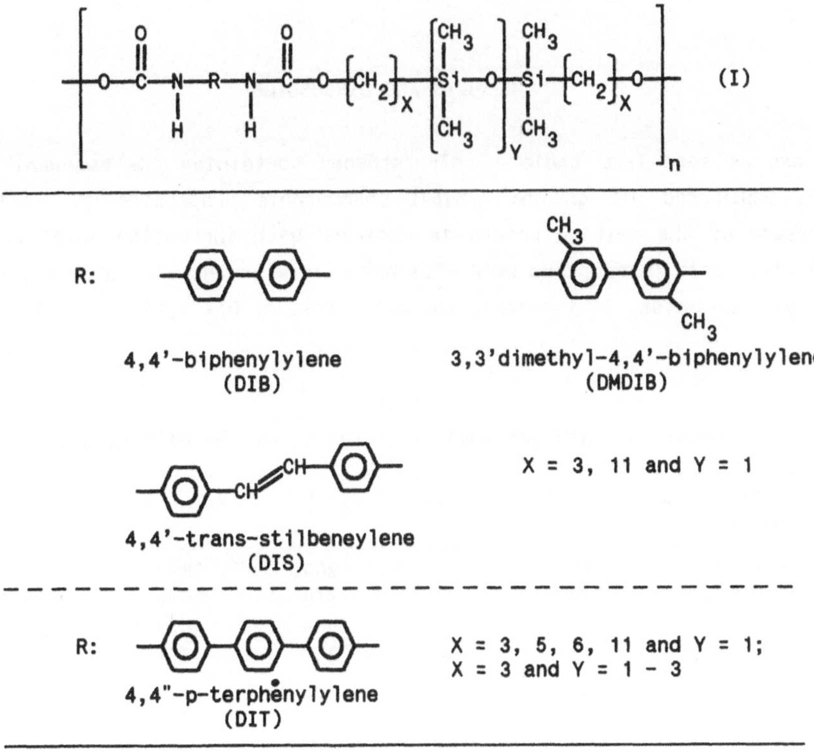

$$\left[\begin{array}{c} O \\ \| \\ O-C-N-R-N-C-O-[CH_2]_X \\ | \quad\quad | \\ H \quad\quad H \end{array} \left[\begin{array}{c} CH_3 \\ | \\ Si-O \\ | \\ CH_3 \end{array} \right]_Y \begin{array}{c} CH_3 \\ | \\ Si-[CH_2]_X-O \\ | \\ CH_3 \end{array} \right]_n \quad (I)$$

R:

4,4'-biphenylylene
(DIB)

3,3'dimethyl-4,4'-biphenylylene
(DMDIB)

4,4'-trans-stilbeneylene
(DIS)

X = 3, 11 and Y = 1

R:

4,4"-p-terphenylylene
(DIT)

X = 3, 5, 6, 11 and Y = 1;
X = 3 and Y = 1 - 3

Scheme 1

MATERIALS AND METHODS

The polyurethanes were prepared by addition reaction of aromatic diisocyanates and bis(ω-hydroxyalkyl)oligodimethylsiloxanes for 3 h at 100 ^{0}C in

dimethylacetamide. Dibutyltindilaurate was used as a catalyst. The syntheses
of the monomers are described in previous papers [2, 6, 7]. The composition
of the polyurethanes were checked by elemental analysis and the proposed
structures are in agreement with IR- and NMR-spectra. The mesophase behaviour was studied using a Perkin Elmer DSC 2 differential scanning calorimeter
and a Leitz orthoplan polarizing microscope equipped with a Leitz hotstage -
model 350. The thermal stabilities were investigated by means of a Perkin
Elmer thermogravimetric system - model TGS 2.

RESULTS AND DISCUSSION

As can be seen from table 1 polyurethanes containing the mesogenic units
DIB, DMDIB and DIS do not exhibit mesomorphic properties. A significant
increase of the melting points is observed with increasing axial ratio of
the rigid core. Mesophases were obtained only when the elongated 4,4"-p-ter-
phenylylene moiety is inserted. The DSC traces of DIT 1,11 (Fig. 1) show two

TABLE 1

Thermal data and intrinsic viscosities of the polyurethanes

mesogenic unit Y,X		$T_m(^0C)^a$	$T_i(^0C)^a$	$T_B(^0C)^b$	$V_{260}(\%)^c$	$[\eta](dl/g)^d$
DIB	1,3	176	—	200	38	0.35
	1,11	176	—	276	4.6	0.24
DMDIB	1,3	81	—	200	39.7	0.27
	1,11	107	—	290	3.2	0.29
DIS	1,3	205	—	200	38.3	0.28
	1,11	204	—	310	3	0.29
DIT	1,3	247	—e	200	3.7	0.24
	1,5	250	265	270	3.9	0.22
	1,6	241	267	270	4	0.19
	1,11	227	254	281	2.4	—f
	2,3	243	—	210	8	0.19
	3,3	217	—	200	18	0.19

[a] Peakmaxima of the DSC endotherms. [b] Beginning of the decomposition on the
TG traces at a heating rate of 80 °C/min. [c] Per cent weight loss on annealing 30 min. at 260 °C. [d] Intrinsic viscosities in DMA at 20 °C, DIB 1,11
at 40 °C and DIS 1,11 at 50 °C. [e] Clearing point could not be detected by
DSC measurements. [f] DIT 1,11 is not clear soluble in common solvents.

transitions on heating as well as on cooling. The transitions correspond to the melting and clearing point confirmed by observation under the polarizing microscope. A phase assignment could not be done because no specific textures occur in the LC state. Therefore further experiments, like X-ray diffraction or miscibility studies, are necessary. The transition temperatures of the LC polyurethanes decrease only slightly with increasing alkyl-chain length. The decreasing of the melting points is more effective by elongation of the siloxane moiety but this leads to the loss of the LC properties. There is no more indication of a mesophase neither in the DSC traces nor under the polarizing microscope. Figure 1b shows the DSC thermogram of DIT 3,3 which is typical for polyurethanes having short alkyl segments (X = 3) as a part of the flexible spacer. The heating trace of the sample exhibits one broad endothermic peak with a shoulder on the low temperature side. The cooling and a subsequent heating trace do not exhibit exo- or endothermic peaks anymore. Apparently, the bulky dimethylsiloxane segment effects the stability of both the crystalline and the LC phase. No glass transition could be assigned with certainty from the DSC measurements for any polyurethane studied.

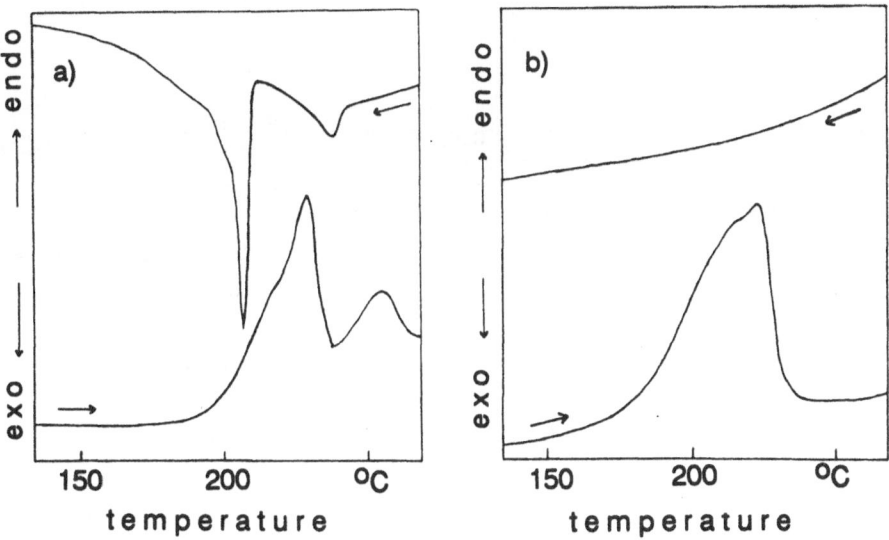

Figure 1) DSC thermograms of polyurethanes: a) DIT 1,11; b) DIT 3,3.

Investigation of the thermal stabilities (T_B and V_{260} in TABLE 1) also shows a significant difference between polyurethanes containing three methylene units (X = 3) and those containing an elongated alkyl chain segment (X = 5, 6, 11). The latter are thermally stable up to approximately 270 °C whereas the decomposition of the former polyurethanes starts in the range of 200 - 210 °C. Due to the lower thermal stability no clearing point could be obtained for compound DIT 1,3 by DSC, although LC behaviour was unambiguously observed by polarizing microscopy. However, a weight loss occurs for all polyurethanes on annealing at 260 °C for 30 minutes.

The results obtained for the polyurethanes are in accordance with those of diurethanes prepared as low molecular weight model compounds [2]. In both cases we have noticed a strong increase of the phase transition temperatures and a reduction of the range of mesophase in comparison with analogues diesters [3] and polyesters [8], respectively.

ACKNOWLEDGEMENT

The authors are indebted to the Deutsche Forschungsgemeinschaft, Bonn, for financial support.

REFERENCES

[1] M. Sato, Porima daijesuto, 1985, 37, 10.

[2] L. Willner, F. Braun, M. Heß and R. Kosfeld, Liq. Crystals, in press.

[3] F. Braun, L. Willner, M. Heß and R. Kosfeld, Makromol. Chem., Rapid Commun., 1989, 10, 51.

[4] B.-W. Jo, J.-J. Jin and R.W. Lenz, Eur. Polym. J., 1982, 18, 233.

[5] H. Ringsdorf and A. Schneller, Brit. Polym. J., 1981, 13, 43.

[6] F. Braun, L. Willner, M. Heß and R. Kosfeld, J. Organomet. Chem., 1987, 332, 63.

[7] F. Braun, L. Willner, M. Heß and R. Kosfeld, J. Organomet. Chem., in press.

[8] F. Braun, L. Willner, M. Heß and R. Kosfeld, to be published.

Mixtures of Thermotropic Nematic Liquid Crystals

Coexistence Curves and Cloud Point Curves

M. Ballauff

Max–Planck–Institut für Polymerforschung

Postfach 3148

65 Mainz, FRG

Abstract

A statistical mechanical treatment of phase equilibria in mixtures of thermotropic nematic liquid crystals with nonuniformly distributed polymers is given. Based on the Flory lattice model of nematic fluids it allows the explicit calculation of coexistence curves, cloud point curves and phase volume ratios without invoking adjustable parameters. It is shown that the strong fractionation taking place in nematic–isotropic phase equilibria has a profound influence on the phase behavior.

Introduction

In a number of recent publications [1–3] a theory of phase equilibria in mixtures of thermotropic nematic liquid crystals and flexible polymers has been given. It is based on the Flory lattice model [5,6] adapted to the description of thermotropic systems through introduction of anisotropic dispersion forces and free volume. The principal result of this treatment is the demixing of the liquid crystal and the coil upon onset of ordering. This "entropic demixing" proceeds in purely athermal systems and is strongly dependent on molecular weight. A comparison with available experimental data [7] showed semi–quantitative agreement [3]. From experimental data it is evident that the pronounced dependence of demixing on molecular weight will strongly determine the phase behavior. Especially, the coexistence curve for nonuniform polymers will not coincide with the cloud point curve, i.e., the point where the first nematic droplet is formed. This is the analogous case to liquid–liquid demixing which has been studied thoroughly in the classical work of Koningsveld and Staverman [8] and Rehage and Möller [9]. To elucidate this point for the nematic–isotropic equilibria under consideration here, the treatment of reference [3] is extended to include mixtures of nematic liquid crystals with nonuniformly distributed polymers.

Theory

The system considered here consists of a rodlike species the axial ratio of which is denoted as x_r and a series of homologous random coil species identified by the subscripts α, β.... The contour length of the latter components is given by the quantities x_α. As in ref.[3] the incomplete occupation of the space available to the molecules is introduced by the device of allowing some of the lattice sites to remain empty. The combinatorial analysis along the lines given in ref. [1–3] is easily adapted for the problem under consideration here. Let n_0 denote the total number of lattice sites, n_r the number of rodlike molecules and $n_p = \Sigma n_\alpha$ the total number of coiled molecules. The evaluation of the configurational partition function Z_M which may be split into its combinatorial part Z_{comb} and its orientational part Z_{orient} [5]:

$$Z_M = Z_{comb} \cdot Z_{orient} \tag{1}$$

may proceeds as follows: Let the quantity y specify as usual [5] the disorientation of the rodlike particle. For a molecule with the long axis at the angle Ψ with respect to the domain axis [5]:

$$y = \frac{4}{\pi} x_r \sin\Psi \tag{2}$$

Going along the lines of reference [6] and [3] we obtain for the partition function Z_M:

$$Z_M = \frac{[n_0 - n_r(x_r - \bar{y})]! \; Z_{orient}}{(n_0 - n_r x_r - n_p \bar{x})! \; n_r! \; \Pi \, n_\alpha!} \cdot n_0^{-n_p(\bar{x} - 1) - n_r(\bar{y} - 1)} \tag{3}$$

The quantity $\bar{x}$ is the number average of the number of segments per coiled species:

$$\bar{x} = \Sigma \, n x_\alpha / n_p \tag{4}$$

and $\bar{y}$ denotes the average value of the disorder index y (cf. ref. [3]). As in ref. [3] the conformation of the coiled species is assumed to be independent of the order in the respective phase. Hence, the orientation–dependent interaction energy between the segments of the nematogen are given by (see ref. [3])

$$\epsilon_\Psi = -(k_B T^* / \tilde{V}) \, s \, v_r (1 - \tfrac{3}{2} \sin\Psi) \tag{5}$$

where s denotes the common order parameter, v_r the volume fraction of the nematogen, T^* the characteristic temperature of anisotropic interaction [5,3] and $\tilde{V}$ the reduced volume defined by $\tilde{V} = V/V^*$, i.e., the ratio of the volume of the system to the hard core volume. Since eq.(5) is identical to the expression given in ref. [3], the evaluation of Z_{orient} can be done as outlined in references [1] and [3]. The volume fraction v_r follows as

$$v_r = \tilde{V}_r \, n_r x_r \, / \, (n_r x_r + n_p \bar{x}) \, \tilde{V} \tag{6}$$

and the volume fraction of the coiled species α as

$$v_\alpha = \tilde{V}_\alpha \, n_\alpha x_\alpha \, / \, (n_r x_r + n_p \bar{x}) \, \tilde{V} \tag{7}$$

with $\tilde{V}_r$ and $\tilde{V}_\alpha$ being the reduced volumes of the nematogen and of component α, respectively. The reduced volume of the system is

$$\tilde{V}^{-1} = v_r / \tilde{V}_r + v_c / \tilde{V}_c \tag{8}$$

with

$$v_c = \Sigma \, v_\alpha \tag{9}$$

and

$$\tilde{V}_c^{-1} = \Sigma \frac{v_\alpha}{\tilde{V}_\alpha} \, / \, \Sigma \, v_\alpha \tag{10}$$

With the definition of the reduced temperature $\theta = T/T^*$ and of the quantity a [1,3]

$$a = -\ln \left[1 - \frac{v_r}{\tilde{V}_r} \left(1 - \frac{\bar{y}}{x_r} \right) \right] \tag{11}$$

we obtain for the chemical potentials at equilibrium

$$\Delta\mu'_r/RT = \ln(v_r'/x_r\tilde{V}_r) + v_r'(\bar{y}-1) + v_c'x_r(1 - \frac{\tilde{V}_r}{\tilde{V}_c\bar{x}})$$

$$+ x_r(\frac{\tilde{V}_r}{V'} - 1) + x_r(\tilde{V}_r - 1)[a+\ln(1-1/\tilde{V}')] - \ln f_1$$

$$-(x_r v_r' \cdot s / \tilde{V}\theta)[1 - s + v_c s + \frac{1}{2} s\, v'_r \frac{\tilde{V}}{\tilde{V}_r}] + \chi v_c'^2 \frac{\tilde{V}_r x_r}{\tilde{V}_c} \qquad (12)$$

and

$$\Delta\mu'_\alpha/RT = \ln(v'_\alpha / \tilde{V}_\alpha x_\alpha) + v'_r \frac{\tilde{V}_\alpha x_\alpha}{\tilde{V}_r x_r}(\bar{y}-1) + v'_c \tilde{V}_\alpha x_\alpha (\frac{1}{\tilde{V}_r} - \frac{1}{\tilde{V}_c\bar{x}'})$$

$$+ \tilde{V}_\alpha x_\alpha a + x_\alpha \tilde{V}_\alpha (\frac{1}{V'} - \frac{1}{\tilde{V}_r}) + x_\alpha(\tilde{V}_\alpha - 1)\ln(1 - \frac{1}{V'})$$

$$+ \frac{s^2 x_\alpha}{2\theta V'} \frac{v_r'^2}{\tilde{V}_r}(2\tilde{V}_\alpha - \tilde{V}') + \chi x_\alpha v'_r (1 - v'_c \frac{\tilde{V}_\alpha}{\tilde{V}_c}) \qquad (13)$$

Here primes are appended to quantities which may differ from their corresponding values in the isotropic phase. The quantity f_1 is given the by usual expression derived in ref.[1] and χ is the common Flory – Huggins interaction parameter (cf. ref. [2]). The corresponding chemical potentials in the isotropic phase become:

$$\Delta\mu_r/RT = \ln(v_r/\tilde{V}_r x_r) + v_r(x_r - 1) + v_c x_r(1 - \frac{\tilde{V}_r}{\tilde{V}_c\bar{x}}) + x_r(\frac{\tilde{V}_r}{V} - 1)$$

$$+ x_r(\tilde{V}_r - 1)\ln(1 - \tilde{V}^{-1}) + \chi v_c^2 \frac{\tilde{V}_r x_r}{\tilde{V}_c} \qquad (14)$$

$$\Delta\mu_\alpha/RT = \ln(v_\alpha /\tilde{V}_\alpha x_\alpha) + v_r \frac{\tilde{V}_\alpha x_\alpha}{\tilde{V}_r}(1 - \frac{1}{x_r}) + v_c \tilde{V}_\alpha x_\alpha (\frac{1}{\tilde{V}_r} - \frac{1}{\tilde{V}_c\bar{x}})$$

$$+ x_\alpha \tilde{V}_\alpha (\frac{1}{V} - \frac{1}{\tilde{V}_r}) + x_\alpha(\tilde{V}_\alpha - 1)\ln(1 - \tilde{V}^{-1}) + \chi x_\alpha v_r (1 - v_c \frac{\tilde{V}_\alpha}{\tilde{V}_c}) \qquad (15)$$

Phase equilibria may be calculated by numerical solution of the equilibrium conditions $\Delta\mu'_r = \Delta\mu_r$ and $\Delta\mu_\alpha' = \Delta\mu_\alpha$. A more detailed account of the derivation of eq.(12) – (15) will be given elsewhere [10].

Results and Discussion

To demonstrate the influence of molecular weight distribution all calculations are done for a Schulz distribution [11] where the weight fraction w_α of the α-th species is given by

$$w_\alpha = \frac{x_\alpha^f}{f!} (-\ln p)^{f+1} p^{x_\alpha} \tag{16}$$

with $p = (\bar{x}_n - 1)/(f - 1 + \bar{x}_n)$. Since $\bar{V}_\alpha$ is assumed to be independent of α, we have $v_\alpha = w_\alpha$.

As an example we discuss here a mixture of the nematic liquid crystal EBBA (p–butyl–N–(p–ethoxybenzylidene)aniline) and oligomeric polystyrene. The respective equation–of–state data have been gathered in reference [1]. To demonstrate the influence of molecular weight distribution (mwd) most clearly, the parameter f in eq. (16) was chosen to be 10, i.e., the nonuniformity $\bar{x}_w/\bar{x}_a - 1$ is to 0.1. This value is often regarded as small and resulting mwd as rather narrow. In the present case we show that this breadth of the mwd already suffices to produce strong alterations of the phase behavior when compared to a molecularly uniform sample.

Fig. (1) shows the ratio Φ of the volume of the nematic phase to the volume of the system. The number average degree of polymerization $\bar{x}_n$ of the polystyrene sample has been set to 25.3. At $\Phi=0$ the intersection with the abscissa gives

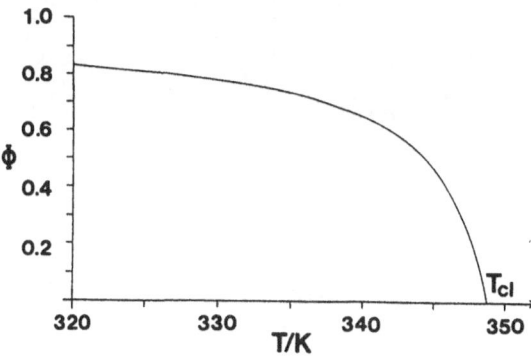

Fig.(1) Calculated phase volume ratio of the system EBBA/polystyrene ($\bar{x}_n = 25.3$).
Volume fraction of polystyrene : 0.05.

the cloud point temperature T_{cl} where the first nematic droplet is formed (cf. ref. [8]).

In this region the curve $\Phi(T)$ is rather steep. Therefore T_{cl} can be determined very accurately. However, the point where $\Phi=1$, i.e., where the whole system consists of a homogeneous nematic phase is obviously very difficult to reach. The $\Phi(T)$ curve is rather flat indicating the point $\Phi = 1$ to be located at much lower temperatures. Here crystallization of the nematic solvent will intervene and render the observation of the homogeneous nematic phase impossible. The reason for this difficulty is clearly located in fractionation. The longer chains are enriched in the isotropic phase. Increasing the volume of the nematic phase is followed by a transfer of more and more chains into the ordered state. As a consequence of this the temperature where $\Phi = 1$ is strongly depressed. Experiments using the nematic liquid crystal PAA (p–azoxy–anisole) as solvent and polystyrene as the coiled component are in qualitative agreement with these deductions [12].

The cloud point curves resulting from theory for the above mwd ($\overline{x}_n = 25.3$, u $= 0.1$) are shown in the following graph (Fig. (2)).

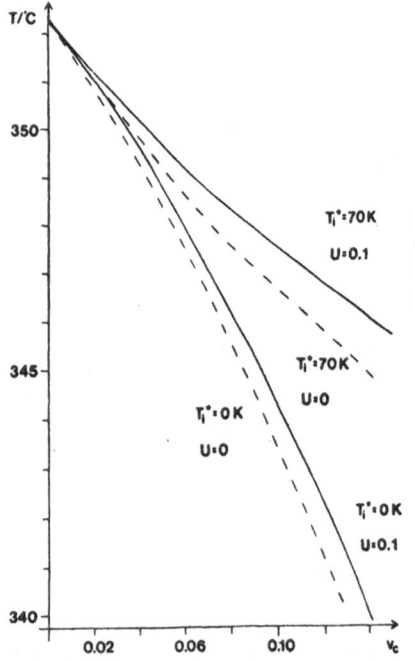

Fig. (2) Cloud point and coexistence curves. Bold lines: Cloud point curves calculated for the system

EBBA/polystyrene ($\overline{x}_n = 25.3$, U $= 0.1$).

T_i^* is the characteristic temperature of isotropic interaction. Dashed lines: Concentration of the polymer in the isotropic phase for a monodisperse solute ($x_\alpha = 25.3$, U=0) for T_i^*=0K and 70K.

From a comparison of the curves referring to $T_i^* = 0K$ it becomes clear that even a narrow mwd with U $= 0.1$ shifts the cloud points when compared to the coexistence curve (U = 0). In both cases the curve corresponding to $\Phi = 1$ is not discernible from the ordinate. The influence of the finite breadth of the mwd is even more pronounced in systems where χ or $T_i^*\neq 0$. For U = 0 the coexistence curve is only shifted slightly, but the cloud point curve exhibits a characteristic upward curvature also observed experimentally (see e.g. ref. [7]). Thus the present theory predicts that for nematic–isotropic equilibria of liquid crystals with polymers the mwd of the solute can not be disregarded even when the nonuniformity is small.

Acknowledgement: Financial support of the Deutsche Forschungsgemeinschaft, Schwerpunkt "Thermotrope Flüssigkristalle", is gratefully acknowledged

References

[1] M. Ballauff, Mol. Cryst. Liq. Cryst. 136, 175 (1986)

[2] M. Ballauff, Mol. Cryst. Liq. Cryst. Lett. 4, 15 (1986)

[3] M. Ballauff, Ber. Bunsenges. Phys. Chem. 90, 1053 (1986)

[5] P.J. Flory, Adv. Pol. Sci. 59, 1 (1985) and further references cited therein

[6] P.J. Flory, Macromolecules 1978, 11, 1178

[7] B Kronberg, I. Bassignana, D. Patterson, J. Phys. Chem. 82, 1719 (1978)

[8] R. Koningsveld, A.J. Staverman, J. Pol. Sci. A–2, 6, 349 (1968)

[9] G. Rehage, D. Möller, J. Pol. Sci. Part C, 16, 1787 (1967)

[10] M. Ballauff, in preparation

[11] G.V. Schulz, Z. physik. Chem. B43, 25 (1939)

[12] H. Orendi, M. Ballauff, in preparation

THEORETICAL AND EXPERIMENTAL ASPECTS OF THE REINFORCEMENT OF THERMOPLASTICS BY LIQUID CRYSTALLINE POLYMERS

S.CLAßEN, U. GALLENKAMP, M. WOLF, J. H. WENDORFF
Deutsches Kunststoff-Institut
6100 Darmstadt, Schloßgartenstr. 6, FRG)

ABSTRACT

Thermoplastic may be reinforced by liquid crystalline polymers which are either heterogeneously or homogeneously distributed. The probability of achieving a homogeneous distribution is low since flexible and rigid chain repel each other in blends for entropic reasons.The compatibility may be enhanced by frustrating the tendency of the liquid crystalline polymers to crystallize and by introducing a negative heat of mixing via specific intermolecular interactions. A very effective way of increasing the compatibility of rodlike units and of promoting also the reinforcement consists in constructing macromolecules having the shape of rigid stars or possessing both a rigid backbone and rigid side chains .

INTRODUCTION

Composites of liquid crystalline polymers and nonthermotropic polymers constitute a new class of materials which is expected to play an increasingly important role in the future. Blends containing side chain liquid crystalline polymers, for instance, may be used to fabricate light control films (1,2) and blends containing rigid rod-like liquid crystalline polymers offer favorable mechanical properties such as a high stiffness and a high strengh (3-5).

HETEROGENEOUS VS. HOMOGENEOUS BLENDS

Blends having rodlike polymers as the dispersed phase offer in certain respects advantages as compared to composites containing glass fibers,carbon fibers or metal whiskers. One decisive advantage is that the reinforcing species is in the fluid state during processing and solidifies on cooling. This results in a strong reduction of the viscosity of the composites relative to that of the pure thermoplastic on the one side whereas the solidification of the dispersed phase on cooling causes a strong increase of the stiffness and strength of the composite over that of the thermoplastic (3-5). In addition, the distribution of the dispersed phase may be modified during processing in such a way that either a fiber like, a spherical like morphology or even a three dimensional network arises, each displaying its specific mechanical properties (3).

It is in certain cases advantageous to use homogeneous blends of rigid rod-like polymers and flexible chain thermoplastics (6). One of such cases is the reinforcement of transparent polymers such as polycarbonate. In addition, the expectation is that single rigid molecules should be very effective in reinforcing thermoplastic materials because of the intrinsic strenght and stiffness of individual molecules , because of the large aspect (length to diameter) ratio of high molecular weight polymers and because of the good molecular coupling . (Synergistic effects may furthermore be expected once a "percolation" concentration is surpassed). The requirement is in any case the existence of compatibility between flexible and rigid chain molecules.

BLENDS WITH ROD-LIKE MOLECULES
Lattice calculations performed by Flory (7) have revealed that a strong repulsion exists between flexible and rigid chain molecules for entropy reasons. The isotropic phase may accomodate only a low concentration of rigid chains even in favorable instances, the anisotropic phase usually does not accomodate flexible chains at all. One way of overcoming the strongly limited compatibility at least to a certain extent consists in attaching flexible side groups to the rigid chain backbone. This problem was treated theoretically again using the lattice model by Ballauf (8). Experimental findings and theoretical predictions agree that such modifications increase the compatibility strongly in binary systems as well as in ternary systems, containing also solvents . The disadvantage is that the effectiveness of the chain molecule to reinforce is reduced due to its increased effective diameter. Two other

approaches towards molecular reinforcement, avoiding this
issue to a certain extent are:
- increase of the enthalpic driving force towards mixing
- blending with rigid rods possessing rigid side chains

ENTHALPIC DRIVING FORCES TOWARDS MIXING

One approach is based on designing the liquid crystalline
polymer component in such a way that the compatibility with
commercially available thermoplastics is increased for
enthalpic rather than purely for entropic reason. Figure 1
shows ternary phase diagrams characteristic of rigid rods,
flexible chains and solvents for the particular case of a
negative interaction parameter (9). Large negative values of
the parameter lead to a compatibility which is considerably
enhanced. Some experimental results will be presented in the
following which illustrate the problems involved in working
with such polymers (10,11).

The rigid chain molecule which was synthesized to act as
a reinforcing medium was designed in such a way that it was
not able to crystallize easily. Otherwise the usually large
enthalpy of crystallization would counteract the enthalpy of
mixing. This was achieved by substituting the phenyl rings by
a group leading to a "twist deformation". The lateral
substitute was chosen , in addition , in such a way that the
symmetry was effectively broken ,reducing thus the tendency
towards crystallization even more. The lateral groups
increase the tendency towards miscibility also , of course,
via entropic contributions. Figure 2 shows the chain molecule.
The thermoplastic materials used for the blends were, for
instance, polycarbonate, polystyrene and polymethyl
methacrylate (10,11).

All blends turned out to heterogeneous, despite the
particular choice of the chemical structure of the
thermotropic polymer. This is apparent from light
microscopical and scanning electron microscopical studies. The
design was thus not sucessful, indicating the strong entropy
controlled tendency of rigid and flexible chain molecules to

avoid each other. The average sizes of the dispersed liquid crystalline regions were found to depend strongly on the matrix and to be of the order of 0.1-10 μm.

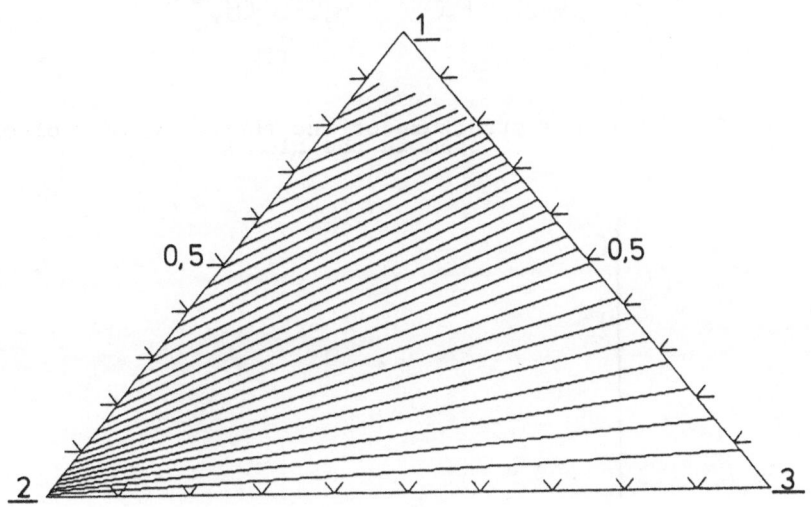

Figure 1 a .   Ternary phase diagram for a system containing a solvent (1), a rod-like chain (2) (length x=100) and a flexible chain (3) (x=100).
The interaction parameter is 0

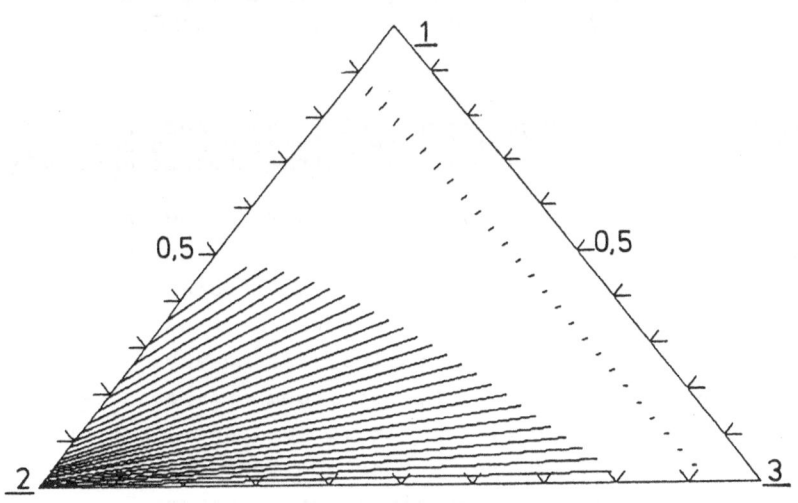

Figure 1 b .   Ternary phase diagram for a system containing a solvent (1), a rod-like chain (2) (length x=100) and a   flexible chain (3)(x=100). The interaction parameter is -0.8

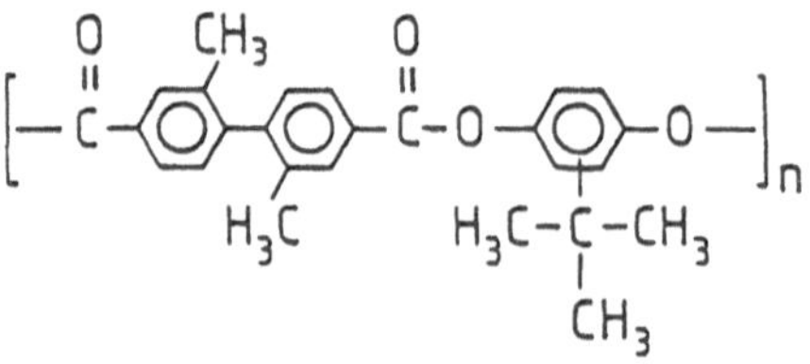

236

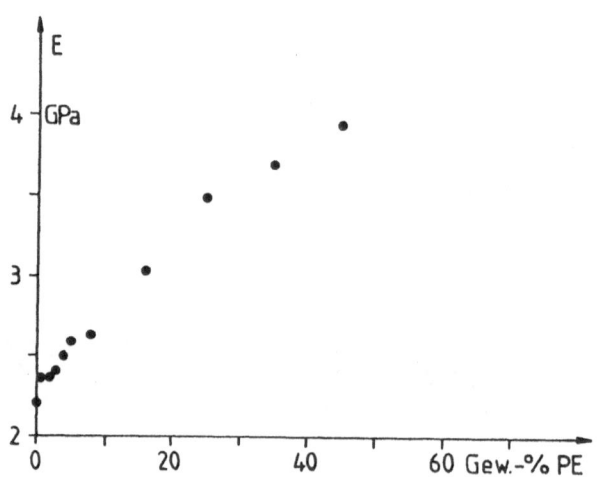

Figure 2 . Chemical structure of the thermotropic polyester
used in the blends

Figure 3 Dependence of the modulus of the
polycarbonate/polyester blend on the
concentration of the thermotropic polyester

Figure 3 reveals in agreement with the expectation that
the effective reinforcement is of limited magnitude.
Significant differences were nevertheless apparent as far as
different matrix polymers are concerned. Figure 3 displays the
dependence of the elastic modulus, as measured dynamically at
a frequency of 100 Hz at room temperature on the
composition of the blend. It is apparent that the liquid
crystalline polymer reinforces polycarbonate considerably.
Similar results were obtained for polystyrene whereas its
impact on polymethylmethacrylate was found to be weak. The

reinforcement activity continues to be large for the polycarbonate matrix even at concentrations as high as 45%. The reinforcement is of the same order as the one by glass spheres, as far as the polycarbonate matrix is concerned. It is apparent that better results will be obtained, if the dispersed phase is elongated and secondly if the molecules are distributed on a molecular level. The effectiveness will nevertheless be limited for macroscopically isotropic samples as apparent from a simple model calculations (11).

BLENDS WITH RIGID STAR MOLECULES AND WITH RIGID RODS HAVING RIGID SIDE CHAINS

The disadvantage of increasing the miscibility of rigid rods by adding flexible side chains is - as pointed out above- that the effectiveness of the rigid backbone to enhance the mechanical properties of the composite is reduced due to the isotropic chain elements which were added. This was the reason why a further structural modification was evaluated by us - based on lattice model calculations (9) and why first steps towards synthesizing such modified molecules were done.

The concept consists in attaching rigid side groups either in a freely jointed way or rigidly to the chain backbone (giving rise to combined main chain/side chain systems (12) in the first case or "barbed wire" kinds of chain molecules with short rigid side groups in the second case .

The central point of our approach is that large sequences of rigid rod-like chain units can be arranged within a macromolecule in such a way that flexible chains do not recognize them as alient species, as far as the combinatorial entropy of mixing is concerned. Lattice calculations show, for instance, that rigid star molecules having arms which point in three orthogonal directions (Figure 4 a) and which have the same length possess a combinatorial entropy of mixing corresponding to that of flexible chain molecules.So they behave in mixtures in the same way as flexible chain molecules do.

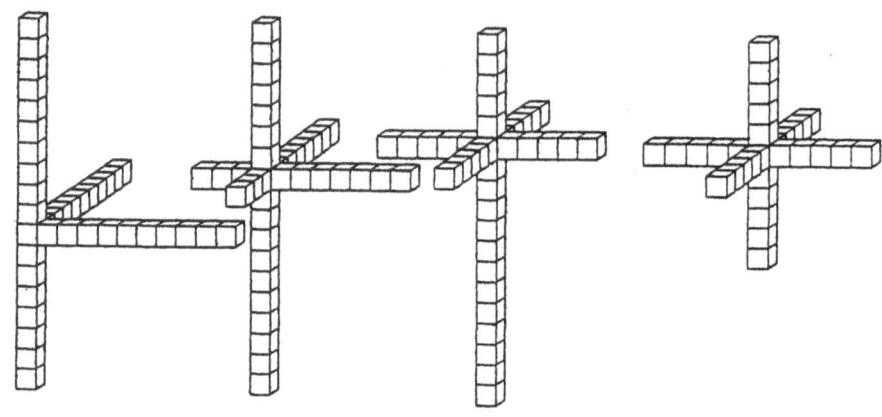

Figure 4 a. Rigid star molecules and rigid molecules having
rigid side groups (schematically)

Modifications of this shape, as shown in Figure 4 a lead
to larger and larger deviations from the effective isotropic
character of the molecule. The strength of its anisotropy
remains , however, much smaller than that of a rod even if
only a limited number of small rigid side chains are attached
to it. Such rigid side chains , on the other side , are much
more effective in reinforcing the matrix than flexible
ones.This is particularly obvious in the case of rigid star
molecules with arms of equal length. Figure 4 b gives an
example of the kind of molecules which can be used.

The lattice calculations actually showed (9) that the
solubility of such types of molecules in flexible chain
polymers both in ternary systems containing solvents and in
binary blends may be surprisingly large . Figures 5 and 6 give
examples of the kind of phase diagrams predicted to occur for
such systems.Displayed are ternary phase diagrams of rigid
rods having various amounts of rigid side groups with flexible

coils in solvents.

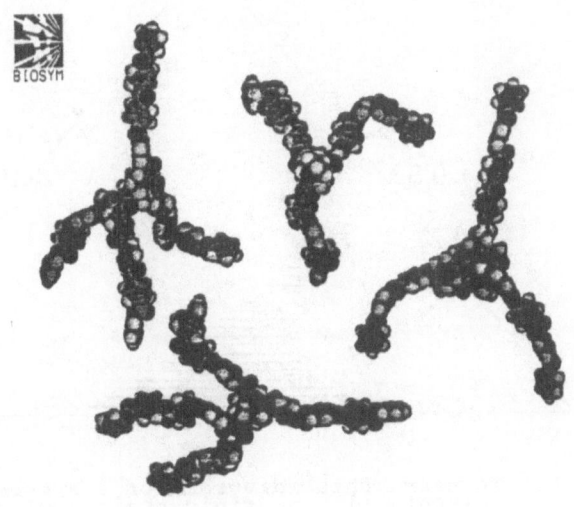

Figure 4 b . Quasi-rigid star molecule (computer simulation)
(core: twisted diphenyl unit, arms:PHB)

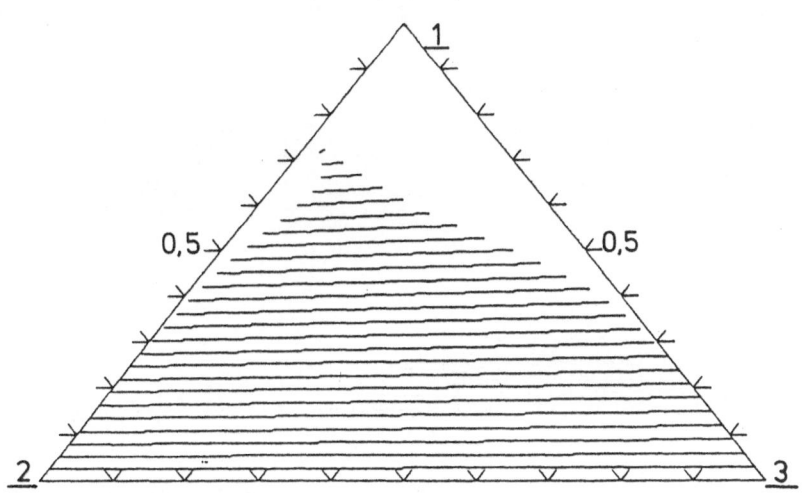

Figure 5. Ternary phase diagram for a system containing a
solvent (1) a rigid chain molecule 2) (chain
length x=158, arm length 58) and a flexible chain
molecule (3) (x=100)

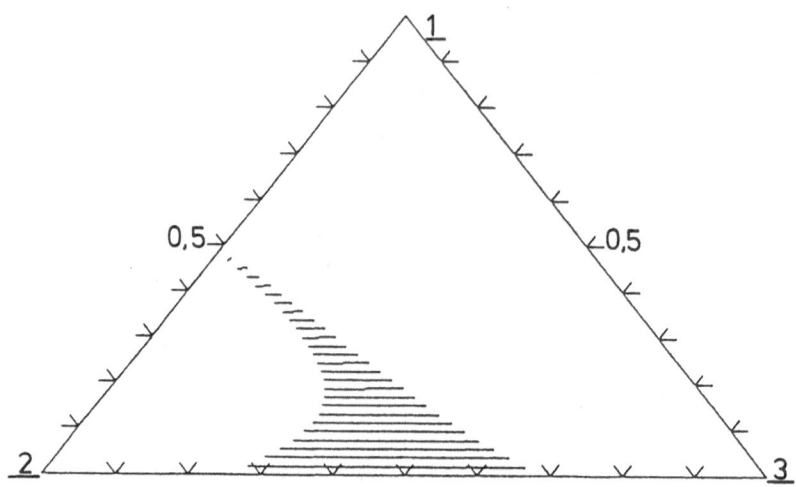

Figure 6. Ternary phase diagram for a system containing a solvent (1) a rigid chain molecule (2) (chain length x=258, arm length 158) and a flexible chain molecule (3) (x=100)

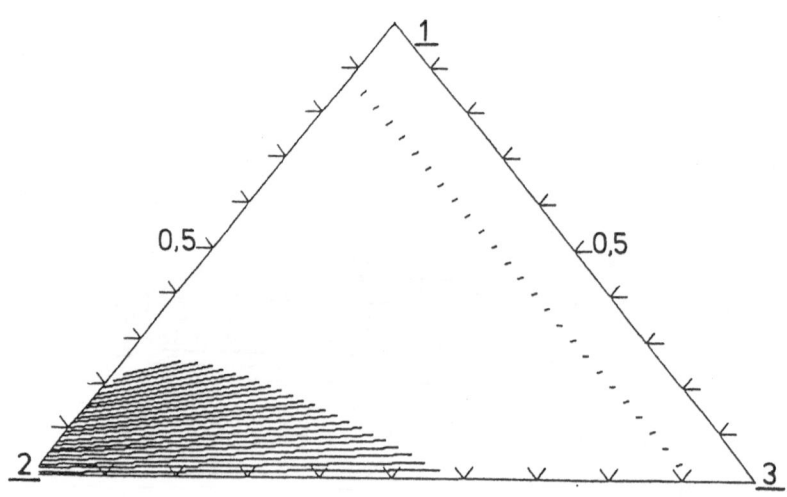

Figure 7. Ternary phase diagram for a system containing a solvent (1) a rigid chain molecule (2) (chain length x=105, arm length 5) and a flexible chain molecule (x=100), the interaction parameter is -0.6 (compare with Figure 1)

The interesting prediction is that a limited compatibility may not only exist in the isotropic but also in the anisotropic state, as apparent from Figures 5 and 6 . The compatibility may furthermore be increased if specific interactions are introduced , giving rise to a negative interaction parameter in the approach used here (Figure 7).

It seems conceivable that such kinds of reinforcing molecules lead to a considerable enhancement of the properties of the composites even at low concentrations and even simultaneously in orthogonal directions.

REFERENCES

1. Vaz, N.A., Smith, G.W. and Montgomery, G.P., A light control film composed of liquid crystalline droplets dispersed in an epoxy matrix. Mol.Cryst.Liq.Cryst. , 1987, 146, 17-34

2. Zumer, S., Light scattering from nematic droplets: anomalous-diffraction approach. Physical Review a, 1988, 37, 4006-4015

3. Takayanagi, M., Polymer Composites of Rigid and Flexible Molecules. Pure & Appl. Chem. , 1983, 55, 819-832

4. Brostow, W., Polymer blends containing polymeric liquid crystals. Kunststoffe, 1988, 78, 411-419

5. Paci, M., Barone, C., and Maganini, P.L., Calorimetric study of blends of poly (butylene terephthalate) and a liquid crystalline polyester. J.Polym.Sci.Polym.Phys.Ed., 1987, 25, 1595-1605

6. Wiff, D.R., Timms, S., Helminiak, T.E., and Hwang, W.F., Molecular Entanglementts for Rigid Rod Molecular Composites. Polym. Eng. Sci., 1987, 27, 424-432

7. Flory, P.J., Molecular Theory of Liquid Crystals. Adv.Polym.Sci., 1984, 59, 1-36

8. Ballauff, M., Phase eqilibria in rodlike systems with flexible side chains. Macromolecules, 1986, 9, 1366-1374

9. Gallenkamp , U., Florysche Gittertheorie und Mischbarkeit: Verallgemeinerung der zugrundegelegten Teilchenform. PhD-thesis, TH Darmstadt 1989

10. Claßen, S., Schmidt, H.W., and Wendorff, J.H., Untersuchungen an Legierungen aus steifkettigen aromatischen LC-Polyestern mit amorphen Polymeren. Frühjahrstagung der DPG, Darmstadt, 1989, manuscript in preparation for Polymers for Advances Technologies

11. Claßen, S., Master Thesis, TH Darmstadt , 1989

12. Ebert, M., Endres, B, Wendorff, J.H., Reck, B., and Ringsdorf,H.,Combined main chain/side chain polymers; A new class of polymers with unusual structural, thermodynamic and dynamical properties, submitted to Liq..Cryst., 1988

Acknowledgement
We would like to thank H.W.Schmidt (University of Marburg) for the samples and the Fonds der Chemie as well as the BMFT who provided us with the work station for computer simulations.

MOLECULAR ORDER, MORPHOLOGY AND PHASE SEPARATION IN BLENDS OF ISOTROPIC AND LIQUID-CRYSTALLINE POLYMERS

H.G. ZACHMANN, D. CHEN, J. NOWACKI, E. OLBRICH, C. SCHULZE
Institut für Technische und Makromolekulare Chemie der
Universität Hamburg, Bundesstr. 45, D-2000 Hamburg 13, FRG

ABSTRACT

Poly(ethylene terephthalate) (PET), the copolyester of PET and
40 mol-% PHB, Poly(ethylene naphtalene-2,6-dicarboxylate)
(PEN), and the copolyester of PEN and 40 mol-% of PHB were
synthesized. Blends of PET and PET/PHB, of PEN and PEN/PHB,
and, finally, of PEN and PET were obtained by coprecipitation
from solution Phase separation, crystallization, melting, and
transesterification in these blends were investigated. A single
phase is obtained in the amorphous blend of PET and PET/PHB if
the PET content is smaller than 60 wt.-% as well as in the
amorphous blend of PET and PEN if the PEN content is smaller
than 43 wt.-%. In the blend of PET and PET/PHB, during melt-
pressing at 280°C, transesterification is smaller than in the
corresponding single component systems. This indicates that in,
the blend, on a molecular scale, miscibility is not perfect. In
PET blended with PET/PHB, the melting point before transesterification is almost the same as in pure PET and decreases
with increasing transesterification. The crystallization peak
is at a lower temperature than in pure PET and shifts to higher
temperatures as a consequence of transesterification.

INTRODUCTION

Blends of liquid crystalline and isotropic polymers are of increasing interest from both a scientific and a technological
point of view.

Copolyesters of poly(ethylene terephthalate) (PET) and poly(p-
hydroxybenzoic acid) (PHB) are examples for liquid crystalline

polymers which can be used in such blends. They were first synthesized by Jackson and Kuhfuss[1,2] who also showed that these copolyesters are liquid crystalline if the PHB-content exceeds 40 mol-%. Another class of liquid crystalline polymers is formed by the copolyesters of poly(ethylene naphthalene-2,6-dicarboxylate) (PEN) and PHB. These polymers were studied in our laboratory[3]. They are liquid crystalline if the PHB content exceeds 30 mol-%. The crystallization and melting behavior of the homopolymer PEN, which like PET is not liquid crystalline, has been extensively studied, too[4,5].

In the investigations of blends of polyesters, transesterification reactions are an important effect[6-9]. By such chemical reactions, the blend might be transformed within minutes into a uniform copolymer. Therefore, in many cases miscibility of the two components may be just a consequence of transesterification. Transesterification during blending can be completely avoided if the blend is obtained by coprecipitation of the two components from solution. In this case, however, the problem of fractionation during precipitation may arise.

Different studies were performed on blends of PET and the copolyester PET/PHB containing 40 mol-% PET. A few results indicated some miscibility[10,11]. More detailed investigations, however, seem to prove that the components are not miscible[12-17].

The question arises of how the miscibility is changed if the fraction of PET in the copolyester increases. In addition to the blends of PET with PET/PHB, it seems interesting to investigate the corresponding blends of PEN with PEN/PHB and the blends of PEN with PET. In the following, we want to report some results which we have obtained by studying the phase separation, the crystallization and the melting of such blends formed under different conditions.

STUDIES OF THE PURE COMPONENTS

The components of the blends we have studied are PET, PEN and
copolyesters of the two polymers and of these polymers with
PHB. The properties of the copolyesters PET/PHB, PEN/PHB and
PET/PEN/PHB are summarized[3] in Fig. 1. A nematic liquid
crystalline state is only obtained if the PHB content of the
copolymer exceeds 30 to 40 mol-%. In the binary and ternary
copolyesters, PEN and PET crystallize if more than 50 mol-% of
one of these materials is present. PHB only crystallizes if
more than 80 mol-% is present, and even at this high concentration the less ordered high temperature modification is only
formed[3].

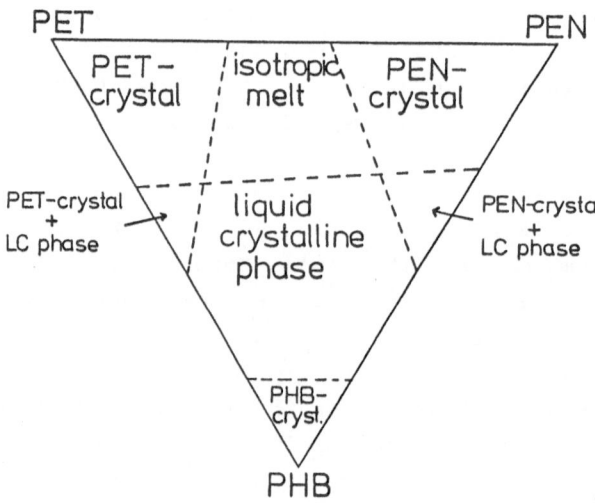

Figure 1. Phase diagram of binary and ternary copolyesters of
PET, PEN and PHB

The glass transition temperatures were measured by means of
dynamic mechanical analysis (DMA), using the instrument DMA 983

from the company DuPont. The instrument was used in the
resonance frequency mode. Fig. 2 shows the loss modulus of the
copolyester PET/PHB (60/40) as a function of temperature. It is
well known from microscopic examinations that this material is
heterogenous having liquid crystalline domains and isotropic
domains. Correspondingly, one observes two peaks. The one at
the lower temperature corresponds to the glass transition of
the liquid crystalline domains, the other one to that of the
isotropic domains.

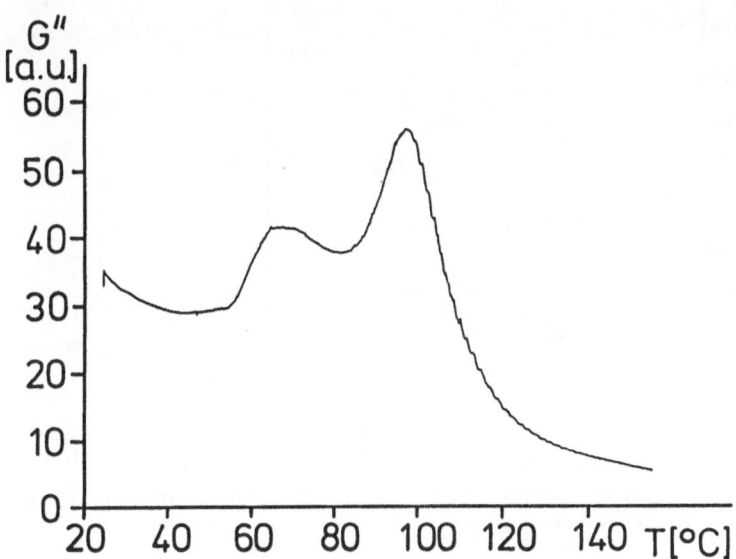

Figure 2. Loss modulus G" of the copolyester PET/PHB (60/40)
as a function of temperature T

Fig. 3a shows the glass transition temperature T_g of the
PET/PHB copolyesters as a function of the PET content. One
clearly recognizes that the liquid crystalline material has

lower values of T_g than the isotropic material. By extrapolation one obtains an hypothetical glass transition temperature of 125°C for the PHB. This is in good agreement with results obtained by Rosenau-Eichin et al.[19] on copolyesters of p-hydroxybenzoic acid and m-hydroxybenzoic acid and by Kricheldorf and Döring[20].

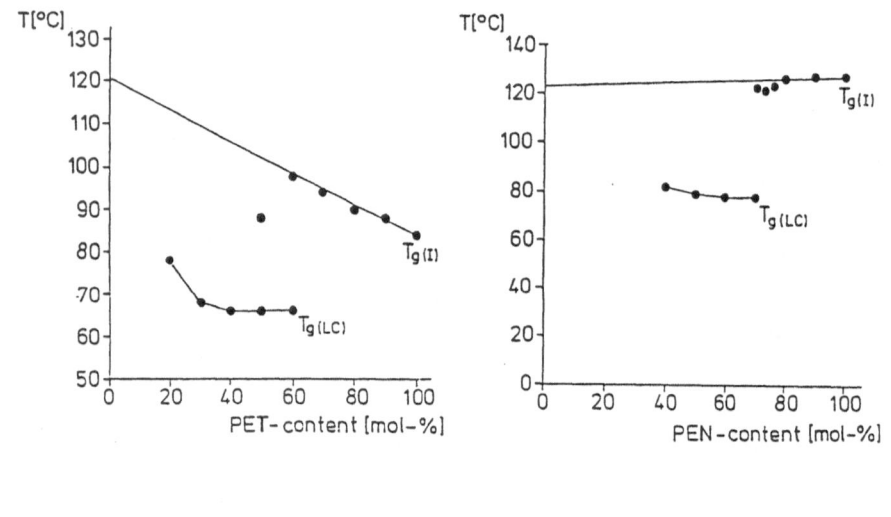

a   b

Figure 3. Temperature position of the peak of the loss modulus
G" as a function of the PET content in the copolyester PET/PHB (a) and as a function of the PEN
content in the copolyester PEN/PHB

Fig. 3b represents the corresponding results for the PEN/PHB copolyesters. The hypothetic T_g-value for PHB is the same as that obtained from the PET/PHB copolyesters.

STUDIES OF THE BLENDS OF PET AND PET/PHB

PET and the copolyesters PET/PHB were synthesized as described
earlier[3,21]. The two polymers were dissolved in hexafluoro-
isopropanol and coprecipitated by ethanol. After drying for
24 h at 50°C in vacuo, amorphous films were obtained by melt
pressing for 30 s at 280°C followed by quenching in ice water.
As it was shown before, not only the copolyester but also PET
is molten at this temperature if it is precipitated from
solution as described above. Probably, the crystals formed
during precipitation are very small and thermally unstable.

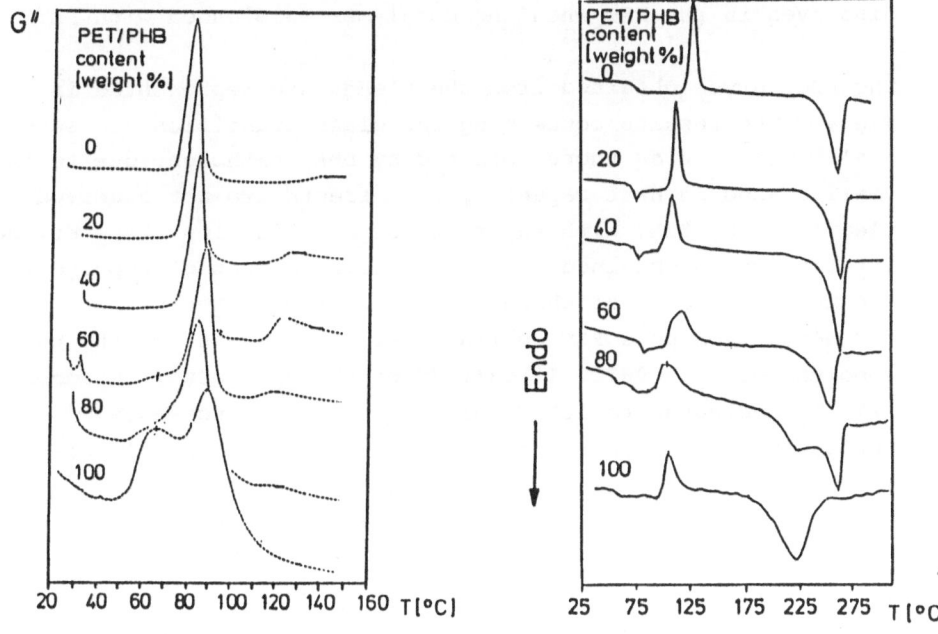

a   b

Figure 4. Loss modulus G" as a function of temperature (a) and
DSC curves (b) for blends of PET and PET/PHB (60/40)

Fig. 4a shows the loss modulus as a function of temperature for
the blends with the copolyester PET/PHB (60/40). This copolyester has two glass transitions as shown in Fig. 2, one at 66°C
arising from the liquid crystalline regions and one at 98°C
arising from the isotropic regions. PET has a glass transition
temperature at 78°C if measured by DMA at 15 Hz. When the two
materials are blended, up to 40 wt.-% copolymer, a single glass
transition temperature is observed which gradually increases
from 78 to 84°C. Obviously, the two components form a single,
isotropic phase. If the amount of copolymer is 60 wt.-% and
more, a second peak appears at about 66°C, probably arising
from a liquid crystalline phase. The relative intensity of this
peak is comparatively small. Probably a part of the chain
forming a liquid crystalline phase in the pure copolymer is
dissolved in the PET when the copolymer is blended with PET.

The DSC-curves obtained from the blends are represented in
Fig.4b. The results concerning the glass transition are essentially the same as those obtained by DMA, although, due to the
small change in heat capacity, the effects are not observed so
clearly as by DMA. With regard to crystallization, however, new
information is obtained. With increasing amount of copolyester,
the temperature T_c at which crystallization takes place
decreases. For pure PET we find T_c=128°C whereas for the pure
copolyester T_c = 98°C. If only 20 wt.-% copolyester is added to
PET, T_c decreases to 112°C while T_g slightly increases.
Obviously, the tendency of the PET units in the copolyester to
crystallize at temperatures which are close to the glass
transition temperature is transferred to the PET units in the
homopolymer. This might become useful for technical applications where an increased tendency of crystallization of PET is
wanted. The total degree of crystallinity per gram PET in the
homopolymer and in the copolyester also increases when the
copolyester is added (see Table 1).

TABLE 1
Melting Enthalpy ΔH_m and change of
Enthalpy during cold crystallization, ΔH_c,
of blends of PET and PET/PHB (60/40)

PET/PHB wt.-%	ΔH_c J/g	ΔH_c J/gPET	ΔH_m J/g	ΔH_m J/gPET
0	29.6	29.6	46.1	46.1
20	31.5	33.5	57.8	61.4
40	28.3	32.1	50.9	57.7
60	8.0	9.7	31.6	38.3
80	11.1	14.5	36.6	47.8
100	5.8	8.2	21.4	30.3

While the addition of the copolyester considerably changes T_c, a much smaller effect on the melting temperature T_m is observed. The end of the melting range and the position of the melting peak only decreases by a few degrees if PET is blended with the copolyester. The effect is much smaller than in the case of a copolyester of PET and PHB of the same composition as that of the blend. This is explained by the fact that the entropy of mixing is negligible in the blend and that, the enthalpy of mixing is comparatively small obviously.

From the results obtained we have concluded that a single phase is obtained if PET is blended with PET/PHB (60/40) up to 40 wt.-%. However, although the blend was only molten for 30 s, transesterification during melting cannot be completely excluded. In order to investigate any possible transesterification effect we performed another series of DSC experiments on samples melt-pressed for different times. The results are represented in Fig. 5. The upper curve shows the DSC curve of the powder of the blend obtained by coprecipitation without melt-pressing. One can clearly distinguish two crystallization peaks showing that two phases are present, probably the PET and the copolyester. The next curves, obtained after melt-pressing for 12s, 30s, etc., show a single crystallization peak as already reported above. With increasing melting time this peak shifts to higher temperatures indicating that transesterification occurs. After 45 min T_C lies at 136°C. This temperature corresponds to the crystallization temperature found with a copolyester having a composition corresponding to the one the molecules in the blended material would assume after trans-esterification is completed (lowest curve in Fig. 5). Therefore, we believe that during melt-pressing for a longer period of time, transesterification occurs resulting in an almost random copolyesters. The assumption that transesterification takes place during longer annealing is supported by the observed decrease of the melting point with increasing annealing time. The small blocks in the copolyester contribute a considerable mixing term to the melting entropy thus decreasing the melting point, while the mixing of the long molecules contribute only an almost negligible term.

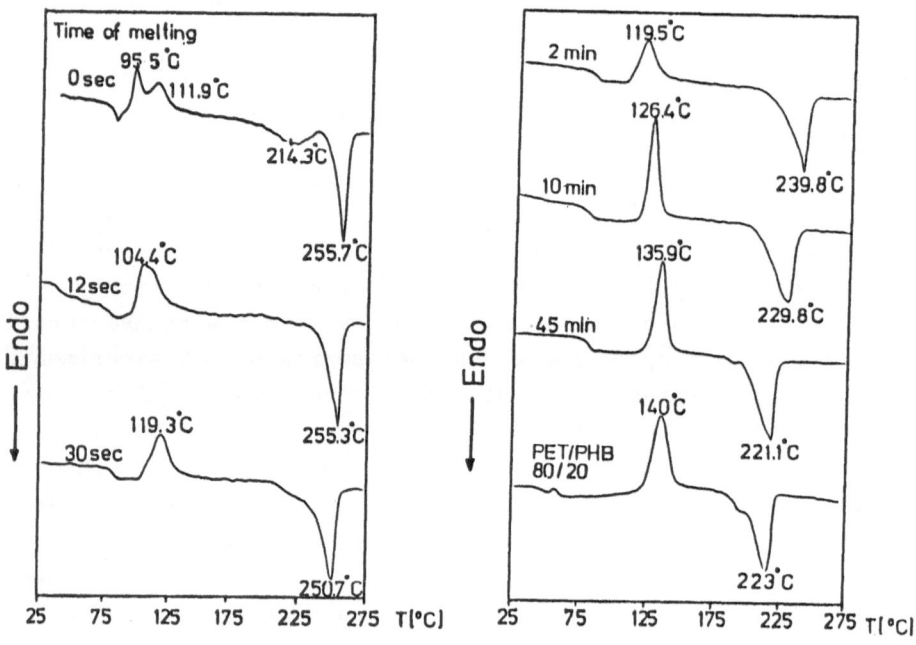

Figure 5. DSC curves of blends of PET with PET/PHB (60/40), containing 40 wt.-% of PET, after different times of melting

However, we have to ask the following question: Do the two crystallization peaks in the upper diagramm in Fig. 5 prove that
the components are not miscible in the melt and that miscibility is only a consequence of transesterification? The answer
is no. Firstly, the two components forming a single phase
before crystallization may form two different phases after
crystallization, thus giving rise to two crystallization peaks.
Secondly, and more important, the following has to be
considered: As can be easily shown, the solubility of the
copolyester in the solvent is not as good as that of the PET.
Therefore, phase separation obtained by coprecipitation may be
a result of a fractionation during precipitation.

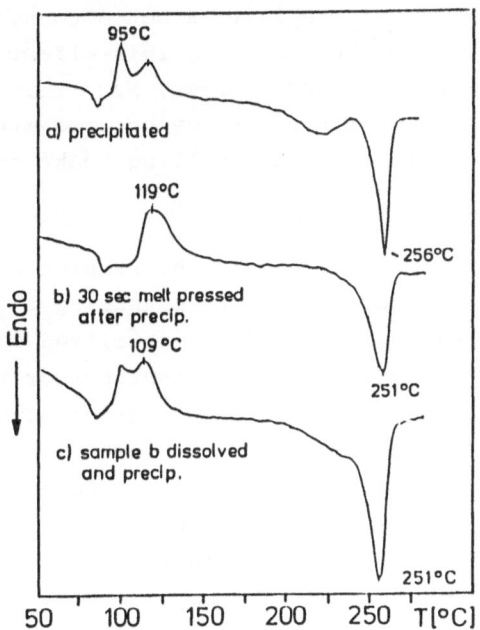

Figure 6. DSC curves of blends of PET with PET/PHB (60/40)
containing 40 wt.-% PET

In order to find out if this is the case, the blend which had been melt-pressed for 30 s was dissolved again and then precipitated. Fig.6 shows the DSC-curve of the sample after the first precipitation (a), the DSC-curve of the sample precipitated and melt-pressed for 30 s (b) and the DSC-curve of a sample precipitated, melt-pressed, dissolved and again precipitated (c). In the curve of the last sample one can again clearly distinguish two crystallization peaks, although they are not as well separated as after the first precipitation.

Therefore, we conclude the following: During coprecipitation of the two components, due to fractionation, phase separation occurs. When afterwards the sample is melt-pressed, a single phase is formed within the first 30s. In this phase, within about 40 min, transesterification leads to a more or less statistical copolymer. Miscibility is not a consequence of transesterification; on the contrary, this effect takes place after the single phase has been formed. Note that a small transesterification effect has already occured after 30 s manifested by the decrease of the melting peaks from 256°C to 251°C.

Interesting results are obtained if the samples from Fig. 5 are crystallized at 200°C for 1 h before measuring the DSC curve. The DSC curves obtained from these crystallized samples are shown in Fig. 7. In contrast to the results from the samples which were not annealed (Fig. 5), two melting peaks are obtained. Obviously, during the longer lasting crystallization process at 200°C two types of crystals are formed, one probably consisting of the PET units in the copolyester PET/PHB (60/40) and the other one of the PET units in the homopolymer which is gradually transesterified with the copolyester. During the comparatively rapid heating process applied to the samples of Fig. 5, the copolyester forming a uniform molecular mixture with the PET was either not at all able to crystallize or it was not able to form a single phase during crystallization.

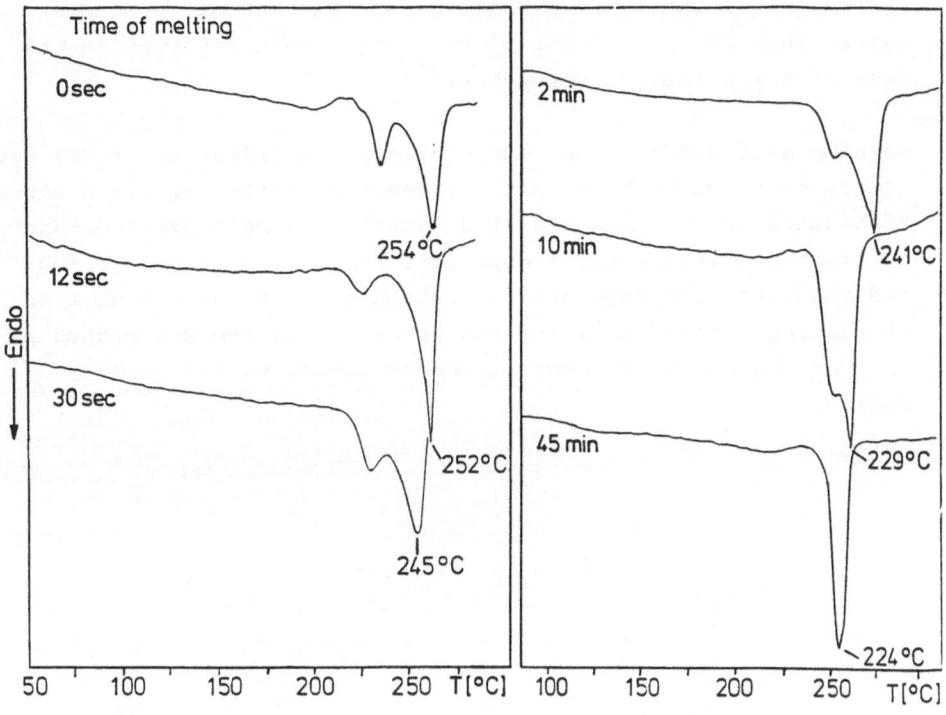

Figure 7. DSC curves of blends of PET with PET/PHB (60/40)
containing 40 wt.-% PET annealed at 200°C for 1h
after different times of melting

The DSC diagramms shown in Fig. 5 also demonstrate how
increasing blockiness of the copolyester influences the
crystallization behavior. The longer the blocks, the lower T_c.
And they also suggest a simple technical process to manufacture
copolyesters of PET having different block lengths: One has to
precipitate the PET and the copolyester from solution and to
keep the blend a certain amount of time at 275°C. From neutron
scattering experiments we know that, in the "statistical"
copolyester to which the DSC curve at the bottom of Fig. 5

refers, consists in the average a PET-sequence of 10 monomer
units. This is by a factor of two larger than expected in the
case of a statistical copolymer.[22]

We have also studied the rate of transesterification in PET and
copolyesters PET/PHB by means of neutron scattering. If a blend
of deuterated and nondeuterated material is melt-pressed, due
to transesterification, a copolymer consisting of deuterated
and nondeuterated sequences is obtained. With increasing time
of melting, transesterification proceeds and the deuterated as
well as the non-deuterated sequences become smaller and
smaller.

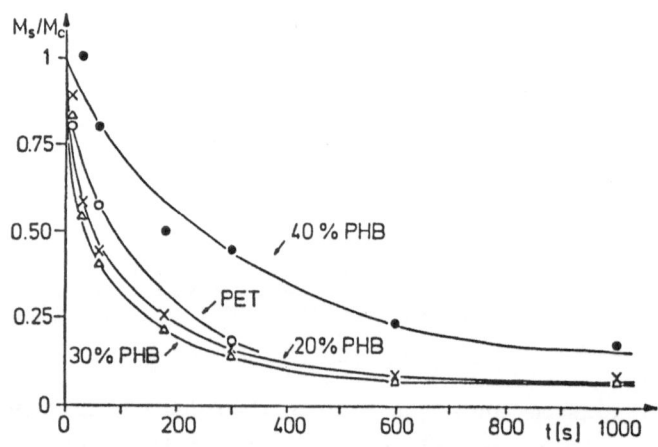

Figure 8. Average molecular weight of deuterated and nondeuterated sequences in PET/PHB copolyesters M_s, divided by
the initial molecular weight M_0, as a function of
melt-pressing time t

By means of neutron scattering, the average molecular weight of
the deuterated and nondeuterated sequences is measured. From
the time decay of the average molecular weight, the rate of
transesterification can be deduced[7,23]. Fig. 8 shows the

decrease of this average molecular weight with time for PET and for some copolyesters PET/PHB. In the case of PET and copolyesters containing up to 30 % PHB, one clearly recognizes that the molecular weight of the sequences decreases within three minutes to a fourth of the molecular weight of the material, indicating that, on the average, 3 transesterification reactions between deuterated and nondeuterated molecules have occurred within this time. In the case of the copolyester containing 40 mol-% PHB, the reaction is considerably slower. This may due to the fact that this copolyesters is liquid crystalline and, therefore, the mobility of the molecules is reduced.

According to Fig. 5, the transesterification reaction in the blend of PET with PET/PHB does not proceed so rapidly as in the pure components. A reason for that may be that, on a molecular scale, miscibility is not perfect.

Similar results were obtained for the blend of PET with the co-polyesters PET/PHB (70/30) and (80/20).

STUDIES OF THE BLENDS OF PET AND PEN

PEN has a glass transition temperature of 120°C and a melting point of 270°C. The corresponding values for PET are 78°C and 268°C. Fig. 9a shows the temperature dependence of the loss modulus of the blends of PET and PEN. Up to 27 wt.-% PEN, a single maximum appears which shifts to higher temperatures with increasing amount of PEN. Obviously, in this range of composition the two polymers form one phase. Materials containing more than 43 wt.-% PEN show an additional maximum at larger temperatures. The position of the two maxima are not identical with those of pure PET and pure PEN respectively. This indicates that in the blends separation into two phases occurs each of which is a mixture of the two components.

G''

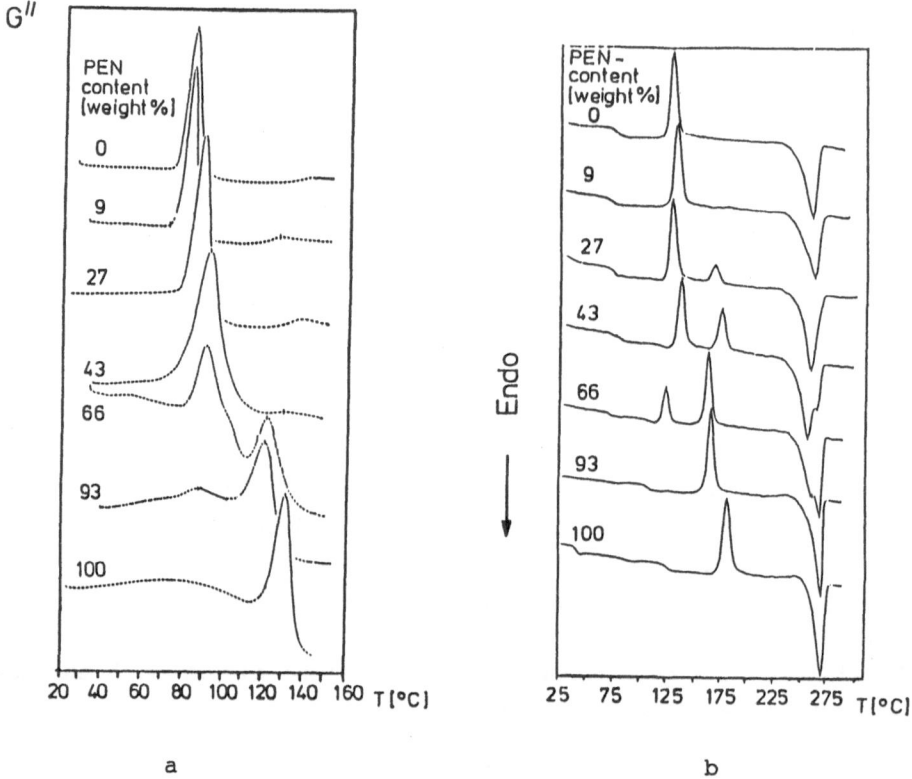

a

b

Figure 9. Loss modulus G" as a function of temperature (a) and
DSC curves (b) for blends of PET and PEN

Fig. 9b shows the DSC curves of the different blends. As one
can recognize from the crystallization peak in the blend
containing 9 wt.-% PEN only the PET crystallizes. In the blend
containing 93 % PEN, only the PEN crystallizes. Obviously, if
one component is present in a small concentration, crystal
nucleation and crystal growth becomes difficult.

In the intermediate concentration range, both components
crystallize giving rise to two crystallization and two melting
peaks. The two temperatures of crystallization, T_{c1} and T_{c2},
oscillate with increasing amount of PEN. This seams to be the
result of the superposition of the influence of the change of

the glass transition temperature and of the change of concentration.

STUDIES OF THE BLENDS OF PEN WITH PEN/PHB (60/40)

Fig. 10 shows some DSC curve obtained on blends of PEN and the copolyester PEN/PHB (60/40) which is liquid crystalline. The blend contains 30 wt.-% PEN. The powder precipitated from the solution shows two crystallization peaks, one at 128°C the other at 158°C. These temperature are somewhat different from those of the pure components. After 12 s of melting, one crystallization peak appears at 160°C and another one at 242°C. With increasing melting time, these peaks shift to smaller temperatures.

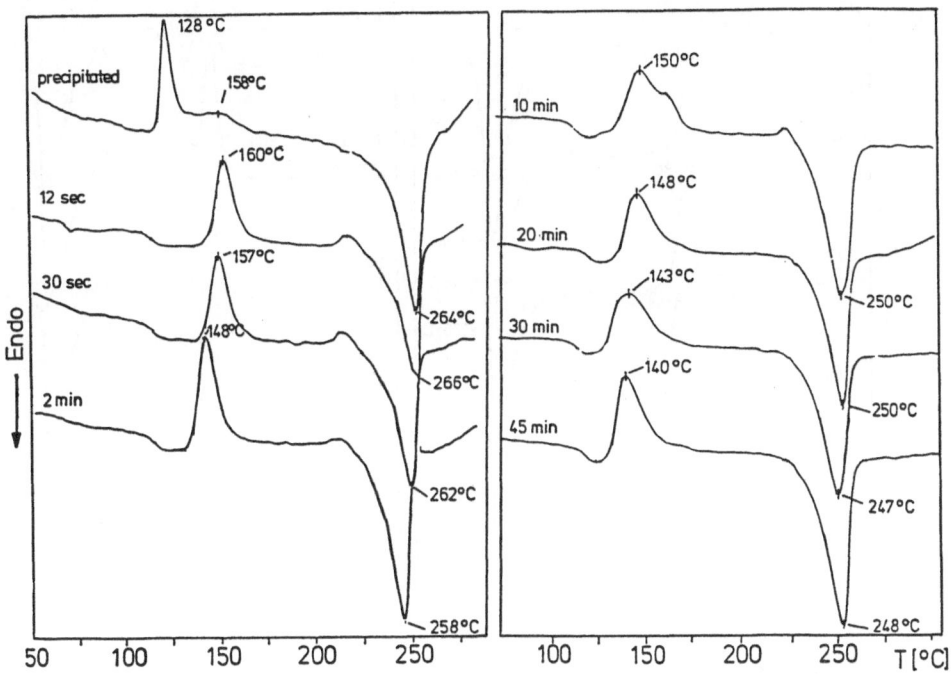

Figure 10. DSC curves of blends of PEN with PEN/PHB (60/40) containing 30 wt.-% PEN after different times of melting

We think that after precipitation, due to fractionation, a two-phase system is present. During melt pressing for 12 s, a single phase consisting of the two components is formed by diffusion. The rate of crystallization is slower in the single phase system, therefore the crystallization peaks are shifted to higher temperatures. With increasing melting time transesterification takes place. Obviously the copolyester formed by transesterification is able to crystallize at lower temperature than the PEN units in the single phase blend. As in the case of the blends of PET with PET/PHB, the assumption that transesterification takes place during longer annealing is supported by the observed decrease of the melting point with increasing annealing time.

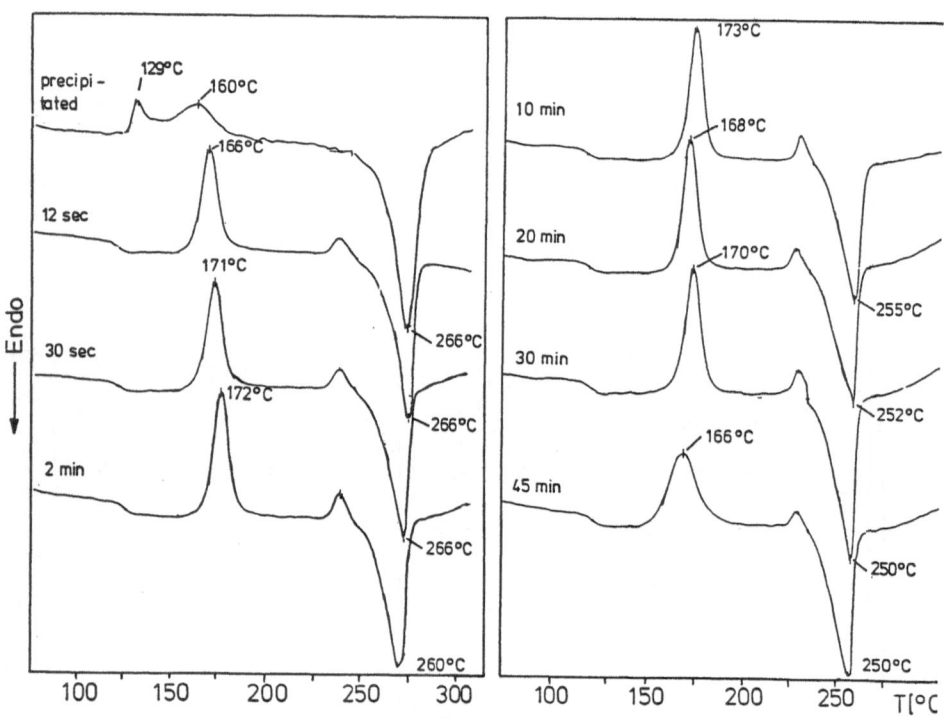

Figure 11. DSC curves of blends of PEN with PEN/PHB (60/40) containing 50 wt.-% PEN after different times of melting

Similar results are obtained from a blend containing 50 wt.-% PEN (Fig. 11). The crystallization peak obtained at the higher temperatures is smaller than in Fig. 10. This may indicate that the peak at higher temperature arises from the PEN units in the homopolymer.

ACKNOWLEDGMENT

The authors wish to thank the company DU PONT DE NEMOURS for making available the DMA 983 for performing the dynamic mechanical measurements. The neutron scattering investigations were funded by the German Federal Minister for Research and Technology (BMFT) under the Contract No. 03-ZA1HAM-3.

REFERENCES

1) Jackson, W.J. and Kuhfuss, H.F., J. Polym. Sci., Polym. Chem. Ed. 1976, 14, 2043

2) Jackson, W.J., Macromolecules 1983, 16, 1027

3) Buchner, S. Chen Di, Gehrke, R. Zachmann, H.G., Mol. Cryst. Liq. Cryst., 1988, 155, 357

4) Buchner, S., Wiswe, D., Zachmann, H.G., Polymer, 1989, 30, 480

5) Cheng, S.Z.D. and Wunderlich, B., Macromolecules, 1988, 21, 789

6) Gilmer, J.W., Wiswe, D., Zachmann, H.G., Kugler, J., Fischer, E.W., in L.A. Kleintjens, P.J. Lemstra: Integration of Fundamental Science and Technology, Elsevier Applied Science Publisher 1986. p. 563

7) Kugler, J., Gilmer, J.W., Wiswe, D., Zachmann, H.G., Hahn, K., Fischer, E.W., Macromolecules, 1987, 20, 1116

8) McAlea, K.P., Schultz, J.M., Gardner, K.H., Wignall, G.D., Macromol. 1985, 18, 447; Polymer 1986, 27, 1582

9) Kosfeld, R., Hess, M., Friedrich, K., Material Chem. and Phys., 1987, 18, 93

10) Joseph, E., Wilkes, G.L., Baird, D.G., Polymer Prepr.,
    1983, 24, No. 2, 304

11) Bhattacharya, S.K., Tendolkar, A., Misra, A., Mol. Cryst.
    Liq. Cryst., 1987, 153, 501

12) Joseph, E., Wilkes, G.L., Baird, D.G., Polymer Prepr.,
    1984, 25, No. 2, 94

13) Friedrich, K., Hess, M., Kosfeld, R., Makromol. Chem.,
    Macromol. Symp., 1988, 16, 251

14) Amano, M. and Nakagawa, K., Polymer, 1987, 28, 263

15) Hasegawa, H., Shiwaku, T., Nakai, A., Hashimoto, T.,in
    Dynamics of Ordering Processes in Condensed Matter, ed. by
    S. Komura and H. Furukawa, Plenum Publishing Corporation,
    1988, p. 457

16) Nakai, A., Shiwaku, T., Hasegawa, H. and Hashimoto, T.,
    Macromolecules, 1986, 19, 3008

17) Hashimoto, T., Nakai, A. Shiwaku, T., Hasegawa, H.,
    Rojstaczer, S. and Stein, R.S., Macromolecules, 1989, 22,
    422

18) Chen, D. and Zachmann, H.G., to be published

19) Rosenau-Eichin, R., Ballauff, M., Grebowicz, J. and
    Fischer, E.W., Polymer 1988, 29, 518

20) Kricheldorf, H.R. and Döring, V., Makromolekulare Chem.
    1988, 189, 1425

21) Günther, B., Zachmann, H.G., Polymer, 1983, 24, 1008

22) Olbrich, E., Chen, D., Lindner, P., Zachmann, H.G. and
    Benoit, H., Physica 1989, 156 & 157, 420

23) Benoit, H.C., Fischer, E.W., Zachmann, H.G., Polymer, 1989,
    30, 379

X-RAY DIFFRACTION STUDY OF NETWORKS FORMED BY LIQUID CRYSTALLINE DIACRYLATES

R.A.M. Hikmet and D.J. Broer
Philips Research Laboratories
PO Box 80.000, 5600 JA Eindhoven, The Netherlands

ABSTRACT

Highly oriented structures are obtained by photopolymerisation of liquid crystalline diacrylates. Using X-ray diffraction it is shown that by photopolymerisation, apart from the macroscopic structure the microstructure of the molecules also becomes frozen in. Consequently with the aid of photopolymerised samples it is deduced that a cybotactic phase is formed before the transformation into the monotropic smectic C phase. Further it is concluded that molecules in the liquid crystalline state assume an all-trans conformation.

INTRODUCTION

Oriented structures are of great academic as well as industrial interest due to their anisotropic properties. Such oriented structures possess anisotropic electrical, optical and mechanical properties. Most of the methods developed in order to obtain oriented structures rely on deformation of long chain molecules (1,2). Recently it was shown that oriented systems can also be obtained by in-situ photopolymerisation of liquid crystalline (LC) acrylates(3-5). The main advantages of this method is that any orientation within the system can easily be achieved in electric and magnetic fields. It is also possible to use common surface treatments currently applied in display units to obtain a desired orientation. With LC diacrylates this orientation can be permanently fixed by photopolymerisation of the system as a cross-linked network. In this article, with the aid of X-ray diffraction we would like to show that by photopolymerisation macrostructure as well as the microstructure of molecules in the LC state become frozen in. Using photopolymerised samples we shall be showing the effect of temperature on the grouping of the molecules and try to correlate these results with molecular conformation in order to explain the formation of so called cybotactic nematic phases (7). Further we would like to explain some of the diffraction peaks in terms of intra-molecular effects.

EXPERIMENTAL

The structure of the LC diacrylate used in this study is shown below.

$$CH_2 = CH-COO + CH_2 \overline{)_6} O - \langle \rangle - COO - \langle \rangle - OOC - \langle \rangle - O + CH_2 \overline{)_6} OOC-CH = CH_2$$

Figure 1. The LC diacrylate monomer.

The monomer was provided with 2%w/w 2,2-dimethoxy 2-phenyl actophenone which acts as a photoinitiator. Oriented networks were obtained by sandwiching the monomer with spacers between microscope slides provided with uniaxially rubbed polymer coatings and initiating the polymerisation using a high intensity UV source. X-Ray diffraction patterns were recorded by a Statton camera using Ni filtered Cu_α radiation. The camera was provided with a magnetic heating cell so that the temperature of the samples could be thermostatically controlled as the molecules were uniaxially oriented under the influence of the magnetic field.

RESULTS AND DISCUSSION

At room temperature the monomer is crystalline and melts at around 108 °C to get into a nematic phase before becoming isotropic at 155 °C. Upon cooling the material shows a monotropic smectic phase below 88 °C before crystallisation occurs at around 70 °C. Upon inclusion of 2% w/w photoinitiator all transition temperatures are lowered by about 4 ° C. Mesomorphic behaviour of the LC diacrylate is described in detail in ref (5).

The effect of the freezing in the macroscopic orientation is demonstrated in Fig. 2.

Figure 2. A photograph of a frozen-in LC display.

In Fig. 2 a photograph of a LC display unit, placed between crossed polarisers is shown.The display was made by applying an electric field to reorient LC molecules between the electrodes of the display and photopolymerising the system. In this way the information on the display became permanently locked and it can be made visible between crossed polarisers as birefringent and non birefringent (areas where the molecules are oriented perpendicular to the substrate under the influence of the electric field) areas as demonstrated in Fig. 2. The effect of the polymerisation temperature on the refractive indices of the network can be found in reference(6).

Figure 3(a) shows an x-ray diffraction pattern of the LC monomer in the nematic phase at 90 ° C. In Fig. 3(b) a schematic drawing of the pattern is shown.

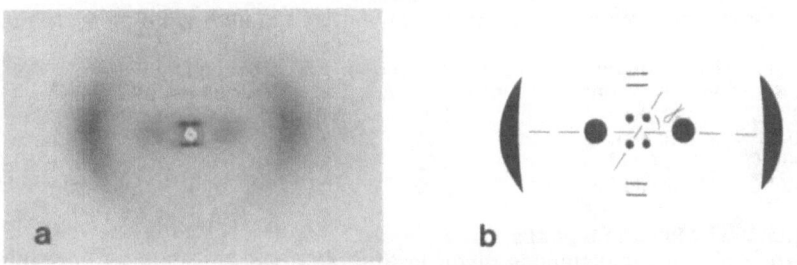

Figure 3. a) Diffraction pattern of the monomer at 90 ° C.
b) Schematic drawing of the pattern.

In this picture apart from broad equatorial peaks,a four point small angle pattern and meridional peaks at higher angles can be seen. This pattern is very typical of the so-called cybotactic nematic phase(7).

Figure 4 shows diffraction patterns of samples obtained at room temperature after polymerisation at different temperatures.

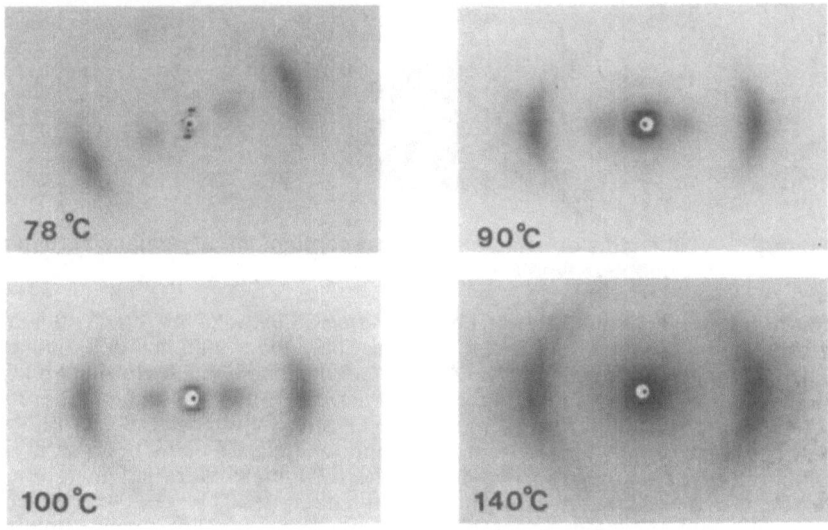

Figure 4. Room temperature diffraction patterns of samples polymerised at the temperatures shown.

The diffraction pattern shown in Fig. 4(b) was obtained after polymerisation at 90 °C and it is almost the same as that obtained before polymerisation (Fig. 3) indicating that the microstructure of the molecules in the LC state becomes frozen in after polymerisation. Bragg angles corresponding to equatorial wide angle peaks and small angle peaks together with calculated molecular parameters for samples polymerised at various temperatures are given in Table1.

Table 1
Various parameters for samples polymerised at different temperatures

Pol. Temp.	Bragg angle for equatorial peaks	Nearest neighbour distance (Å)	α	Bragg angle for four point pattern	Mol. length (Å)
78 ° C	9.955 °	4.98	61 °	1.2939 °	39.06
90 ° C	10.040 °	4.94	64 °	1.1355 °	43.28
100 ° C	9.894 °	5.02	-	-	-
140 ° C	9.820 °	5.05	-	-	-

Equatorial wide angle peaks
Nearest neighbour distances given in this table were calculated for samples polymerised at different temperatures assuming random arrangement for the molecules (7) using the equation $2d_n \sin \theta = 1.117\lambda$ where θ is the Bragg angle and d_n is the nearest neighbour distance. Here it can be seen that the calculated nearest neighbour distance for polymerised samples are in the order of 5 Å and in the nematic state it increases with increasing polymerisation temperature. In order to pack the aromatic rings with a width of 6.4 Å and a thickness of 3.6 Å so that the nearest mean neighbour distance of 5 Å is realized, a herringbone type of arrangement such as shown in Fig. 5 can be considered. Such an arrangement means that free rotation of the aromatic units about their long axis is greatly hindered.

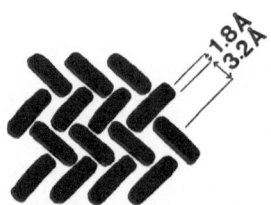

Figure 5. A schematic drawing of the cross section of the aromatic core in a herringbone type arrangement.

Furthermore, since the distance between units attached to the main chain is two carbon atoms (about 2.5 Å) and the nearest neighbour distance is 5Å ,adjacent mesogenic groups attached to the main chain must be lying in alternating layers so that the parallelism of the mesogenic units are preserved at high degrees of conversion observed for these materials. Meaning that possibly a syndiotactic network is formed. When the nearest neighbour distance of the polymer at room temperature was compared with that of the monomer before polymerisation at 90 °C it was found that after polymerisation it decreased by about 4%. Since the linear thermal expansion in the direction perpendicular to the mesogenic units in the units is about $3 \times 10^{-4} °C^{-1}$ giving rise to cooling shrinkage of about 2%, the rest must be due to the polmerisation shrinkage in the direction perpendicular to the mesogenic units and it conforms well with the measured value.

Small angle peaks

The small angle peaks in Fig. 4(a) show that at 78 °C molecules are in the smectic C phase where the mesogenic units are tilted about 39 ° with respect to the layers. Here it is interesting to point out that the observation of this montropic phase in a monodomain became possible only after photopolymerisation since at such low temperatures the sample crystallises during the x-ray measurements. At higher temperatures the sample becomes nematic and the four point pattern at small angles appear. This is considered to be as a result of critical fluctuations between the nematic and the smectic phase and it is observed at temperatures up to 90 °C, 6 °C above the nematic to smectic transition temperature. At higher temperatures the four point pattern becomes replaced by meridional streaks which become more and more diffuse with increasing temperature.

Table 1 also shows the calculated molecular length using the equation $l \sin \alpha = d$ where l is the molecular length α is the angle shown in fig 3(b) and d is the distance corresponding to the small angle Bragg peaks. Here it can be seen that there is a large difference between the two values. The value of 41 Å obtained for the end to end distance of the molecule (L) in all trans conformation using a molecular model however lies in between the measured values of 39 and 44 Å for samples polymerised at 78 and 90 °C, respectively. We would like to correlate this difference with the conformation of the molecules and illustrate it with the schematic drawing below.

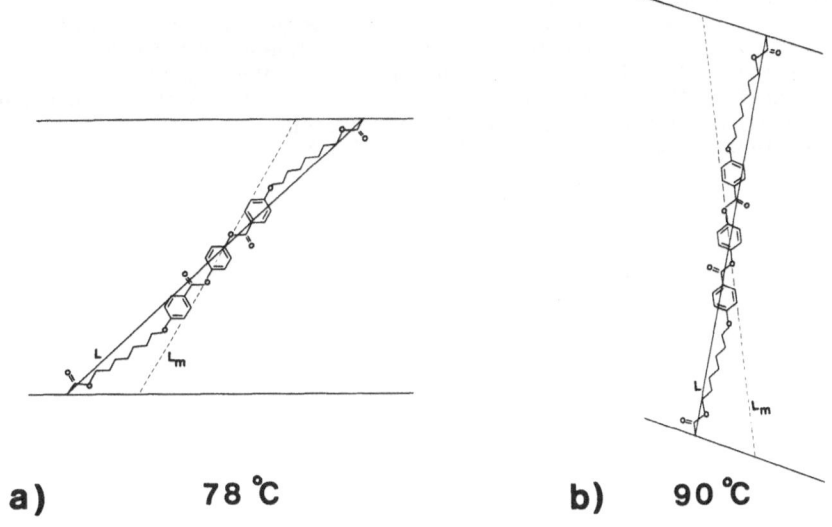

a)　　78 °C　　　　　　　b)　　90 °C

Figure 6. Schematic drawings of possible arrangement of the molecules at the temperatures shown.

In fig 6(a) we show that if the molecules assume all trans conformation and make an angle with respect to the smectic layers, the calculated molecular end to end distance (L_m) can be an under estimation and can explain the observed value for the sample polymerised at 78 °C. In the same way in fig 6(b) we show that over estimation of the end to end distance can also occur for all-trans confor- mation of the molecules (i.e. the sample polymerised at 90 °C).

Other off-equatorial peaks

In order to explain the origin of these peaks a computer program was written to calculate peak positions due to intra molecular scattering from molecules in all-trans conformation. It was found that the calculated peak positions correlated well with the experimentally observed values. This indicates that indeed the origin of these peaks must be intra molecular scattering.

CONCLUSIONS

Here it was shown that by photopolymerisation the macroscopic as well as the micoscopic structure of LC acrylates can be frozen in. In these systems the packing of the aromatic unit is quite tight and the rotation of the molecules is hindered. A cybotactic nematic phase was observed to form before the molecules transform into the smectic C phase. From molecular length calculations it was concluded that molecules assume an all-trans conformation.

REFERENCES

1. Ward, I.M., Mechanical Properties Of Solid Polymers, John Wiley & Sons, Chichester,1983.
2. Smith, P., Lemstra, P.J., J. Polym. Sci. Polym. Phys. Ed. ,1981 ,19, 877.
3. Broer,D.J., Finkelman,F., Kondo,K., Makromol. Chem. 1988,189,185.
4. Broer,D.J., Mol,G.N., Challa,G., Makromol. Chem. 1989,190,19.
5. Broer,D.J., Mol,G.,N., Challa,G., accepted for publication in Makromol. Chem.
6. Broer,D.J., Hikmet,R.A.M., Challa,G., in preparation.
7. De Vries,A., Mol. Cryst. Liq. Cryst.,1985,131,125.

TIME SCALES OF STRUCTURAL CHANGES DURING FLOW IN LIQUID CRYSTALLINE POLYMERS

JAN MEWIS and PAULA MOLDENAERS
Department of Chemical Engineering,
K.U.Leuven, 3030 Leuven, Belgium

ABSTRACT

The transient rheological behaviour of liquid crystalline polymers (LCPs) is very complex and differs considerably from that observed in normal polymer liquids. The transients reflect, to a certain extent, the complex structural changes which are induced by the flow in LCPs. Rheological measurements reveal a wide range of times scales, these can also be observed in optical measurements. Parameters as shear rate and temperature are systematically varied. In this manner similarities between the structural mechanisms of the various transient phenomena are investigated. Two uncoupled time scales have been distinguished. Molecular orientation and domains, or defects, are logical choices for the structural interpretation of the rheological transients.

INTRODUCTION

The mechanisms underlying the flow of LCPs are not well understood. Within the framework of rational continuum mechanics the Leslie-Ericksen theory [1] provides a basis for studying liquid crystals. However the basic theory is of limited value for polymeric materials as it only describes equilibrium behaviour in the Newtonian region. Following a molecular approach, Doi [2] developed a theory based on structural arguments. It simulates qualitatively some of the experimentally observed features [3] but fails to generate an adequate description [4]. The major differences with ordinary polymer melts and solutions are: the limiting low shear behaviour of normal forces and storage moduli [4], the eventual presence of a shear rate region with negative normal forces [4,5] and the oscillating transients [3,6,7].

From microscopic observations and from rheo-optical measurements it is known [8,9] that flow induces complex changes in the microstructure of LCPs. The interaction between flow and structure is considered responsible for the complex mechanical behaviour of the materials under consideration.

A structural interpretation has been hypothesized for the rheo-optical data, based on the presence of domains and disclination points [9]. To guide further structural and rheological modelling, systematic experiments have been performed on model systems under various conditions, with the objective of identifying eventual correlations and scaling laws. Here, results of different types of transient experiments on several lyotropic LCPs will be reviewed and compared.

STRESS TRANSIENTS DURING FLOW

A first series of transient experiments consists of suddenly changing the shear rate and following the resulting evolution of stress in time [7]. If the initial shear rate has been applied during a sufficiently long period, i.e. for a strain of a few hundred units, an equilibrium situation develops. It is this structure which is then probed in the subsequent transient by applying a stepwise change in shear rate.

Stepwise increase in shear rate: For isotropic polymer fluids in the Newtonian region the reduced stress will describe a unique, monotonous curve, as given by the theory of linear viscoelasticity [10]. Figure 1 shows a typical response for an LCP, in this case a solution of 12% by weight of poly(γ-benzyl-L-glutamate) (PBLG) with MW of 250000 in m-cresol [4,7]. Although both the initial and final shear rates are within the Newtonian region, no monotonous evolution but a damped oscillation is obtained for the stress. In addition the transient depends on the final

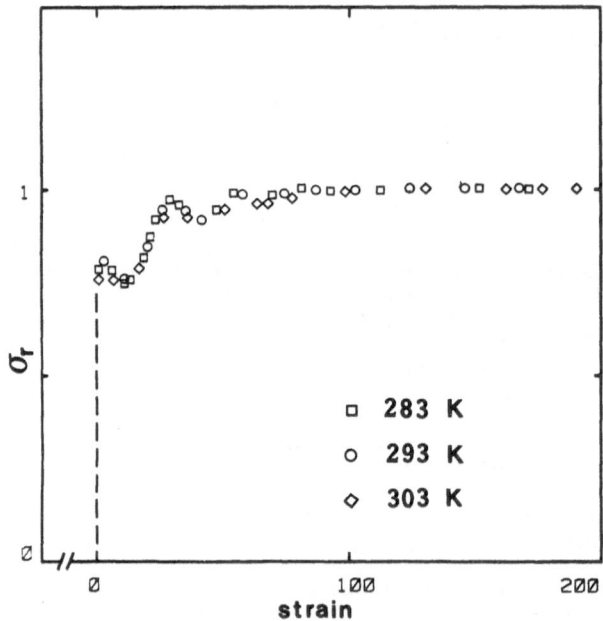

Figure 1. Stress transient after stepwise increase in shear rate

shear rate. Both phenomena constitute deviations from ordinary polymers.

The time scale for the transients has ueen shown to change with the inverse of the final shear rate, i.e. it scales with strain [7]. This scaling cannot be explained on the basis of diffusional phemonena, it rather suggests a convective mechanism like rotation of some structural elements. Linear dichroism measurements [11] confirm this picture, suggesting in particular a possible rotation of domains.

Changing the initial shear rate, within the Newtonian region, affects the reduced stress transient only slightly, the period of oscillation remains around 30 strain units. If the final shear rate is within the shear thinning region, the period becomes shorter and the initial peak for the reduced stress increases above the equilibrium value. This might suggest some structural breakdown, which is assumed to take place at sufficiently high shear rates. Decreasing the temperature over 20 K causes a change of 350% in steady state viscosity but does not affect, within measuring accuracy, the shape of the stress transients (fig. 1) [12]. This proves that time effects do not scale with the zero shear viscosity as is often the case in other polymer fluids.

Stepwise decrease in shear rate: Instead of a sudden increase, a stepwise decrease in shear rate can be applied (stepdown experiments). For ordinary polymers this should result in a stress transient closely related to stress relaxation. For LCPs an oscillating curve is obtained, similar to the one resulting from a stepwise increase in shear rate. It has been shown [7] that the transient stress curve for a stepwise decrease consists of two parts. The rapid initial decay displays a time dependency which is independent of shear rate, at least within the Newtonian region. If the initial shear rate is within the shear thinning region, the relaxation accelerates with increasing shear rate; this is also the case in normal polymers. The initial relaxation is quite fast: about 90% of the stress is decayed in 2 seconds for the sample of fig. 1. The subsequent oscillating part scales again with strain rather than with time. It has a similar period as in the stepwise increase in shear rate and requires also about 100 strain units to equilibrate.

As far as data have been published for other LCPs, the strain scaling for stepwise changes in shear rate has been confirmed [13-15]. It therefore seems to be an important general feature of LCPs, which should be reproduced by suitable rheological models. The Doi model does not satisfy this requirement.

TRANSIENT BEHAVIOUR UPON CESSATION OF FLOW

A stepdown experiment with the final shear rate equal to zero is an interesting limiting case. There is however a complication for some materials, such as the PBLG sample mentioned above, because they gradually change their nematic structure into a cholesteric one if the flow stops. This is not the case in racemic PBG mixtures. Hence it can be verified whether this structural transition affects the rheological transients after cessation of flow.

How the structure evolves after stopping the flow can be probed

rheologically in three ways. The first and obvious choice is similar to
the experiments described above, i.e. by observing the shear stress relax
[16]. As in stepwise decrease in shear rate, two regions can be distinguished (fig. 2). The fast initial part is independent of shear rate in
the Newtonian region, it coincides with the fast part in stepwise decrease
in shear rate. It is followed by a slow "tail", the time scale of which is
inversely proportional to the previous shear rate.

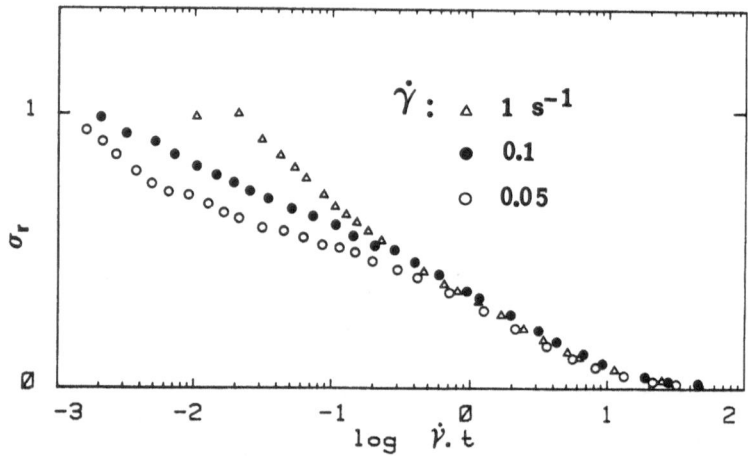

Figure 2. Shear rate scaling for stress relaxation upon cessation of flow
(20% racemic PBG in m-cresol, MW=330000)

The initial part of the curve seems to obey the basic rules for
relaxation upon cessation of flow from the isotropic linear theory of
viscoelasticity [10]. The integral of the curve approximates quite well
the viscosity as predicted by theory. When changing the temperature this
part also scales, within measuring accuracy, with viscosity, as expected
for isotropic fluids. The normal stresses cannot be calculated adequately
from the relaxation of the shear stresses. In this calculation the second
part of the curve, which shows a totally different behaviour, dominates.

As part of the stress relaxation depends on shear rate, the same
should hold for the equilibrium structure during flow, even in the Newtonian region. This shear rate dependence of the final part of the stress
relaxation is similar to that observed for the relaxation of the birefringence in similar materials [9]. The structural change during relaxation cannot be explained by a nematic-cholesteric transition as fig. 2
refers to a racemic mixture of PBG which remains nematic at rest [17]. .

Two other transient experiments can be performed to detect the structural changes upon cessation of flow. Oscillatory testing at small ampli-
tudes offers one nondestructive method to follow changes in microstructure
[4]. Applied to the sample of fig. 1, the moduli decrease with time during
rest rather than increase, as is eventually encountered for normal polymeric fluids in the non-Newtonian region [18]. However this is not a

general rule for LCPs. For a liquid crystalline solution of hydroxypropyl-cellulose (HPC) in acetic acid an increase is found [15] and with other PBG samples an initial increase was followed by a subsequent decrease [19]. The results for these three materials are shown in fig. 3. In all cases the average time scale for the change in moduli is orders of magnitude larger than that for stress relaxation. It seems to be associated with the tail of the relaxation curve: it is also proportional to the inverse of shear rate and does not depend on temperature [4].

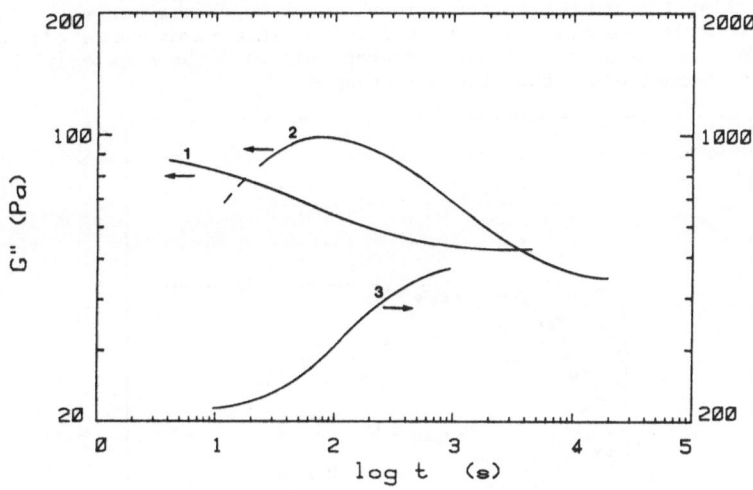

Figure 3. Diverging shapes of the curves for the dynamic moduli after cessation of flow (1 = sample of fig. 1; 2 = 25% PBDG, MW=310000; 3 = 40% HPC in acetic acid, MW=100000)

The basic theories for liquid crystals do not describe relaxation curves with two different time scales. These theories do not take into account defects in the liquid crystalline structure. It is well known that real LCPs contain bands and defects or domains, which are assumed to be responsible for the complex behaviour [8,9,20]. In solutions of hydroxy-propylcellulose the formation of band structures after cessation of flow follows the same shear rate dependence [20] as the present mechanical measurements. A quantitative model on this basis [9] has been mentioned above for the rheo-optical transients. It has not been extended to the rheological experiments discussed here.

A problem arises when discussing oscillatory behaviour because there is no suitable theory yet for dynamic moduli of LCPs. Larson has modified the Doi theory to correct this shortcoming [21]. He integrates over inde-pendent domains of varying orientation to predict moduli that increase or decrease with time, depending on concentration. Domain orientation could be an important factor but domain interfaces should also be substantial sources of stresses in LCPs. The scaling of the time constant with the previous shear rate could reflect a domain size effect according to the

272

dichroism measurements [11], but the results are not definitive yet.

A third technique to study the evolution of structure in time consists in measuring the stress transient when restarting the flow after a variable rest period (intermittent shear flow)(fig. 4). These stress transients display a rather complex evolution with the duration of the rest period. After very short times, the stress rapidly recovers its equilibrium value again but even so a small oscillation is visible. The initial amplitude of the damped oscillation becomes gradually larger, thereby causing the minima in the stress signal to deepen out with time. The stress curve reaches a constant profile after approximately 10^4 seconds. This very long time scale compares well with the slow equilibration of the moduli after the flow has stopped.

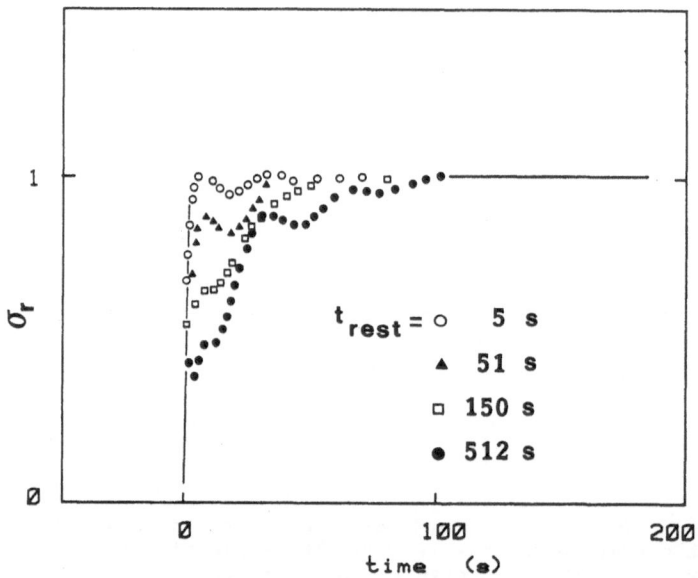

Figure 4. Effect of rest period in intermittent shear flow, sample as in fig. 2 ($\dot{\gamma} = 1s^{-1}$).

The period of the oscillatory part (fig. 4) increases only slightly with time. If this feature is associated with a tumbling element, the latter must have nearly a constant aspect ratio. The second peak of the damped oscillation seems to split in two separate peaks after a few hundred seconds, which does not confirm the explanation of a tumbling element. In addition a separate sharp peak develops at small strains (a few units). The time scales of these phenomena are hardly affected by temperature, as was the case for those of the moduli after cessation of flow. Hence there seems to be a strong similarity between the time-dependent linear moduli and the start-up curves, suggesting that they probe similar structural features.

DISCUSSION AND CONCLUSIONS

Transient behaviour can be observed in lyotropic LCPs over more than 4 decades of time. In stepwise changes of shear rate, including relaxation experiments, the major part of the stress is adjusted in a few seconds or less. This part seems to correlate with the viscosity, as can be seen by changing the temperature or by calculating the viscosity from the relaxation curve. There seems to be no dichroism effect associated with this response which is attributed to changes in molecular realignment. The damped oscillation, which accompanies the stress transient in some experiments, does not scale with time but with strain.

Upon cessation of flow, changes still occur long after the shear stress has relaxed. This can be observed in the evolution of the moduli, in intermittent shear flow and in the final part of the stress relaxation. This time effect does not scale with viscosity, e.g. it is practically independent of temperature. On the other hand it changes inversely proportional to the preceding shear rate in the three experiments mentioned above. The average time constant for this mechanism is 2-3 orders of magnitude larger than that of the stress relaxation and it is affected differently by the test parameters. Therefore different structural mechanisms seem to be responsible for the two time scales. The longest time scale could reflect some large scale structural rearrangements, without necessitating a nematic-cholesteric transition. In these rearrangements defects or domains can be expected to play an important role. This is further corroborated by the presence of similarly large, shear rate dependent, optical relaxation times.

REFERENCES

1. Leslie, F.M., Theory of flow phenomena in liquid crystals. Adv. Liq. Cryst., 1979, 4, 1-81.

2. Doi, M., Molecular dynamics and rheological properties of concentrated solutions of rodlike polymers in isotropic and liquid crystalline phases. J. Polym. Sci., 1981, 19, 229-243.

3. Metzner, A.B, and Prilutski, G.M., Rheological properties of polymeric liquid crystals. J. Rheol., 1986, 30, 661-691.

4. Moldenaers, P. and Mewis, J., Transient behavior of liquid crystalline solutions of poly(benzylglutamate). J. Rheol., 1986, 30, 567-584.

5. Kiss, G. and Porter, R.S., Rheology of concentrated solutions of poly-(γ-benzylglutamate). J. Polym. Sci. Polym. Symp., 1978, 65, 193-211.

6. Viola, G.G., and Baird, D.G., Studies on the transient shear flow behavior of liquid crystalline polymers. J. Rheol., 1986, 30, 601-628.

7. Mewis, J. and Moldenaers, P., Transient rheological behaviour of a lyotropic polymeric liquid crystal. Mol. Cryst. Liq. Cryst., 1987, 153, 291-300.

8. Kiss, G. and Porter, R.S., Rheo-optical studies of liquid crystalline solutions of helical polypeptides. Mol. Cryst. Liq. Cryst., 1980, 60, 267-280.

9. Asada, T., Onogi, S. and Yanase, H., A rheo-optical study on the reformation of structure in racemic poly(γ-benzyl glutamate) liquid crystals. Polym. Eng. Sci., 1984, 24, 355–360.

10. Ferry, J.D., Viscoelastic Properties of Polymers, J. Wiley, New York, 1980.

11. Moldenaers, P., Fuller, G. and Mewis, J., Mechanical and optical rheometry of polymer liquid-crystal domain structure. Macromolecules, 1989, 22, 960–965.

12. Moldenaers, P. and Mewis, J., Transient rheological behaviour of polymeric liquid crystals. In World Congr. III Chem. Eng., Soc. Chem. Eng. Jap., 1986, vol. IV, pp. 546–549.

13. Wissbrun, K.F., Observations on the melt rheology of thermotropic aromatic polyesters. Brit. Polym. Jl., 1980(12), 163–169.

14. Doppert, H.L. and Picken, S.J., Rheological properties of aramid solutions: transient flow and rheo-optical measurements. Mol. Cryst. Liq. Cryst., 1987, 153, 109–116.

15. Moldenaers, P. and Mewis, J., Time-dependent behaviour of polymeric liquid crystalline solutions. In Proc. Xth Int. Congr. Rheol., Sydney, 1988, vol. 2, 134–136.

16. Moldenaers, P. and Mewis, J., Mechanical relaxational phenomena of lyotropic polymeric liquid crystals, to be published.

17. Berghmans, S., Intermittent shear flow in nematic polymeric liquid crystals. Dissertation, Dept. Chem. Eng., K.U.Leuven, 1989.

18. De Cleyn, G. and Mewis, J., A constitutive equation for polymer liquids: application to shear flow. J. Non-Newtonian Fluid Mech., 1981, 9, 91–105.

19. Moldenaers, P., Yanase H. and Mewis, J., Effect of the shear history on the rheological behaviour of lyotropic liquid crystals. Polymer Preprints, ACS, to be published.

20. Navard, P., Formation of band textures in hydroxypropylcellulose liquid crystals. J. Polym. Sci. Polym. Phys. Ed., 1986, 24, 435–442.

21. Larson, R.G., Linear viscoelasticity of nematic liquid crystalline polymers. Proc. Xth Intern. Congr. Rheol., Sydney, 1988, vol. 2, pp. 64–66.

Part 5

STRUCTURE/MORPHOLOGY

Biaxial-Drawing of Dried Gel Films of
Ultra-High-Molecular-Weight Polyethylene

Y.SAKAI, K.UMETSU, P.D. HONG and K.MIYASAKA
Department of Organic and Polymeric Materials
Tokyo Institute of Technology,Meguro-ku Ookayama Tokyo 152

ABSTRACT

Simultaneous-biaxial drawing of the UHMWPE gel films about 130 μm thick was possible at high temperatures such as 135°C. The original film was required to have high uniformities both in the thickness and structure, particularly to get high draw ratios. The maximum draw ratio ever achieved was 20x20. The structure and properties of biaxially drawn films were studied.

INTRODUCTION

Since Smith and Lemstra found that UHMWPE(Ultra High Molecular Weight Polyethylene) gels had an extraordinarily high drawability and the drawn materials had very high values of modulus and strength, uniaxial drawing of the materials prepared through gelation has been studied a lot.

Biaxial drawing of these materials,however, has hardly been studied[1],in spite that it has many interesting problems such as biaxial drawability, the structure and properties of drawn materials. In this paper are presented some results obtained on biaxial drawability of UHMWPE dried gel films, and the structure and properties of the biaxially-drawn films.

EXPERIMENTAL

Sample preparation: A homogenized decahydronaphthalene solution containing 4wt% polymer (Hizex Million: $Mv = 4.5 \times 10^6$) was prepared in a flask at 160°C, being stabilized by 0.5wt% of an antioxidant di-tert-butyl-p-cresol. The solution was cooled to room temperature to make the gel. The gel was taken out of the flask and pressed between aluminum boards under a pressure of $100 kgcm^{-2}$ at $150^{\circ}C$ for 10min, followed by quenching in water at $20^{\circ}C$. Uniform thick films could not be obtained below $150^{\circ}C$ (temperature of the compression machine). Some partial melting resulting in the decrease in drawability might occur during compression. Subsequently, the gel sheets were dried at room temperature to make films 0.11-0.13mm thick. This sample preparation is different from that of so-called dried gel films of Smith and Lemstra. The WAXD indicates that the c-axis of crystals tends to orient in the thickness direction of the sheet. The SAXS photograph has a meridian two-point pattern giving a 12nm long spacing. These structural aspects are similar to those of single crystal mats of both medium molecular weight (MMW) PE and UHMWPE and to those of dried gels.

Biaxial drawing: Samples of $10x10cm^{2}$ were cut from the film for simultaneous biaxial drawing performed at $135^{\circ}C$ using an Iwamoto biaxial film stretcher.

RESULTS AND DISCUSSION

As seen in Fig.1, the original film consists of particles about $5\mu m$ in diameter. Each particle seems to be connected with its neighbors through many thin short fibrils formed during pressing. These fibrils must play an important role in the subsequent biaxial drawing. The particle structure disappears almost completely at a draw ratio of 10x10, having transformed into fibrils. It should be noted about Fig.1 that most fibrils are not straight but bent, in contrast with uniaxial drawn materials.

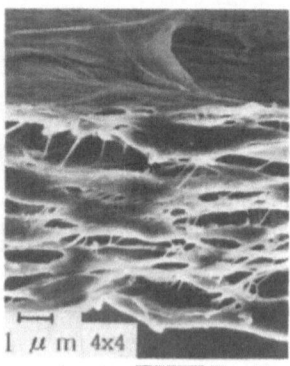

(a)   (b)

Fig.1 SEM photographs of si-multaneously biaxially drawn films: (a)1x1;(b)20x20

Fig.2 FE-SEM photograph of the plane on which the film thickness of drawn film.The plane was prepared breaking in liquid N_2.

SEM of fractured section of a 4x4 drawn film shown in Fig.2 indicates that biaxial drawing is performed making layers. It should be noted that any meaningful fracture surface, reflecting the real structure of the section, could not be obtained at draw ratios larger than 4x4. SEM at high draw ratios indicate that something like melting had happened during fracturing.

The change in the diameter of fibrils was followed by SEM with the results shown in Fig.3. The average diameter of fibrils decreases with drawing, and the distribution of diameter becomes sharper. The thinning of fibrils is due to the drawing of each fibril and also to the splitting of fibril. The results in Fig.3 seem to indicate that each fibril was drawn till high draw ratios of 16x16 at least.

As to the fibrillar network, it is interesting that in films as highly drawn as 16x16, some fibrillar entanglements were made by deeply folded fibrils. This sort of entanglement may be resulted from micro-scale structural irregularities. The entanglement resists further drawing, as is evidenced by the fact that this sort of fibrillar entanglement was hardly observed at 20x20 the highest draw ratio ever achieved.

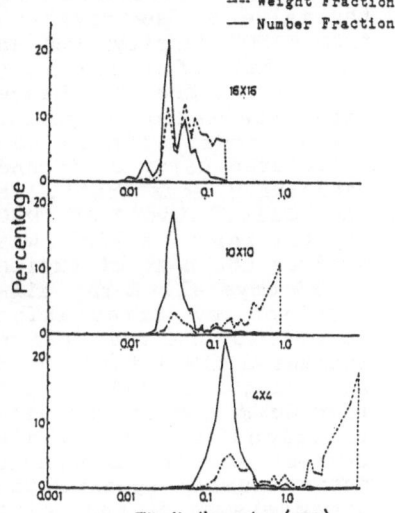

Fig.3 Change in fibrillar diameter with biaxial drawing.

The structural characteristics of the biaxially drawn UHMWPE films are summarized as follows:

Crystal plane orientation was caused in the modes of the (100) and (110) planes parallel to the film plane as shown in Fig.4.

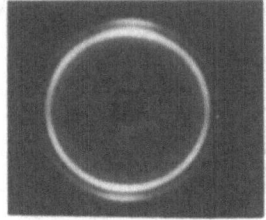 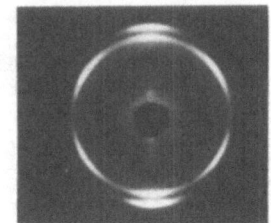

(a)   (b)

Fig.4 WAXD photographs, taken with the incident beam parallel to film plane.The thickness direction is vertical.
(a):4x4 (b):16x16

These modes are the same as those already well known for usual

280

PE drawn films. SAXS peak due to the long spacing disappeared at 10x10, suggesting the transformation from the crystalline-amorphous in-series structure to that of extended chain morphology. However, it is not certain whether this transformation took place or not in reality. In the case of uniaxial drawing, the modulus of drawn material can be a measure to judge this possibility of transformation, by comparison with the value of the crystal modulus. The crystallinity estimated from WAXD, density, DSC and IR shown in Fig.5 is much less than that of uniaxially drawn films and the change with draw ratio is small. The IR crystallinity was calculated using Okada and Mandelkern equation[2]. In the calculation of DSC crystallinity, $293 Jg^{-1}$ was used as the heat of fusion of PE crystal. X-ray diffraction by irradiation normal to the film plane increased the intensity in 2θ range around $15°-20°$ (for Cukα) where the diffractions by monoclinic crystal and the amorphous phase usually appear. The broadness and the high thermal stability of the diffraction suggest the contribution of the amorphous region may be predominant over that of monoclinic crystal. Fig.6 shows Thermoluminescence (TL) measured during heating following x-ray irradiation. This suggests origination of a great number of defects providing the trap-sites for the electrons. The similar increase in TL intensity could be observed for uniaxial drawn UHMWPE. TL glow peaks with their maximum at about $70°C-80°C$ correspond to the so-calledα_c mechanical dispersion related to dispersions within a crystal and at the boundary of a crystal. TL intensity remarkably increases between 6x6 and 10x10, and the enhancement seems to become milder at draw ratios

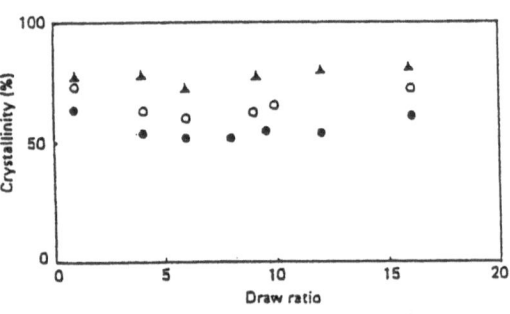

Fig.5 Crystallinity as a function of biaxial draw ratio: ▲ ,from density; ○ ,from IR; ● ,from heat of fusion.

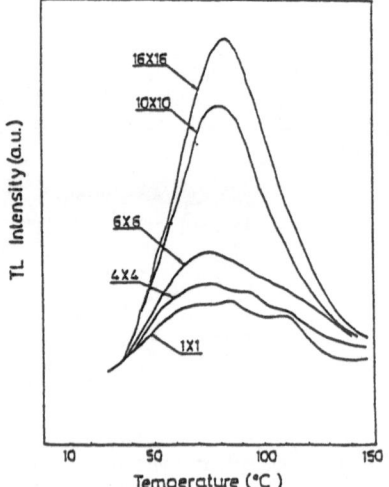

Fig.6 Change in TL glow intensity curve with biaxial drawing.

over 10x10. The increase in TL intensity is due to the increase in the number of defects within a fibril. The milder increase in TL intensity at high draw ratios over 10x10 implies the increase in the contribution of interfibrillar slippage to the drawing.
Properties of biaxially drawn films are summarized as follows:

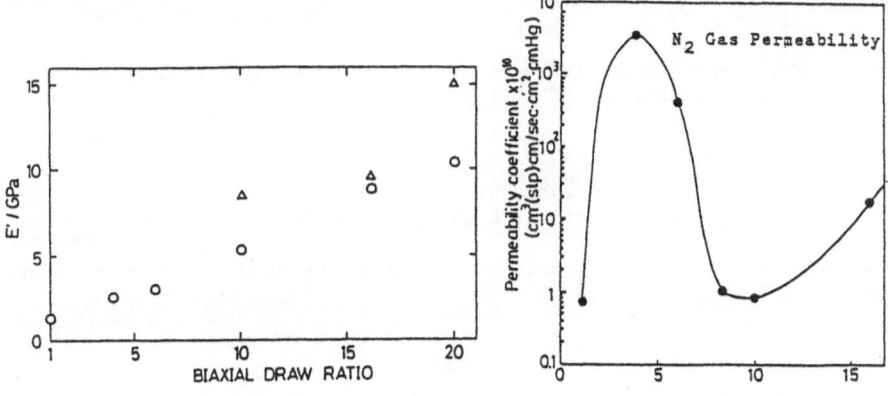

Fig.7 Storage dynamic modulus at 20°C of drawn films as a function of biaxial draw ratio.

Fig.8 N$_2$ gas permeability of drawn films as a function of biaxial drawn ratio

Young's modulus increased with increasing draw ratio, as expected, and the highest value ever achieved is about 12 Gpa at 20x20, as indicated by Fig.7. This value is supposedly close to the upper limit for this sort of films consisting of microfibrils orienting random on the film plane. As already observed, the drawn films had microfibrillar network structure. The theory for modulus of this sort of fibrillar network is desired to establish. N$_2$ gas permeability of biaxially drawn films remarkably changed with draw ratio as indicated by Fig.8. 10^{-11} (cm^3(STP)cm/sec cm^2 cmHg) order before drawing increased to 10^{-7} at 4x4, and then decreased to be restored to 10^{-11}, corresponding well to the SEM observation. The permeability again increased at draw ratio over 10x10. This suggests that the porosity slightly increased due to the origination of microvoids, whereas the modulus continued to increase.

REMARK: This review is prepared on the bases of Ref.3 and 4.

REFERENCES

1) Mimami,S. and Itoyama,K.,Am.Chem.Soc.Polym.Prep., 1985,26,2,245
2) Okata,T. and Mandelkern,L.,J.Polym.Sci.,1967,A-2,5,239
3) Sakai,Y. and Miyasaka,K.,Polymer,1988,29,1609
4) Sakai,Y. and Miyasaka,K.,Polymer,in press

LARGELY UNRECOGNIZED FEATURES OF POLYMER CRYSTALLIZATION UNDER CONDITIONS OF HEAT TRANSFER AND FLOW

H. JANESCHITZ-KRIEGL

Linz University, A-4040 Linz, Austria

Dedicated to Prof. D.W. Van Krevelen on the occasion of his 75th birthday.

ABSTRACT

Basic ideas of polymer crystallization, rheology and heat transfer are integrated into a description of the fundamentals of the industrial solidification processes in thermoplastic materials. Aspects of texturing are included.

INTRODUCTION

A fairly concise review is given of some of the more fundamental results obtained in a national working party on injection moulded articles, which was set up in 1983 by the author. At that time one could find only little support from previous publications except for the wrong textbook advice that classical theory should be used for the calculation of solidification in polymers. This advice, however, ignores the effect of supercooling which is so important in polymers. It is just this effect which opens a whole spectrum of aspects never treated systematically before. Some useful contributions by Astarita, Van Krevelen and Malkin and their cooperators will be cited in due course. An extended review for Progress in Polymer Science is in preparation.

Global Heat transfer

A semi-infinite body is considered for convenience. Before the envisaged experiment is started, the body is thought to be at a uniform temperature well above the melting point. At zero time the temperature of the plane at x = 0 (the "wall") is suddenly dropped to a temperature below the melting point and kept at that temperature. In Fig. 1 the possible processes following such a quench are presented schematically.

In this figure the uppermost curve represents the classical square-root-law (1). This curve shows the progress of a crystallization front if supercooling is neglected (infinite rate of nucleation, <u>heat diffusion</u>

controlled process). However, if the propagation of a crystallization front is treated correctly (2), one obtains the lower curve. One can show that this process is always nucleation rate controlled in the beginning, in contrast to the classical theory. The initial slope represents the growth speed of the crystalline layer at the temperature of the quenched wall. According to this correct treatment, however, the crystallization front (moving phase boundary) is supercooled. As a consequence also the melt in front of the phase boundary is supercooled. So one cannot avoid that disperse nucleation occurs in the supercooled part of the melt. This explains the occurrence of a diffuse crystallization zone (3), which is indicated by the hatched area in the graph. This figure also suggests where the front is superseded by the zone.

If a slab of finite thickness is cooled, one can get almost simultaneous dispersed crystallization over the whole sample, if some critical number is small enough (3)(4)(5). (Ratio of Stefan over Deborah number, kindly proposed by Astarita to be called after the present author.)

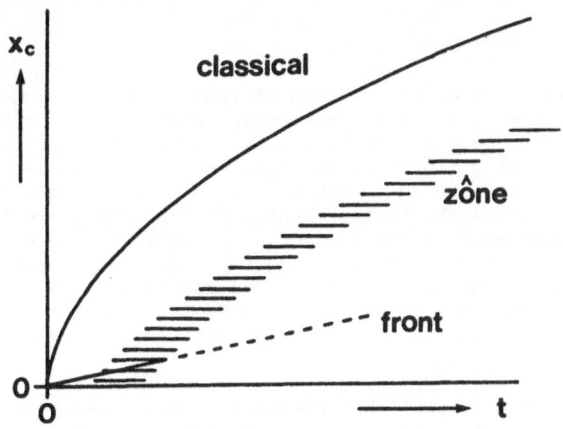

Figure 1. Schematic presentation of the progress of crystallization into a semi-infinite body quenched at zero time at the plane x = 0: x_c ... distance at which crystallization occurs, t ... time.

Spatial Size Distribution of Grains

If H is the latent heat of fusion, c is the heat capacity of the melt, T_i is the initial bulk temperature and T_w is the wall temperature, a Stefan number

$$St = \frac{H}{c(T_i - T_w)}$$

(1)

of the order of unity, as usually observed with polymer processing, means that, with thermal equilibration of a finite piece of material, only about twice the amount of heat must be removed, if crystallization occurs. This fact justifies a global treatment (crystallization as a kind of self-accelerating reaction (6)), as implied in the previous section (3). Along this line, the heat conduction problem is practically uncoupled from the actual nucleation problem: The larger the distance from the wall, the lower the (average) speed of cooling, the larger the size of grains (spherulites). In Fig. 2 this effect is demonstrated for a slab of polypropylene after a quench on one side.

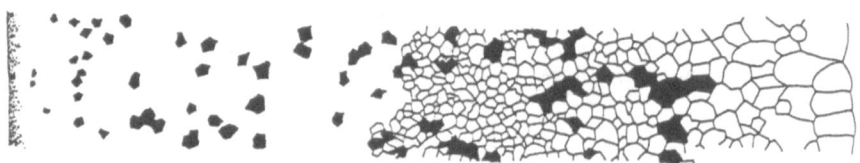

Figure 2. Drawing made after a series of contiguous microphotographs as taken from a cut through a polypropylene tablet quenched on the left side of the figure. Observe the increase of grain (α-spherulite) size from left to right until the line is reached, where crystallization is halted (at the melting point). Black areas indicate where the faster growing β-spherulites are found. At the quenched surface the size of α-spherulites is comparable with the extension of the small dots placed. In the intermediate range α-spherulites are not drawn. The distance between the boundaries on the left and on the right sides is about 7 mm. ($T_i = 200°C$, $T_w = 40°C$, Ratajski, still unpublished.)

It goes without saying that progress of heat transfer, nucleation and grain growth can also be treated simultaneously (7) (8), with the aid of numerical techniques. These techniques become mandatory if in addition to heat conduction also heat convection is involved. The necessary physical data, however, are still scarce. Pioneering work has been carried out in the latter field by Van Krevelen (9) and his cooperators.

Flow Induced Crystallization, a Relaxation Phenomenon

As explained in refs. (10) and (11), the induction time t_i for shear induced crystallization is governed by the following equation:

$$\int_0^{t_s} \phi\,(\dot{\gamma},\,t)\,d\,t + \int_{t_s}^{t_i} \phi_r\,(\dot{\gamma},\,t)\,d\,t = t_\infty \quad , \tag{2}$$

where φ is a probability function for the occurrence of lamellar (shear induced) crystallization (where φ is increasing with time from zero to a saturation value $\phi_\infty \leqslant 1$. Parameters: τ ... relaxation time, $\dot{\gamma}_a$ ... critical shear rate of activation), ϕ_r is the relaxing probability function (after shear flow is stopped), t_s is the shearing time, t_i is the induction time ($t_i \geqslant t_s$) and t_∞ is the induction time at optimum conditions ($\phi \equiv 1$, $t_s = t_i = t_\infty$, t_∞ being the third parameter of the model).

Experimental results corroborate this theory. Pertinent results are shown in Fig. 3. During shearing one observes flow birefringence. Together with the shear stress this birefringence goes to zero, when flow is stopped. The shear induced crystallization coming up belatedly is indicated by the second rise of the birefringence. The shorter the shearing time t_s, the longer is the prolonged induction time t_{ip}. With decreasing t_s the prolonged induction time goes to infinity before t_s goes to zero: There is a minimum shearing time t_{sm}, below which no shear induced crystallization comes up. Certainly, t_{sm} is a function of shear rate and temperature. This explains the sharp boundaries between shear induced boundary layers and spherulitic cores in injection moulded articles. At a temperature of 158°C, as used for the described experiment, normal spherulitic crystallization cannot be observed even after many hours in a quiescent polypropylene melt.

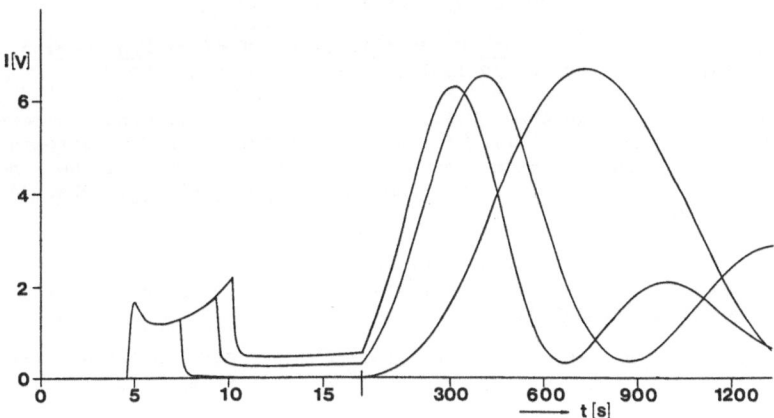

Figure 3. Measure of birefringence (light intensity) vs. time for inter-
mitted shear of a polypropylene melt at 158°C after Liedauer
(in preparation). Observe the different time scales for shearing
and for prolonged observation (shear rate ... $\dot{\gamma} = 100$ s^{-1}).

The probability function φ seems proportional to the total surface area of lamellae developing apparently also at (higher) temperatures where they decay by relaxation when flow is stopped. This can be concluded from elaborate quenching experiments (11), (12).

For engineering purposes, convection of previously nucleated material will be a new subject of research.

286

ACKNOWLEDGEMENTS

The author is very much indebted to Prof. W. Schneider, Vienna University
of Technology, for his great cooperativeness in the development of parts
of the theory and to the Austrian "Fonds zur Förderung der wissenschaft-
lichen Forschung" for sponsoring this research under project nr. S 33 02.

REFERENCES

1. Carslaw, H.S. and Jaeger, J.C., Conduction of Heat in Solids, 2nd ed.,
   Clarendon, Oxford 1959, p. 281-296.

2. Eder, G. and Janeschitz-Kriegl, H., Stefan Problem and Polymer Processing, Polym.Bull. 1984, 11, 93-94.

3. Berger, J. and Schneider, W., A Zone Model of Rate Controlled Solidification, Plastics and Rubber Processing and Applications 1986,
   6, 127-133.

4. Astarita, G. and Kenny, J.M., The Stefan and Deborah Numbers in Polymer Crystallization, Chem.Eng.Comm. 1987, 53, 69. See also: Astarita,
   G. and Nicolais, L., Physics and Mathematics of Heat and Mass Transfer
   in Polymers, Pure & Appl.Chem. 1983, 55, 727-736.

5. Janeschitz-Kriegl, H., One Hundred Years of Chemical Engineering,
   N.A. Peppas, ed., Kluwer/Reidel, Dordrecht, spring 1989.

6. Malkin, A.Ya., Beghishev, V.P., Keapin, I.A. et al., General Treatment of Polymer Crystallization Kinetics - Part 1. A New Macrokine-
   tic Equation and its Experimental Verification, - Part 2. The Kinetics of Non-Isothermal Crystallization, Polym.Engng.Sci. 1984, 24,
   1396 - 1401, 1402 - 1408.

7. Schneider, W., Köppl, A. and Berger, J., Non-Isothermal Crystallization: Crystallization of Polymers, System of Rate Equations,
   Intern.Polym.Processing 1988, 2, 151-154.

8. Berger, J., Erstarrung von Kunststoffen unter dem Einfluß von Wärme-
   leitung und Kristallisationskinetik, doctoral thesis, Vienna Univ.
   Techn. 1988.

9. Van Krevelen, D.W., Crystallinity of Polymers and the Means to
   Influence the Crystallization Process, Chimia 1978, 32, 279-294.

10. Eder, G. and Janeschitz-Kriegl, H., Theory of Shear-Induced Cry-
    stallization of Polymer Melts, Colloid & Polym.Sci. 1988, 266,
    1087-1094.

11. Eder, G., Janeschitz-Kriegl, H. and Krobath, G., Shear Induced
    Crystallization, a Relaxation Phenomenon in Polymer Melts,
    Colloid & Polym.Sci. 1989, in press.

12. Janeschitz-Kriegl, H., Wimberger-Friedl, R., Krobath, G. and Liedau-
    er, S., Über die Ausbildung von Schichtstrukturen in Kunststoff-Form-
    teilen, Kautschuk + Gummi, Kunststoffe 1987, 40, 301-307.

AN ELECTRON MICROSCOPIC STUDY ON THE FORMATION OF POLYETHYLENE FIBRILS

Martin Kunz, Raúl E. De Micheli[2], and Martin Möller[1],[*]

Institut für Makromolekulare Chemie der Universität Freiburg
Hermann-Staudinger-Haus, Stefan-Meier-Straße 31
D-7800 Freiburg, Federal Republic of Germany

[1] Department of Chemical Technology, University of Twente
P.O. Box 217, NL-7500 AE Enschede, Nederland

[2] Centro de Investigaciones Technologicas Para la Industria Plastica-INTI
1650 San Martin, Argentina

ABSTRACT

Crystallite deformation and chain rearrangment in high density polyethylene samples of different molecular weights have been studied by transmission electron microscopy for solid-state-extrusion as well as for ultra-drawing of high molecular weight gels. The crystallite rearrangement is documented at different stages of the deformation process and discussed with regard to the Peterlin model [6]. The destruction of the lamellae at low temperatures and their plastic deformation below the melting transition of the primary material indicates the occurence of chain slipping above the α-transition. In the necking region of drawn gels, the draw ratios weré found to increase from the center to the surface of the tapes. The deformation process in gel drawing and solid state extrusion appear to be the same within the early stages and at elevated temperature. A rather irregular fibrillar morphology was observed for highly drawn gels. Substitution of the solvent by methacrylate monomers and subsequent curing at low temperatures allowed to observe the polyethylene gels in statu nascendi. The gels are formed by lamellar crystals of large width which are linked by amorphous interlayers formed within localized contacts of the lamellar surfaces.

INTRODUCTION

Solid-State-Extrusion and gel drawing of polyethylenes are different processes which were developed to obtain high modulus and high strength polyethylene fibres. Both techniques are based on chain extension in elongational flow fields. However, the results as well as the primary PE-materials
are different [1,2]. In the case of the extrusion technique, low molecular
weight polyethylenes with non oriented crystallites are pressed through a
conical die, in which an extensional flow field yields nominal draw ratios typically below 30. In the gel drawing or gel spinning process, highly oriented
ultrahigh molecular weight polyethylene crystals are uniaxially drawn by a
factor of 70-200. Numerous work has been done to investigate the thermal
and mechanical properties of the resulting fibres and to elucidate the complex molecular reorientation processes which result in the transformation
from a lamellar to an oriented fibrous structure [3]. The most widely accepted model to describe the formation of a microfibrillar structure has been
proposed by Peterlin [6]. Crystal blocks are broken off the lamellae and
microfibrills are pulled out at the cracks. The model has been developed to
explain the fiber formation on drawing of spun or extruded flexible polymers,
thus, to describe the transformation of nearly isotropically oriented lamellae.
Confirming information about the morphology could be obtained from X-ray,
infrared-, and Raman-spectroscopy as well as from electron microscopy on
drawing melt crystallized polyethylenes [7,8]. Recently van Aerle and Braam
proposed a three stage deformation model for the gel drawing process in
which the transformation of lamellae into microfibrills during the initial stages
of drawing is described according to the Peterlin model [9]. Tautening of the
tie molecules between the crystal blocks which are incorporated into the
microfibrills and subsequent unfolding of these crystallites on increase of
extended chain inter-microfibrillar crystals represent the second and the third
step. This three stage model is based on careful small and wide angle X-ray
scattering experiments.

In the present paper we compare transmission electron microscopic studies
on the deformation and morphology changes of polyethylene crystallites in
the solid state extrusion and the gel drawing process. The plastic deformation process was followed up from the starting polymer, e.g the polyethylen-
gel as crystallized, to the fully drawn fibrous material. Newly developed
staining [10] and electron microscopic [11] techniques were employed to
achieve TEM images of superior contrast and detailed resolution.

MATERIALS AND METHODS

HDPE (M_w = 59.000, M_n = 19.000, Alathon, Du Pont) was crystallized at different temperatures and pressures [12]. Solid-state-extrusion was performed by pressing the sample through a tapered die with 20° entrance angle according to Porter et. al. [1].

Polyethylene gels were prepared from solutions of UHMW PE (Hizex 240 M, $M_w \approx 1.5*10^6$) in xylene according to Smith and Lemstra [13]. Concentrations were varied between 0.5 and 6 % wt while the crystallization conditions were kept constant. The dried gels were compressed and annealed at 70°C under vacuum at a pressure of 20 MPa to remove voids and subsequently cut into tapes. The tapes were calibrated by ink marks in length and drawn at 110°C to various draw ratios.

Extruded samples as well as dried gels were embedded in epoxy (Epon 812, Serva) which was cured for 48 hours at 65°C [14]. The structure of the originally formed gels was fixed replacing the solvent by a mixture of methacrylate monomers (Lowicryl HM 20, Chemische Werke Lowi GmbH) which was polymerized by photoinitiation for 36 hours under UV irradiation at -40°C [15]. The embedded samples were stained with RuO_4 [10]. Ultrathin section were cut at ambient temperature with an Ultracut E ultramicrotome (Reichert & Jung) which was equipped with a diamond knife. Specimen with a thickness of 40 nm or less were selected by the silvery to colorless interference. The drawn in epoxy embedded fibres were cut paralell to the extrusion or drawing direction. Special care was taken in the case of gel drawn tapes to have the section plane perpendicular to the tape surface and parallel to the drawing direction.

Electronmicrographs were recorded by means of a ZEISS EM 902 transmission electron microscope with an integrated electron energy loss spectrometer and a Philips 400 electron microscope working with an acceleration voltage of 80 kV and 100 kV respectively. Several micrographs were used for the determination of the crystal thickness. The distance between the stained crystal surfaces was measured on the negatives. Calibration was done with cross grating replicas and catalase crystals.

Melting was studied at a a standard heating rate of 10°/min using a Perkin Elmer DSC-7 calorimeter on samples of 1,53 mg. The instrument was calibrated with high purity gallium and indium standards. Transition enthalpies as determined by numerical integration of the transition peaks show a variation of less than 5%. The maximum of the endotherms was taken as the melting temperature. The crystallinity was calculated by assuming a melting enthalpy of 293 J/g for 100% crystalline material.

RESULTS AND DISSCUSION

<u>a) Solid-state-extruded Polyethylenes</u>

Polyethylenes were extruded to different draw ratios either below or above the α-transition which is associated with the onset of translational processes within the crystallites [16]. Samples of series A were crystallized under isotropic pressure at 120°C and extruded at 76°C. Samples of series C were crystallized at 130°C and extruded at 125°C. Table 1 represents the nominal draw ratios and the resulting tensile modulus in comparison to the degree of crystallinity from DSC experiments and the lamellar thickness as obtained from the electron micrographs. In both series the tensile modulus increased with the nominal draw ratio (NDR). Extrusion at 76°C resulted in a small decrease in crystallinity which was independent of the NDR while the crystal thickness decreased from 18 to 7 nm. Thinning of the lamellae was also observed on extrusion at 125°C. However, in this case the crystallinity increased from 80 to 86 % with increasing nominal draw ratio.

Table 1: Solid-State Extruded PE Samples

sample	T_{EX} [1] (°C)	NDR [2]	E [3] (GPa)	d [4] (nm)	T_m [5] (°C)	w_c [6] (%)
A		1,0		18	139	80
A1	76	2,95	6	11	136	78
A2	76	5,75	11	9	140	76
A3	76	13,5		7	140	78
C		1,0		17	140	80
C1	125	4,0	8	10	140	81
C2	125	11,7	17	7	141	84
C3	125	30,1	49	5	143	86

[1] extrusion temperature, [2] nominal draw ratio, [3] tensile modulus, [4] crystal thickness, [5] melting point, [6] degree of crystallinity (% weight/volume)

The observed changes of the morphology on extrusion depend significantly on the temperature and the draw ratio. Representative electron micrographs for samples of the series A are shown in Figure 1. The nonoriented sample is shown in Figure 1a. Lamellae of 18 nm thickness are clearly separated by amorphous layers which appear dark due to the preferential staining of strained chains in the interface [11]. Figure 1b demonstrates that at a NDR of 2,95 most lamellae are broken up and only a few remain intact. Both types of crystallites are reduced in thickness: 12 nm for the intact ones and about

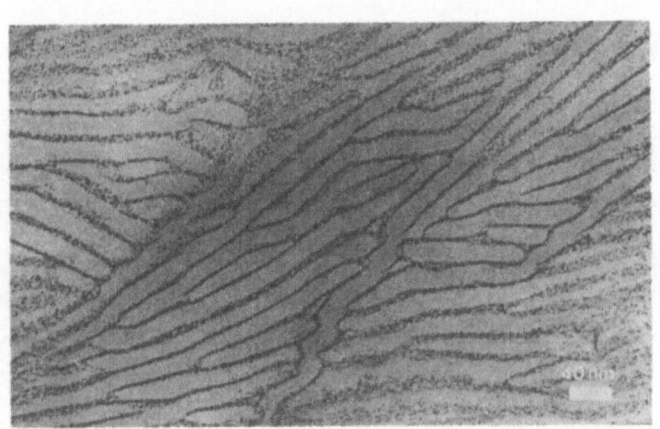

a

b

Figure 1: Electron micrographs of solid-state-extruded poly-ethylene, T_{ex} =76°C; a) sample A, NDR=0; b) sample A1, NDR=2.95.

c

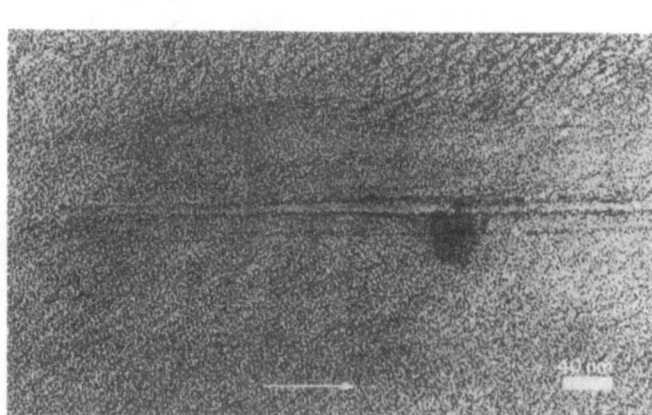

d

Figure 1: (Continued form the previous page) Electron micro-graphs of solid-state-extruded polyethylene, T_{ex} =76°C; c) sample A2, NDR=5,75; d) sample A3, NDR=13,5.

10 nm for those which broke up into smaller blocks. Further elongation to a NDR of 5,75 yielded the morphology shown in Figure 1c. The remaining lamellae were oriented parallel to the drawing direction and reduced in thickness to a value of 10 nm. The crystal fragments tend to orient their surfaces perpendicular to the extrusion direction. This is shown more clearly in Figure 1d for the highest NDR. Hence, the deformation below the α-transition can be described as a process where most of the lamellae are broken up into small blocks which are rotated thus that their surface is oriented perpendicular to the drawing direction. At the same time a relatively small fraction of lamellae is oriented parallel to the drawing direction and remains intact. The reduced thickness of both types of crystallites may be éxplained by tilting of the c-axis due to shearing.

A different situation was found for the samples which were extruded at temperatures above the α-transition although the morphology of the primary material was practically identical to that shown in Figure 1a. Figure 2a represents a herring bone pattern morphology which is the typical morphology found for a NDR of 4. The thickness of the lamellae decreased by a factor of 2-3 due to considerable tilting of the chain axis. The backbone of the herring bone is aligned into the extrusion direction. The lateral width of the lamellae in connection with the highly ordered formation appears remarkable. It may indicate that the deformation is not homogenous but that the original lamellae were broken up into stacks which were sheared against each other. Increasing the draw ratio results in a further drastic convertion of the morpholgy. In figure 2c thin crystallites of small width are perfectly aligned with their surfaces perpendicular to the extrusion direction for the C3 sample with NDR = 30,1. This can be explained by the formation of shish-kebab structures which have been observed typically at melt extruded polyethylenes [4,5]. In some of the micrographs obtained on these samples, the formation of shishs was indicated more clearly. The decrease in the thickness (along the c-axis) of the crystallites is in agreement with partial melting and epitactic recrystallization of low molecular weight fractions on the fibrous core. Figure 2b shows the intermediate morphology as observed for a nominal draw ratio of NDR=11,7. Thus, the lamellae are gradually transformed.

The deformation behavior below the α-transition can be explained by the model of Peterlin [6-8]. In the flow field the lamellae are sheared and the chains become tilted. The thickness of the crystals decreases. Lamellae

294

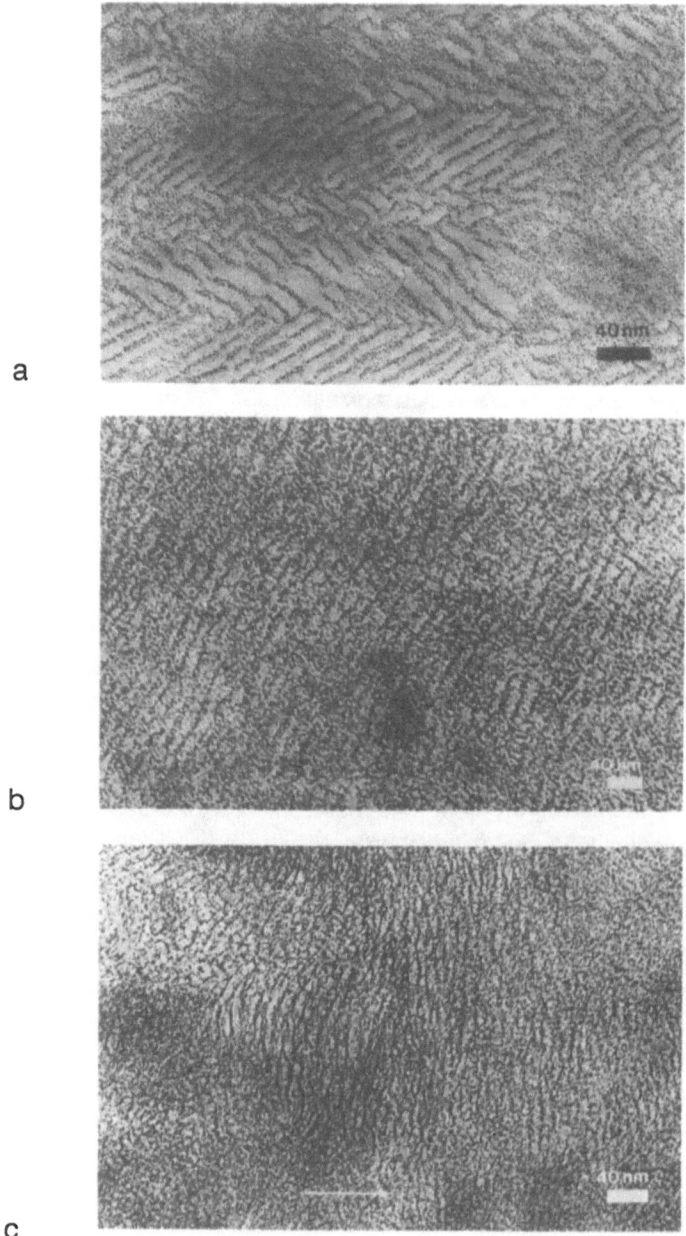

Figure 2: Electron micrographs of solid-state-extruded poly-ethylene, T_{ex} = 125°C; a) sample C1, NDR=4,0; b) sample C2, NDR 11,7; c) sample C3, NDR=30,5.

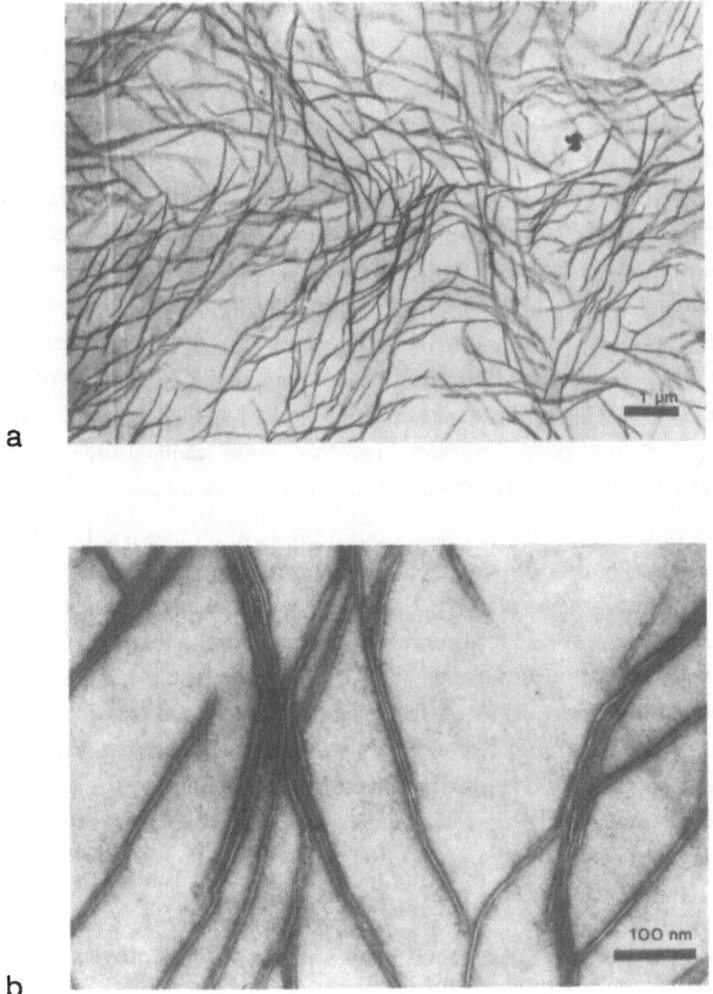

Figure 3: Electron micrographs of embeded UHMW-PE gels which were quenched to room temperature from 3% solution in xylene; a) network; b) connected lamellae.

which are inclined to the deformation direction break into smaller blocks which are rotated perpendicular on further deformation so that they become aligned perpendiculary to the drawing direction..

A different deformation and orientation behavior was observed for the second series, when the samples were extruded at a temperature near the melting point of the primary material (136°C). The observed variations in deformation at 76°C and 125°C can be assigned to the onset of molecular motion within the crystals above the α-transition. While extrusion at low temperatures yields decreased crystallinities, the degree of crystallinity is increased on extrusion at higher temperatures. The first case can be explained by the destruction of crystallites, the second case by annealing effects due to molecular relaxation processes above the α-transition. In addition, the shish-kebab structures indicate partial melting, which may occur in a late stage of the deformation because of the low melting point of smaller crystallites formed at high strains. Typically, large nominal draw ratios are necessary to observe the formation of shish-kebab structures.

At lower draw ratios, the lamellae do not break apart into tiny blocks like in the A-series. Larger segments of the lamellae become aligned in stacks which are tilted simultaneously. Chain tilting and slipping appear to be the main processes involved. A possible explanation may be that first, shearing leads to inclination of the chains until a maximum tilt angle is reached and that the lamellae are tilted in a second step. As a result of this double tilt, the lamellae are inclined and the chains are oriented parallel towards the extrusion direction. Further deformation either by chain slip or breakage of the lamellae along the draw direction leads to the interlocked tapered lamellae observed at high draw ratios [17].

b) Gel spinning

Barham and Keller [18] distinguish three classes of thermoreversible gels, which can be formed from solutions of crystallizable polymers, on the basis of the relation between the gelation and the clearing point. Either the gel will be dissolved (i) below the clearing point, (ii) at the clearing point, or (iii) above the clearing temperature. Gels of the second class are formed, if a semidilute solution of an ultrahigh molecular weight polyethylene is cooled without prior stirring, which is the case in the gel drawing process [13] described by Smith and Lemstra. As the network junctions are dissolved on melting of the lamellae, they are either formed by the lamellae themselves or by entanglement which are locked by incorporation of the chains into the lamellae.

To observe the structure of the swollen gels directly, we replaced the solvent by an amorphous resin. Hence, it was possible to cut thin sections, which were investigated by transmission electron microscopy. This attempt is based on the idea that the three dimensional, ramificated structure of the PE-gel can be stabilized unaltered, if the "solvent" is cured very slowly at low temperatures. The reliablity of this technique has been proved by broad use in electronmicroscopy on biological specimen.

First, the gels were formed by cooling a solution of a UHMW-PE in a high boiling solvent. Afterwards, the solvent was replaced in several extraction steps by methacrylate monomers, thus, that the gel was always covered by the liquid. This step is equivalent to aging of the swollen gel for several hours at low temperature. Finally, the methacrylate was polymerized below room temperature by UV radiation.

Table 2: DSC-data of polyethylene gels
prepared from xylene solutions of different concentrations.

sample	$c^{1)}$ (%)	$T_m^{2)}$ (°C)	$\Delta H^{3)}$ (J/g)	$w_c^{4)}$ (%)
G-0,5	0,5%	135	206	70
G-1,5	1,5%	138	203	69
G-3,0	3,0%	135	215	73
G-6,0	6,0%	136	203	69

[1] PE conc.in the orig. xylene solution; [2] melting point, [3] melting enthalpy, and [4] crystallinity of the dry gel.

The structure of the gels which could be observed this way, is demonstrated by the micrographs in Figure 3. Long lamellae are directly linked by alignment of short segments. A highly branched, isotropic topology results. At the higher magnification in Figure 3b, it becomes obvious that the lamellae are connected to each other via dark interlayers. These interlayers are formed by noncrystalline material which is stained preferentially by the RuO_4 [11]. Checking these gel structures carefully on a large number of micrographs, we could not observe a single branching point, where the lamellae itself ramifies. The existence of tie molecules, which are still in solution, cannot be excluded as it is impossible to image a single polyethylene chain. It must be noted that the lamellae are platelets and that the wormlike appearence in the micrographs results from the fact that only

Figure 4: Electron micrographs of a sedimented polyethylene gel which was prepared from a 1,5% xylene solution and cooled to room temperature with 0,7 K/min.

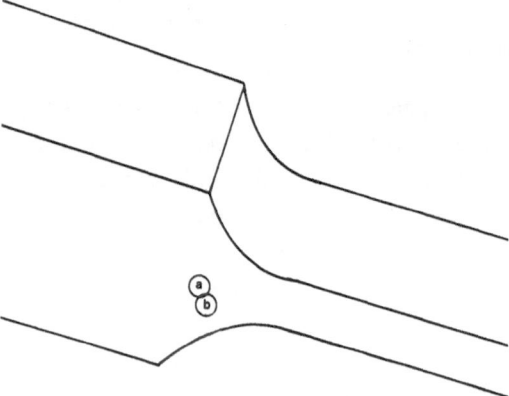

Figure 5: assignment of the micrographs in Figure 6 to different locations in the necking region.

lamellae which are oriented in an angle close to 90° with respect to the plane of the thin section are shown clearly in the two dimensional projection. In principle, the gel can be described as a loose interlocked array of extended, thin lamellar crystallites in which small amorphous interlayers are formed by local contacts which represent the actual crosslinks.

When the concentration of the original polyethylene solution was increased, the network density increased, while the degree of crystallinity and the melting points did not change (Table 2).

When the gels are dried, the lamellae crystallites are sedimented and become oriented parallel to the surface on which the gel ist cast [19]. Figure 4 shows the high orientation of the crystals in such a gel. Like in a mat of single crystals, the orientation is almost perfect with respect to the crystal c-axis but isotropic in the a-b plane. As a consequence, drawing of the tapes is performed perpendicular to the c-axis of the crystallites. It had been shown by X-ray scattering experiments that the chains in the drawn fibres are oriented parallel to the drawing direction [9].

Due to the necking and the fact that the major part of the elongation takes place in the small necking front which proceeds along the drawn tape, it is practically impossible to study the deformation on differently drawn samples. However, various stages of the drawing process should be found in the necking region itself. In order to observe the different stages of the deformation process, we took specimens for electron microscopy at different locations within the necking region as it is shown in Figure 5. Unfortunately it is not possible to determine the actual draw ratio for the microscopic samples. It is only possible to order the specimens with respect to whether they are drawn to a larger or lesser extent.

Figure 6a shows the morphology which was found at the center of the tape. It represents the structure of the unstretched gel similar to Figure 4. The morphology which is found at a location more close to the periphery corresponds to a herringbone pattern (Fig. 6b) No clear and oriented structure could be observed for the specimen which was taken directly at the surface on the outer edge. As one would expect, the actual draw ratio increases from the middle to the surface of the tape. A herringbone pattern was also observed for an extruded sample as shown in Figure 2a. The practically identical morphologies indicate that similar processes are involved in the early stages of deformation in solid-state-extrusion at elevated temperature and in the gel drawing.

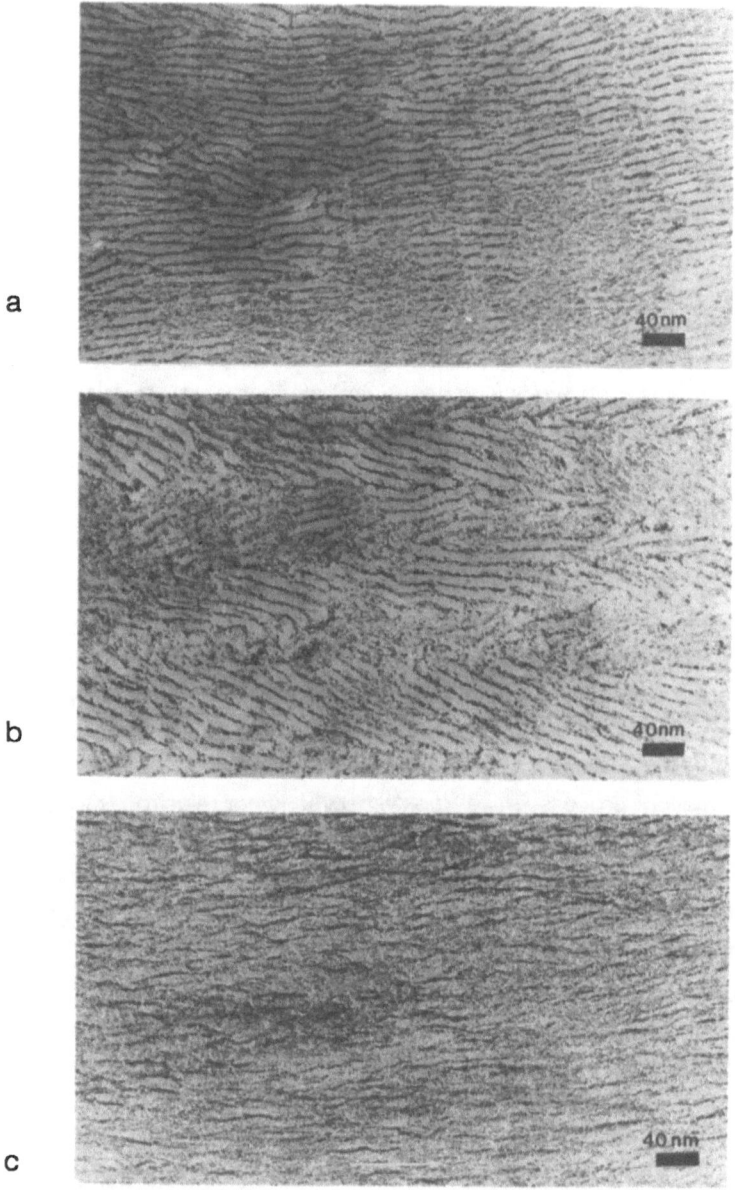

Figure 6: Electron micrographs from the necking region of a gel drawn polyethylene which was prepared from a 1,5% xylene solution and drawn at
110°C. The locations from which the specimens of a) and b) were taken are
given in Figure 5; c) thin section through a polyethylene fibre which was
drawn to $\lambda = 90$.

However, at higher draw ratios the differences in the starting material and the processing itself become evident within the observed morphologies. Figure 6c shows an electronmicrograph which we obtained from a fibre which was drawn at 110°C to a draw ratio of $\lambda = 90$. Dark stripes of a length between 30 and 50 nm are aligned parallel to the drawing direction. These stripes are separated from each other by bright regions with a width of 5-15 nm. In contrast to the structure of the undrawn gel in Figure 4a, the direction of the c-axis is parallel to the dark stripes. The picture represents an image of the microfibrillar structure within a drawn filament [11, 18-24].

RuO$_4$ proved to attack strained chain segments on the crystallite surfaces preferentially compared to completely disordered amorphous PE domains [11]. Because of the restricted diffusion of the staining agent into the crystallite, oxidation of crystalline material is not only slow but also restricted to the crystallite surfaces. Thus, the dark stripes can be assigned to the interface between the oriented crystal blocks which are connected by taut tie molecules [7-9]. Two points may be emphasized. First, the variation in the distances of the stripes is rather large and the stripes are undulating. This is equivalent to a short coherence length of the periodicity. Second, it appears that the crystalline unstained domains do not exceed a length of 100 nm within the drawing direction. Thus, it is questionable whether they should be classified as extended chain crystals. The fact, that SAXS experiments which have been reported on highly drawn PE samples did not give much evidence on a structured morphology of the microfibrils [9] is explained by the irregularity and the small electron diffraction differences.

CONCLUSIONS

Considerable differences have been observed between the solid-state extrusion of polyethylene below and above the α-transition. Chain slippage which allows plastic deformation of the primary lamellae appears to occur only at elevated temperatures.

Solid state extrusion and gel drawing are similar at temperatures near the melting point and for small deformations. The herring bone patterns which have been observed in both cases are consistent with the four point patterns found for small draw ratios in SAXS experiments [9]. In this stage, the deformation process in solid-state extrusion and gel spinning can be described according to Peterlin [6-8]. The lamellae break and the the resulting smaller crystallites are rotated so that their surfaces become inclined with respect to the drawing direction. Due to chain tilting on shearing, the thickness of the lamellae decreases with increasing elongation.

Images of the initially formed gels show that they consist of single lamellae which are interconnected by small amorphous contact layers. The number of interlamellar tie molecules should be small if not even negligible. In contrast, a large fraction of interlamellar tie molecules are assumed for the melt crystallized material [25]. Thus it may be concluded that interlamellar tie molecules or trapped entanglements are of little influence on the rearrangement within the first stage of drawing. In agreement with the relatively large lateral crystallite width between 50 and 150 nm, the deformation can be described as a shearing of lamellae stacks.

Differences between the extrusion and the gel drawing process were observed at higher draw ratios. Most of the material which was oriented by extrusion consisted of chain folded lamellae. The observation of shish-kebab structures at 125°C indicated partial melting. The lateral dimensions as well as the thickness of the lamellae are drastically reduced compared to the primary structure. Drawn gels exhibit a fibrillar structure. Yet we have not been able to obtain an image of the drawn gel at the stage directly after the necking at draw ratios of $\lambda = 10\text{-}30$, where SAXS-experiments [9] have indicated that a fraction of the original lamellae is still present. More detailed information on the morphology of this intermediate stage and its evolution out of the herring bone pattern shown in Figure 6b should be helpful for the understanding of the ultradrawing.

ACKNOWLEDGEMENT: Financial support was granted by the Deutsche Forschungsgemeinschaft, Sonderforschungsbereich 60, Teilprojekt A-1/F-5

REFERENCES

1. Zachariades, A.E.; Mead, W.T.; Porter, R.S.; Chem. Rev., 1980, 80, 351

2. Lemstra, P.J.; Kirschbaum, R.; Polymer 1985, 26, 1372

3. "Development in Oriented Polymers"; Ward, I. M., Ed.; Elsevier Appl. Sci. Publ., Vol 1: New York 1982, Vol 2: New York 1987

4. Bashir, Z.; Odell, J.A.; Keller, A.; J. Mater. Sci. 1984, 19,3713

5. Bashir, Z.; Odell, J.A.; Keller, A.; J. Mater. Sci. 1986, 21,3993

6. Peterlin, A.; Colloid & Polym. Sci. 1987, 265, 357

7.   Peterlin, A.; Ingram, P.; Kiho, H.; Makromol. Chem. 1965, 86, 294

8.   Peterlin, A.; J. Mater. Sci. 1971, 6, 490

9.   van Aerle, N. A. J. M.;Braam, C. W. M.; J. Mater. Sci. 1988, 23, 4429

10.   Montezinos, D.; Wells, G.B.; Burns, J.L.; J. Polym. Sci., Polym. Lett. Ed., 1985, 23, 421

11.   Kunz, M.; Möller, M.; Heinrich, U.-R.; Cantow, H.-J.; Makromol. Chem.,Macromol. Symp. 1988, 20/21, 147

12.   De Micheli, R.E.; Vidal, H.M.; Macchi, E.M.; Polym. Bull. 1985, 14, 61

13.   Smith, P., Lemstra, P.J.; Makromol. Chem., 1979, 180, 2983

14.   Luft, J.H.; J. Biophys. Biochem. Cytol., 1961, 9, 409

15.   Carlemalm, E.; Gravito, R.M.; Villiger, W.; J. Microscopy 1982, 126, 123

16.   Takayanagi, M.; J. Macromol. Sci. Phys. 1970, B4(1), 161

17.   Odell, J.A.; Grubb, D.T.; Keller,A.; Polymer 1978, 19, 617

18.   Barham, P.J.; Keller, A.; J. Mater. Sci. 1985, 20, 2281

19.   Smith, P; Lemstra, P. J.; Pijpers J. P. L.; Kiel, A. M.; Colloid & Polym. Sci. 1981, 259, 1070

20.   Peterlin, A.; J. Polym. Sci. Part C 1965, 9, 61

21.   Peterlin, A.; Polym. Eng. Sci. 1978, 18, 488

22.   Prevorsek, D.C.; Harget, P.J.; Sharma, R.K.; Reimscheussel, A.C.J.; J. Macromol. Sci., Phys. 1973, B-8, 127

23.   Fischer, E.W.; Goddar, H.J.; J. Polym. Sci. 1969, C16, 4405

24.   Clark, E.S.; Scott, L.S.; Polym. Eng. Sci. 1974, 14, 682

25.   Fischer, E.W.; in Integration of Fundamental Polymer Science and Technology I, ed. L. A. Kleintjens & P.J. Lemstra, Elsevier Applied Science Publishers, London, 1987, pp 456

STRUCTURE OF DRAWN FIBRES AS REVEALED BY NEUTRON SCATTERING STUDIES

D.M. SADLER and P.J. BARHAM
H.H. Wills Physics Laboratory, University of Bristol,
Tyndall Avenue, Bristol BS8 1TL, UK

ABSTRACT

A range of fibres prepared from blends of hydrogenated and deuterated polyethylenes have been examined using neutron scattering techniques. It is shown that on drawing through a neck at elevated temperatures (above 70°C depending on molecular weight) a segregation signal appears - this is interpreted as being due to local melting occurring within the neck. On drawing beyond the neck it is shown that the molecular draw ratio increases with the overall fibre draw ratio, and that the final fibre modulus is uniquely determined by the molecular draw ratio.

INTRODUCTION

Whilst there is a large body of literature on fibre drawing including many models of the necking, and subsequent 'ultra-drawing' processes there is very little actual data concerning the changes that occur on the molecular scale during drawing. Neutron scattering techniques using deuterated polymers as an additive permit direct observation of changes at the molecular scale. In particular we can measure the radii of gyration of the molecules along, and normal to, the draw direction and hence determine molecular draw ratios. On those occasions when a segregation signal appears we can deduce that molecular mobility on a large scale occurred in the neck - this we associate with partial melting.

EXPERIMENTAL AND RESULTS

Materials and Methods

HPE samples of various molecular weights from 50,000 to 300,000 were blended in solution with 2% by weight of DPE of various molecular weights between 30,000 and 200,000.

The dried blended polymer was then prepared in sheet form either by melting at 160°C and quenching or as a single crystal mat. The sheets were drawn in contact with a bar at the required drawing temperature.

Neutron scattering measurements were performed using the D11 instrument at ILL, Grenoble. All the usual background subtraction and correction procedures were followed.

Interpretation of Neutron Scattering Data

Figure 1, shows contour plots of the neutron scattering from two fibres drawn through a neck, one at 81°C and the other at 89°C. The fibre drawn at the lower temperature shows high anisotropy indicating that the molecules have become extended in the draw direction. The fibre drawn at the higher temperature shows little anisotropy but displays a large isotropic segregation signal indicating that the molecules have moved over a large distance in the neck. This we assert is symptomatic of local melting. We can determine the radii of gyration of fibres (in the x-direction - normal to the draw direction) using the Zimm relation:

$$I(0,0)/I(q_x,0) = 1 + q_x^2 R_x^2 \qquad (1)$$

Such Zimm plots for the fibres in figure 1 and for the initial undrawn sheet are shown in figure 2.

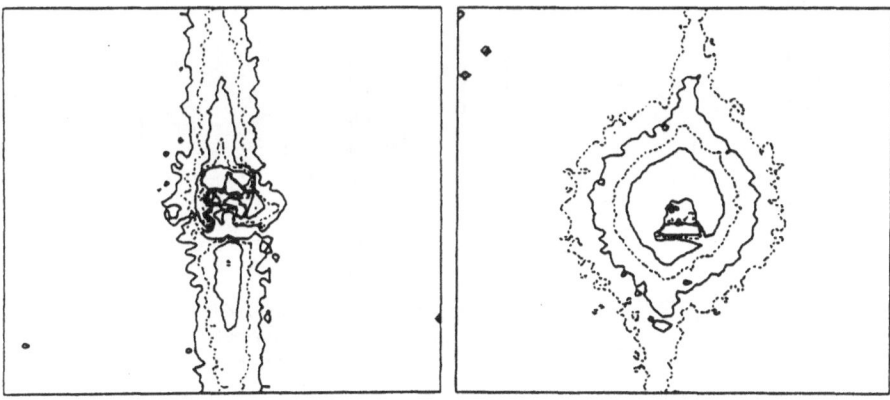

Figure 1 - Contour plots of two polyethylene fibres left drawn x 7.2 at 81°C right drawn x 7.1 at 89°C.

306

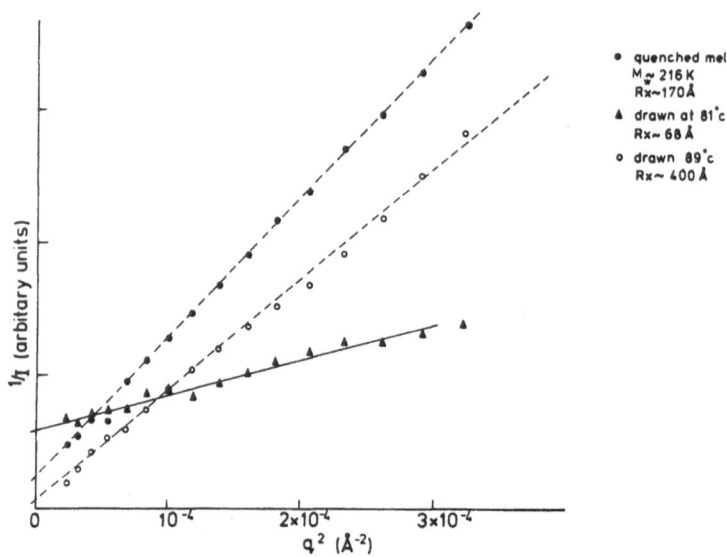

Figure 2. Zimm plots (along the vertical direction) for both
fibres in figure 1, and the initial sheet from which they
were drawn.

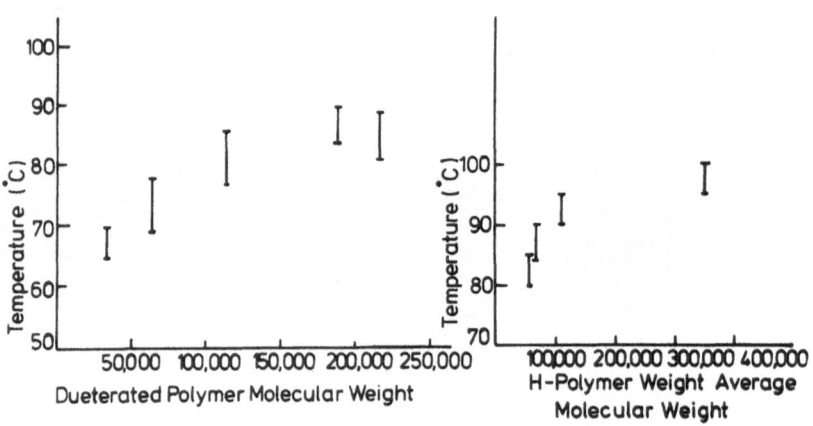

Figure 3. Two diagrams showing how the 'transition'
temperature above which local melting occurs during necking
depends on molecular weight of dopant and host polymers.

Local Melting in the Neck

It is straightforward, from data such as that in fig. 1,
to determine whether or not local melting occurs during
necking. We find, for melt crystallized material, local
melting occurs on drawing above some 'transition temperature'
which itself depends on the molecular weights of both the
host and dopant polymers. This is illustrated by figure 3
(from ref. 2) when drawing single crystal mats we did not
observe a sudden change such as that shown by fig. 1, rather
we found a gradual increase in the degree of segregation (as
assessed by the extrapolated $I(0,0)$ values) as the drawing
temperature was raised above $90^{\circ}C$ (3).

The Molecular Draw Ratio

For melt crystallized samples which showed no local
melting we found that the molecular draw ratio across the
neck was very close to the macrosopic neck draw ratio. For
solution grown crystals in which the molecules are arranged
in 'superfolded' sheets it is not possible to define a
molecular draw ratio. However after necking it is a
straightforward matter to determine both macroscopic and
molecular draw ratios.

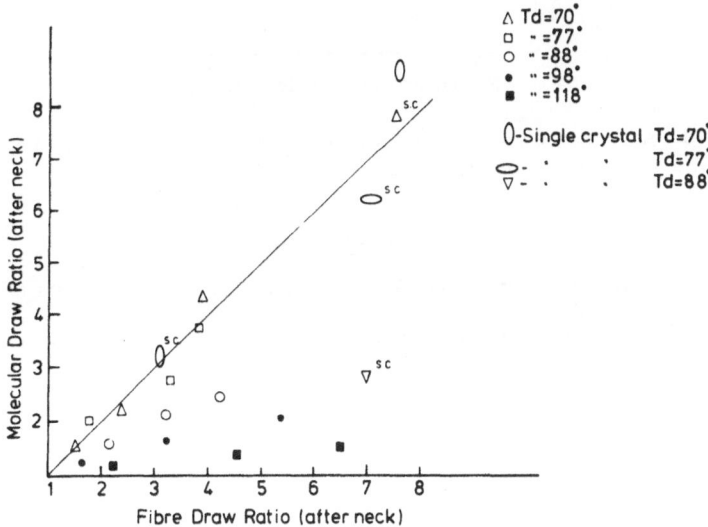

Figure 4. A graph showing the memasured molecular draw ratio
as a function of the macroscopic draw ratio (after necking).

In figure 4 we have plotted molecular draw ratios (after
necking) as a function of macroscopic draw ratios. It can
clearly be seen that as the temperature of redrawing is

raised above ca. 80°C so the effectiveness of the drawing is reduced. In figure 5 we have plotted the final modulus of all the fibres as a function of the molecular (rather than macroscopic) draw ratio (after necking). This shows that the final modulus is uniquely dependent on the molecular draw ratio (4).

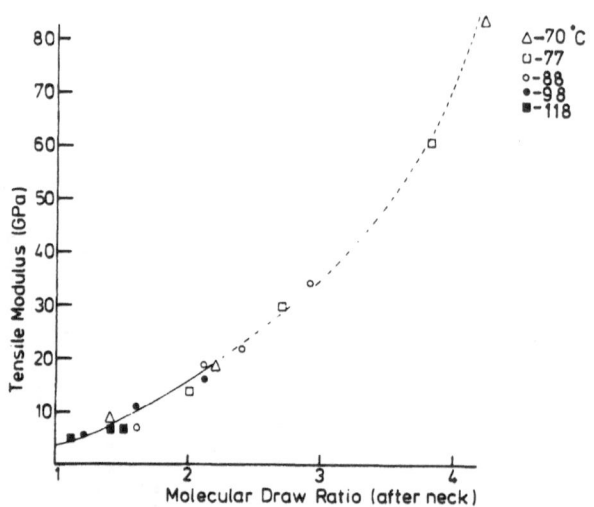

Figure 5. A graph showing how the fibre tensile modulus depends on the molecular draw rate after the neck.

REFERENCES

1. D.M. Sadler and P.J. Barham, J. Polym. Sci. Polym. Phys. Ed. 21 (1983) 309.
2. D.M. Sadler and P.J. Barham, Polymer in Press.
3. D.M. Sadler and P.J. Barham   "   "   "
4. P.J. Barham and D.M. Sadler   "   "   "

FORMATION AND APPLICATION OF HIGHLY ORIENTED SEMICRYSTALLINE POLYMERS

MIROSLAV RAAB, VLADIMÍR HNÁT, MILOŠ KREJČÍ[+] and ECKHARD SCHULZ[++]

Institute of Macromolecular Chemistry, Czechoslovak Academy
of Sciences, 162 06 Prague 6, Czechoslovakia
[+]State Research Institute of Textiles, Liberec,
Czechoslovakia
and [++]Institute of Polymer Chemistry "Erich Correns", Teltow,
GDR

ABSTRACT

The formation and morphology of highly oriented semicrystalline
polymers is briefly discussed. It is shown that unusual composite materials can be prepared by laminating fibrillated films
between isotropic films of polyethylene. One-polymer composites
with high strength and toughness can also be prepared by laminating parallel arrays of ultra-high-strength polyethylene
fibres between sheets of common-grade polyethylene in a heated
press. The mechanical behaviour of these composite films can be
compared to some natural composites (Broad-leaved plantain).

INTRODUCTION

Soon after the synthesis of the first semicrystalline polymers
it was recognized that their strength and stiffness can be enhanced by orientation. Traditionally, melt drawing and cold
drawing have been used to aligh molecular chains preferentially
along one direction. Thus, the high intrinsic modulus and
strength of the backbone chains can be manifested in the bulk
properties. It is well known that during the cold drawing of
many semicrystalline polymers the original lamellar or spherulitic morphology is transformed into fibrillar structure. The
detailed mechanism of this transformation is, however, still
an object of discussion and controversy.

Recently, an alternative to the classical Peterlin and
Prevorsec concepts has been proposed [1], in which the transformation from spherulite to fibre is solely dependent on phase
transition. The model suggests that the stored mechanical energy
is sufficient to cause a local "melting" of the semicrystalline
polymer at the draw temperature. In the second step the

amorphous "melt" undergoes immediately rapid extension and strain induced crystallization. The high stiffness and strength of the oriented polymer is then ascribed to covalent bonds in the extended chain core.

Semicrystalline polymers oriented by the traditional techniques still contain some chain-folded lamellae and amorphous regions between them. However, during the last decade some very effective orientation processes have been developed producing polymer structures with high portion of fully extended and perfectly aligned chains. Some of the techniques are based on solid state processing at a temperature between the alpha crystallization temperature and the melting point [2]. The most perfect orientation and the highest strength and stiffness, however, can be achieved by surface spinning from solution, and particularly by drawing so-called gel fibres [3].

ORIENTATION AND TOUGHNESS

Toughness of polymeric materials is related either to the mechanical work up to break or to the resistance of the material to growing crack. Correspondingly, toughness could be enhanced either by increasing the energy dissipated during straining of the polymer or by introducing barriers against crack propagation. By the latter mechanism the toughness of polymers anisotropically increases with orientation.

A mechanism controlling crack propagation in anisotropic brittle materials was proposed by Cook and Gordon [4]. They analyzed two stress fields aroung a Griffith-type crack in a uniaxially loaded plate: The stresses parallel to the external tensile force showed a maximum at the crack tip. Moreover tensile stresses acting in the cross direction were found with a maximum located somewhat ahead of the crack tip. In the elastic case the maximum of the cross stresses was found to be one fifth of the longitudinal stress concentration. Consequently a plane of weakness parallel to the external stress and perpendicular to the crack path could be opened by the cross tensile stresses ahead of the main crack tip. This would blunt and deflect the growing crack (Fig. 1). The Cook-Gordon mechanism can explain the anisotropical increase of toughness in highly oriented semicrystalline polymers and composites with anisometrical reinforcement, e.g. platy filler [5].

FIBRILLATED FILMS

During cold drawing and orientational strengthening of semicrystalline polymers their cohesion in the cross direction decreases considerably. Due to mechanical straining, but also due to ageing and degradation the highly drawn films of some polymers split into parallel fibrils. This phenomenon is called fibrillation and has been observed particularly for isotactic polypropylene; with other polymers the tendency towards fibrillation decreases in the order: linear polyethylene, polyamides,

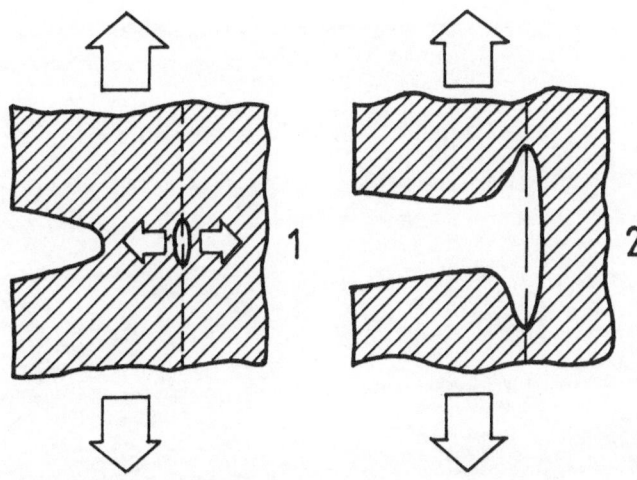

Figure 1. Schematic representation of crack blunting and de-
flection at a plane of weakness [4]

polyesters. Fibrillation has been ascribed to differences in
the natural draw ratios of adjacent fibrils [6]. The "auto-
phobic" behaviour of a polymer as manifested during fibrillation
might also be related to its surface energy.

Fibrillation may cause difficulties in the production of
films by tubular blowing process but, on the other hand, it
could be exploited technologically in polymer film production.
Fibers from fibrillated films are interesting from both the
technical and economic point of view. The process of their
production has markedly lower investment and energy demands in
comparison to classical fiber spinning. Moreover, the resulting
fibres with uneven shape and cross-section can be used as substitute for some natural plant fibres (jute, hemp).

The State Research Institute of Textiles in Liberec,
Czechoslovakia in cooperation with the Institute of Macromolecular Chemistry, Czechoslovak Academy of Sciences, has developed a continuous process and designed a production line
with productivity of about 100 kg/h. The continuous and computer controlled line consists of a usual film blowing operation
which is followed by three drawing zones with individually controlled temperatures (Fig.2). Subject to type of the polymer,
the final drawn film shows a spontaneous tendency to fibrillation. As revealed by the sound propagation method, the films
of isotactic polypropylene drawn 6.5 times already contained
a latent system of parallel splits or crazes [7]. The splitting
is then completed in the next step, when the stressed film is
punctured by rows of tiny needles mounted on a rotating
cylinder (Fig. 3). The resulting fibers have the form of an end-
less network (Fig. 4) and can be further processed by various
textile technologies. Particularly advantageous is their continuous conversion into nonwoven textiles by the "FIBRIL"

Figure 2. One drawing section of the line producing continuously fibres and non-woven textiles of fibrillated films

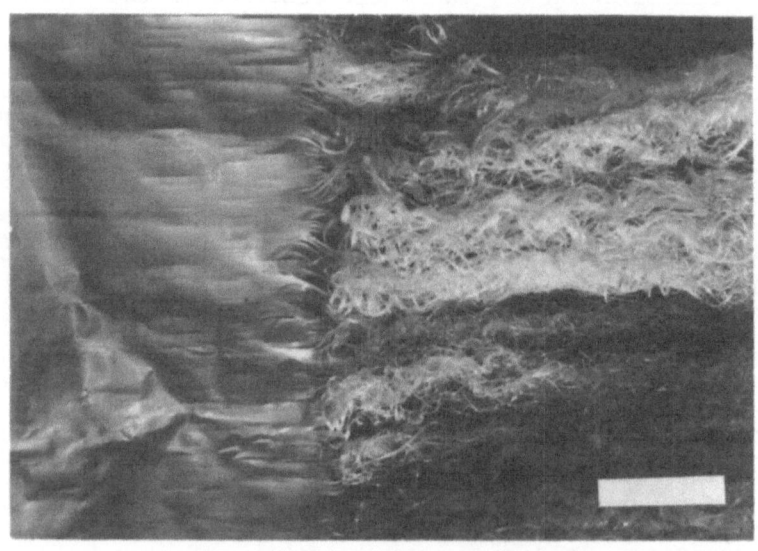

Figure 3. Complete fibrillation of a drawn polypropylene film
after puncturing by a row of needles. Bar = 1 cm

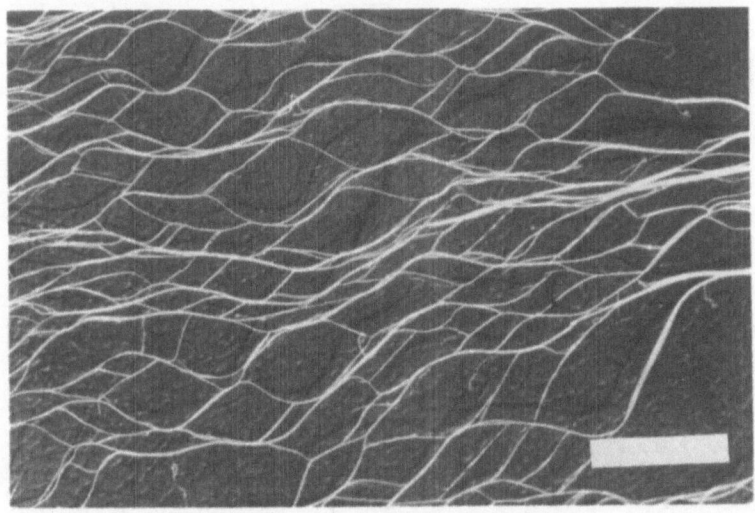

Figure 4. Fibre network resulting from a fibrillated and
punctured polypropylene film. Bar = 1 cm

technology. The non-woven fabrics have found a number of applications such as geotextiles, insulating plates or reinforcements
for composites.

LAMINATED COMPOSITE FILMS

The fibrillated but not punctured films can be laminated
together with suitable interlayers to form flat composite
sheets [8]. The laminating temperature must be chosen above
the melting temperature of the matrix interlayers, but under
the shrinkage temperature of the reinforcing fibrillated sheets.
The shrinkage, however, can be suppresed by constraining the
fibrillated layer during the lamination.
   An example of assemblage of such laminated film is shown
in Fig. 5. In this case a fibrillated film of high density
polyethylene was laminated between two layers of low density
polyethylene films. The lamination was carried out at 136°C,
pressure 10.9 MPa, and during 60 min. Direction of stress during
mechanical testing is also indicated in the schematical sketch.
The stress-strain traces of the composite together with the
starting materials are shown in Fig. 6, both for defect-free
specimens and for specimens with single-edge razor cuts 0.5 mm
deep. It can be seen that even the low volume content of the
reinforcement (6%) raised markedly the yield point of the matrix
material and the energy up to the crack instability (indicated
by arrows) increased more then 6 x. Obviously, this pronounced
increase in toughness could be ascribed to the Cook-Gordon
mechanism. Further balancing of ultimate mechanical properties
can be achieved by laminating several fibrillated films with

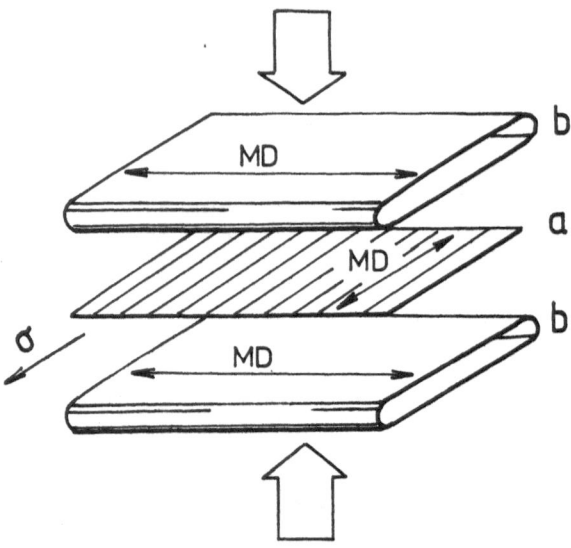

Figure 5. A scheme showing an assemblage of a simple composite film with three layers. MD denotes the machine direction of the film and σ direction of tensile mechanical testing. a - fibrillated film of high-density polyethylene, b - blown tubular film of low-density polyethylene

varying relative angles between orientation directions or by alternative laminating of filled and unfilled polymer films.

THE PLANTAIN EFFECT

One-polymer composite films can also be obtained by laminating parallel arrays of ultra high strength polyethylene fibres between sheets of common-grade high-density polyethylene [9]. High-strength polyethylene fibres with a nearly perfect extended-chain structure show a substantially better thermal stability than common high-density polyethylene and can be overheated for a short time up to 175°C. Thus, time, temperature and pressure of the lamination can be chosen in such a way that firm fusion of the film layers is obtained while the structure and properties of the reinforcing fibers remain preserved. Again, best results are obtained with pre-stressed fibers during lamination. High-strength fibers prepared by surface spinning or drawing of gel fibres can be used as reinforcement.

Despite the large differences in the absolute stiffness and strength values, the polyethylene matrix and the polyethylene reinforcing fibres show some similarities in their mechanical properties. This, in turn, imparts unusual properties to the composite. The yield strains of the matrix and the fibers are very close to each other and both the matrix

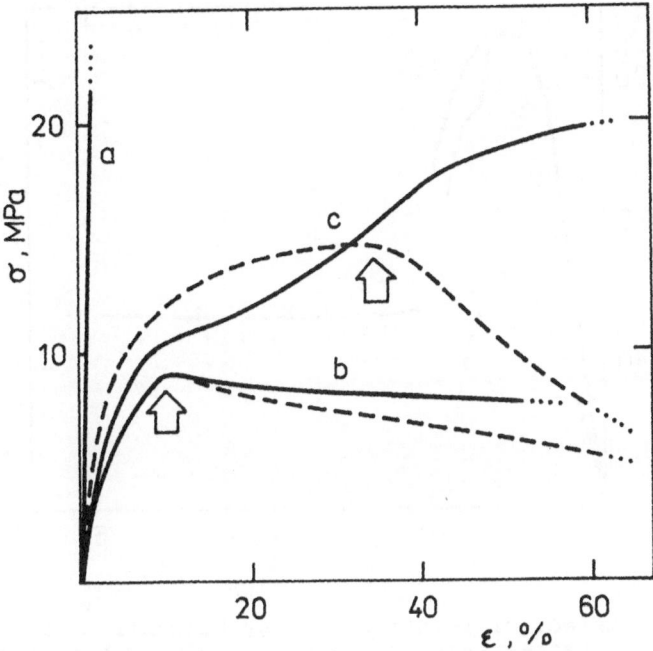

Figure 6. Stress-strain curves of films from Fig.4.
a,b - see Fig. 4, c - composite film. Full curves are for
defect-free specimens, dash curves are for specimens with
single-edge razor cuts 0.5 mm deep. Crack instability is indicated by arrows

and the fibre show a marked strain-rate dependence of strength.
As a result, at low strain values the deformation behaviour of
the one-polymer composite is nearly isostrain, i.e. strains in
the matrix and fibres are approximately equal. Therefore, even
one single fibre has a positive reinforcing effect on the film.
From Fig. 7 it can be seen that a fiber concentration as low
as 0.7 vol.% brings about a dramatic increase in the yield
point and toughness measured as mechanical work up to the crack
instability. It is interesting to note that some natural composites show a similar behaviour. An example is a leaf of the
broad-leaved plantain (Plantago major), which is reinforced with
only 7 strong fibres   but the extensibility of the fibres even
exceeds that of the matrix material. By analogy to plantain
leaves we denote the behaviour of a composite with extensible
fibres as the plantain effect.

316

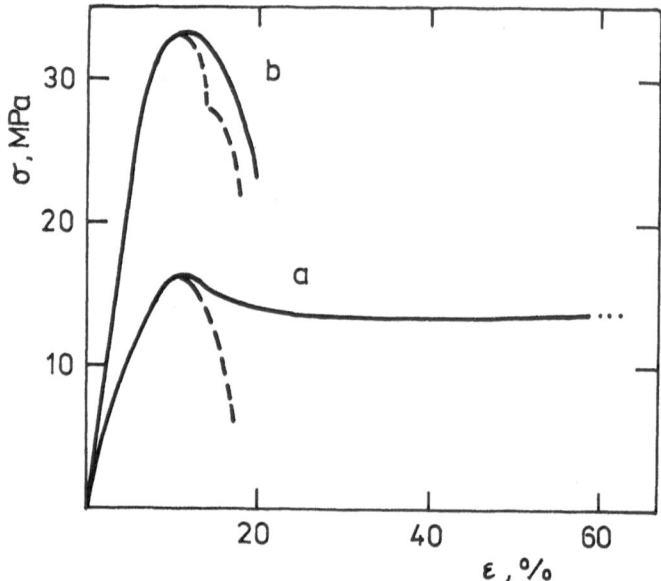

Figure 7. Effect of reinforcement by high-strength polyethylene
fibres on stress-strain curves of high-density polyethylene
films. a - high-density polyethylene film (matrix). b - matrix
film reinforced by 0.7 vol.% of high-strength polyethylene
fibres. Full curves are for defect-free specimens, dash curves
 are for specimens with single edge razor cuts 0.5 mm deep

REFERENCES

1. Jordon, M.E., Juska, T.D., and Harrison, I.R., Shrinkage as
a measure of extended chain conformation. Polym. Eng. Sci.
1986, 26, 690-694.

2. Bigg, D.M., Mechanical property enhancement of semicrystal-
lined polymers - A review. Polym. Eng. Sci., 1988, 28
830-841.

3. Hikmet, R., Lemstra, P.J. and Keller, A., X-linked ultra
high strength polyethylene fibres. Colloid and Polym. Sci.,
1987, 265, 185-193.

4. Cook, J. and Gordon, J.E., A mechanism for the control of
crack propagation in all-brittle systems. Proc. Roy. Soc.
Lond., 1964, A282, 508-520.

5. Eldred, R.J., Effect of oriented platy filler on the
fracture mechanism of elastomers. Rubber Chem. Technol.,
1988, 61, 619-629.

6. Peterlin, A:, Molecular model of drawing polyethylene and polypropylene, J. Mater. Sci., 1971, 6, 490-508.

7. Raab, M., Hnát, V. and Kudrna, M., Fibrillation of polypropylene film as monitored by the sound propagation method. Polymer Testing, 1986, 6, 447-461.

8. Raab, M., Schulz, E., Hnát, V., Pelzbauer, Z., Krejčí, M. and Makovský, J., A polymer composite material based on fibrillated films and a method of its preparation, Czechoslovak Patent Appl. 4165-88.

9. Hirte, R., Hnát, V., Melior, J.P., Pelzbauer, Z., Raab, M., Schulz, E., and Zenke, D., A one-polymer composite polyethylene film: Failure morphology. In Morphology of Polymers, ed. B.Sedláček, Walter de Gruyter & Co., Berlin, 1986, pp. 527-539.

Part 6

BLENDS/COMPOSITES

STRUCTURING COPOLYMERIC BLENDS VIA COMPOSITION VARIATION AND TEMPERATURE JUMP EXPERIMENTS

S. Klotz[*], C. Wendland, V. Krieger and H.-J. Cantow
Institut für Makromolekulare Chemie der Albert-Ludwigs-Universität Freiburg, University of Freiburg, Stefan-Meier-Str. 31, D-7800 Freiburg, Federal Republic of Germany

ABSTRACT

The phase behaviour of homopolymer and blockcopolymer blends is well understood in terms of 1) intermolecular attractive and repulsive forces between the chemically different segments and 2) differences in the equation-of-state properties of the pure components. In the present paper the phase behaviour of statistical poly(styrene-co-maleic acid anhydride)/poly(vinyl methyl ether) blends is discussed with particular respect to intramolecular interaction between styrene and maleic acid anhydride segments. Experimental results from differential scanning calorimetry, turbidity measurements and samll angle light scattering will be given.

INTRODUCTION

Recent theories of statistical copolymer/homopolymer compatibility predict that the phase behaviour is influenced by inter- and intramolecular repulsive forces (1-3). This may lead to homogeneous blends or a negative excess free energy of mixing even when all the binary Flory-Huggins interaction parameters exhibit positive values. Experimental evidence for such behaviour has been given by various systems, i.e. poly(styrene-co-acrylnitril)/poly(methyl methacrylate) (4) a.s.o.(5). In the present paper copolymer/homopolymer blends of poly(styrene-co-maleic acid anhydride) (PS-CO-MAA) and poly(vinyl methyl ether) (PVME) have been investigated (6). The comonomer content of MAA have been varied between 0 and 20 wt%. From solubility parameter theory the incompatibility of the styrene and the maleic acid anhydride segments is known (7). Consequently, the copolymerization of styrene with the repulsive comonomer MAA is expected to cause a miscibility enhancement in blends with PVME.

Usually, the occurence of one glass transition located between those of the pure components is discussed as a criterion for compatibility. Also an optically transparent film cast from a common solution of the polymers gives evidence for a homogeneous mixture. Thus, in the present experimental study differential scanning calorimetry (DSC) and small angle light scattering have been used to characterize the phase behaviour of PS-CO-MAA and PVME.

MATERIALS AND METHODS

PS-CO-MAA/PVME blends have been prepared by solving an appropriate amount of the polymers in reagent grade dioxane and slowly evaporating the solvent under dry nitrogen atmosphere. After evaporation of the solvent the blends have been dried in high vacuum at room temperature for at least two weeks. To avoid the uptake of moisture the dry samples have been stored in vaccum. The molecular weights and the polydispersities of the pure comonents have been measured by gel permeation chromatography and membrane osmosis. Data are listed in TABLE 1.

TABLE 1

Comonomer content MAA in PS-CO-MAA, weight average molecular weights M_w and molecular weight distributions M_w/M_n of the pure components.

polymer	MAA content	M_w (g/mol)	M_w/M_n
PVME	-	80,000	1.3
PS-CO-MAA-5	4.7%	250,000	1.6
PS-CO-MAA-10	9.8%	270,000	1.6
PS-CO-MAA-20	19.5%	220,000	1.6

DSC measurements have been performed on a Perkin Elmer DSC-7 at 20°C/min heating rate. Cloud points have been recorded by measuring the scattered light at samll angles and different heating rates. The cloud points were extrapolated to zero heating rate to obtain the turbidity curve of the particular copolymer/homopolymer blend. Spinodial decomposition have been studied by temperature jump experiments. The temperature equilibrium after the quench lasted about 2 minutes. The developing structure factor for a constant quenching temperature was measured as a function of scattering angle and of time.

RESULTS AND DISCUSSION

Figure 1 shows the DSC traces of the system PS-CO-MAA-5/PVME

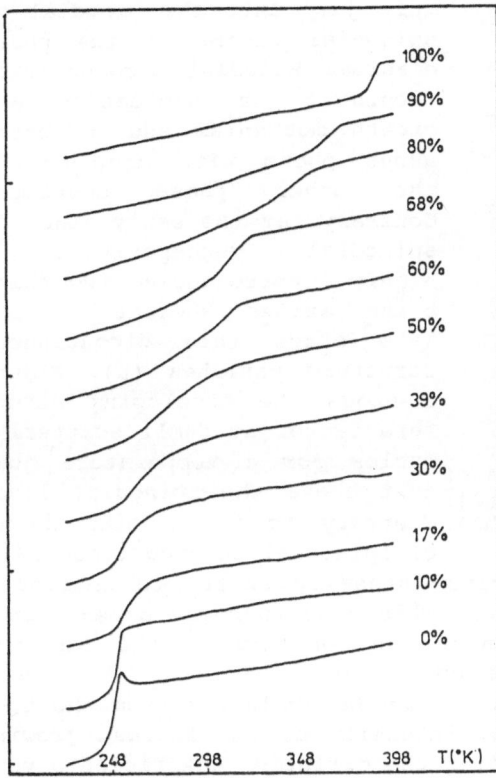

100%
90%
80%
68%
60%
50%
39%
30%
17%
10%
0%

248   298   348   398 T(°K)

Figure 1. DSC-traces of PS-CO-MAA-5/PVME blends as a function of blend composition.

for various blend compositions (6).Over the complete concentration range only one glass transition occurs and thus, blends are compatible. Nevertheless, in the mid concentration regime the glass relaxation starts at rather low temperatures and spread over 50 to 60°C. This behaviour may be explained by broad relaxation time spectra of the blends compared to the pure polymers, and eventually indicates microheterogeneity on a segmental level. In mixtures of PS-CO-MAA-10/PVME this effect is even more pronounced complicating the evaluation of T_g's from those DSC traces. Contrary, in blends of PS-CO-MAA-20/PVME two separated and narrow glass transitions indicated incompatibility. Except, mixtures of 90 wt-%PS-CO-MAA-20 and 10 wt-% PVME were found to be compatible.

For cloud point measurements transparent films of PS-CO-MAA-5/PVME and PS-CO-MAA-10/PVME blends have been prepared by casting thin films from dioxane solution. Blending of PS-CO-MAA-20 with PVME resulted in turbid films except the mixing ratio 90 wt% PS-CO-MAA/10 wt% PVME. In order to get quasi-equilibrium conditions cloud points have been measured at very slow heating rates (0.03°C/min). In Figure 2 the resulting phase diagrams are summarized.

It is evident from DSC and turbidity measurements that by increasing the comonomer content MAA the lower critical solution temperature (LCST) may be shifted towards lower temperature. Variation of the comonomer content from 0 wt% (pure PS) to 12 wt% MAA decreases the LCST from 130°C (pure PS/PVME) to room temperature and either an one phase or a two phase blend may be obtained by a small variation of the MAA/S ratio.

Homogeneous blends may also be structured by a sudden tempera-

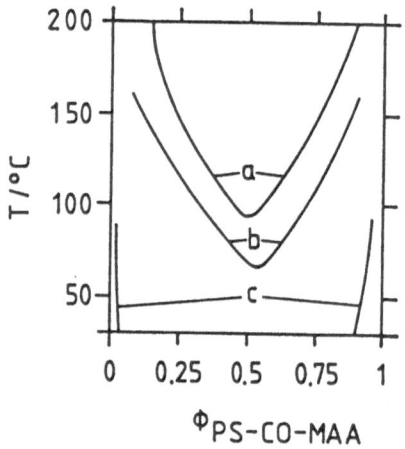

Figure 2. Cloud point curves of blends PS-CO-MAA/PVME with (a)5 wt-%,(b)10 wt-%,(c)20 wt-% MAA.

ture jump into the binodial or spinodial regime of the phase diagram. Binodial decomposition occurs by a nucleation and growth mechanism and a continuous phase with droplets of the other phase develops. Contrary, at the early stage of spinodial decomposition a highly interconnected two phase blend arises whereas in the late stage this bicontinuous structure vanishes (8). Figure 3 shows the developing structure factor at samll scattering angles for a temperature jump just above the spinodial line. Contrary to the linear theory of spinodial decomposition (8) the maximum of the scattering intensity shifts to smaller q vectors as the time increases. This behaviour may be explained by considering the deterministic noise term in the Langevin equation of motion of the structure factor (9,10).

Apparent diffusion coefficients may be evaluated from the time dependence of the scattering intensity of the fastest growing q_m-value (11). By extrapolating the diffusion coefficients versus quenching depth one gets the particular spinodial temperature. Heats of demixing may be calculated from those data provided the Flory-Huggins interaction parameters are known. In the present copolymer/homopolymer systems we applied the experimentally known interaction parameters of the pure homopolymer blends PS/PVME (12). Thus, depending on the particular system and the quenching depth heats of demixing between 2 and 5 J/cm^3 have been obtained (13).

CONCLUSIONS

It has been shown by DSC and turbidity measurements that the compatibility in blends of poly(styrene-co-maleic acid anhydride) and poly(vinyl methyl ether) prepared from dioxane solution decreases with rising maleic acid anhydride content. Blending of PS-CO-MAA with 20 wt-% MAA and PVME results in an almost incompatible copolymer/homopolymer mixture. The spinodial decomposition in those blends have been measured by small angle light scattering. The maximum of the scattering

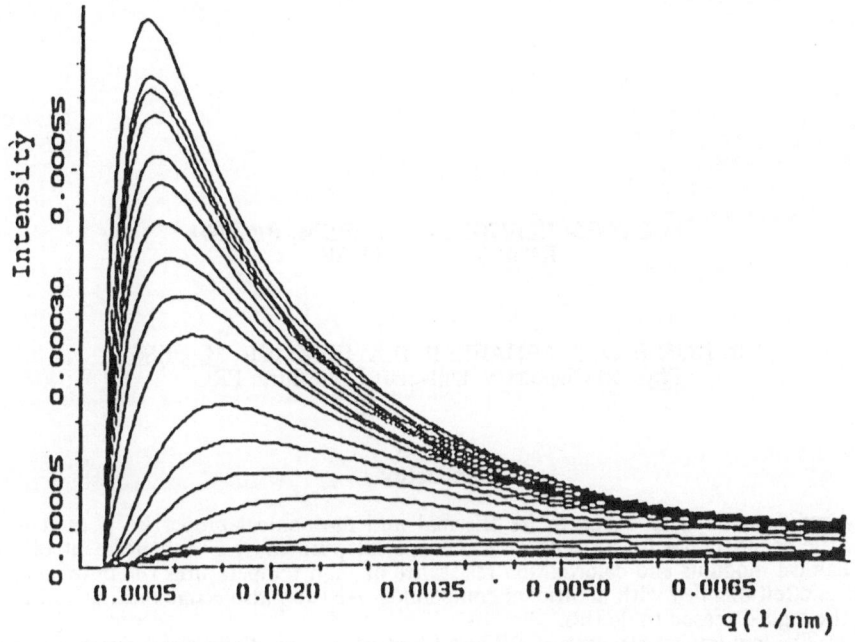

Figure 3. Scattering intensity versus scattering vector q for different waiting times after a quench into the spinodial region from 80°C to 150°C for a 50 wt-% PS-CO-MAA-5/PVME blend.

intensity shifted towards smaller scattering vectors while demixing and thus, exhibited deviation from the linear Cahn-Hilliard theory.

REFERENCES

1. Kambour, R.P., Bendler, J.T., _Macromolecules_, 1983, **16**, 753
2. tenBrinke, G., Karasz, F.G., MacKnight, W.J., _Macromolecules_, 1984, **16**, 1827.
3. Paul, D.R., Barlow, J.W., _Polymer_, 1984, **25**, 487.
4. Bernstein, R.E., Cruz, C.A., Paul, D.R., Barlow, F.W., _Macromolecules_, 1977, **10**, 681.
5. Roe, R.J., Rigby, D., _Adv. Polym. Sci._, 1987, **82**, 103.
6. Wendland, C., Klotz, S., Cantow, H.-J., _submitted_.
7. Hoy, K.L., _J. Paint. Technol._, 1970, **42**, 76.
8. Cahn, J.W., Hilliard, J.E., _J. Chem. Phys._, 1958, **28**, 258.
9. Langer, S., Bar-on, M., Miller, H.D., _Phys. Rev._, 1975, **A11**, 1417.
10. Kawasaki, K., Otha, T., _Prog. Theor. Phys._, 1978, **59**, 362.
11. Hashimoto, T., Izumitani, T., _J. Chem. Phys._, 1985, **83**, 3694.
12. Shibayama, M., Yang, H., Stein, R.S., Han, C.C., _Macromolecules_, 1985, **18**, 2179.
13. Wendland, C., Klotz, S., Krieger, V., Cantow, H.-J., _in preparation_.

POLYMER BLENDS FROM EPDM, PP AND INORGANIC FILLER

R. KOSFELD, K. SCHAEFER, E.A. HEMMER, M. HESS
Physical Chemistry, University Duisburg, FRG

Introduction

Polypropylene (PP) is – due to its thermal and mechanical properties – used in a versatile manner. To fit its properties for special purposes e.g. enhancement of the mechanical modulus and deformation resistance at high temperatures the pure component is often blended with additional compounds. At least also economical advantages may be a good reason to do this.

The low impact strength of PP and filled PP is one of the disadvantages one is often faced with using this material. So it became customary to modify the thermoplast by the well known technique of blending with an elastomeric compound. Semi crystalline ethylene/propylene/dien/terpolymer (EPDM) is well suited for such purposes as it is quite easy to chop it to granules a technique which hardly can be applied to amorphous EPDM. The elastomer granules can easily be extruded together with PP. The increase of impact strength achieved by this procedure unfortunately has to be payed by a decrease of modulus and stiffness.

Lots of papers deal with modification of PP either with filler or with an elastomer. A few authors focused their attention on the properties of ternary systems in order to combine the positive aspects of both modifications.

Stamhuis [1, 2, 3] investigated impact strength and flexural modulus in PP/EPDM blends filled with talcum and short glas fibers.

Pukanszky [4] focused his attention on $CaCO_3$ filled ternary systems.

Fernando [5] described the fatigue mechanism at break in rubber modified PP containing $CaCO_3$, $CaSO_4$ and mica.

Finally Faulkner [6] investigated different types of processing procedures on mica modified PP/EPDM blends.

Further important papers dealing with ternary systems basing also on other thermoplasts than PP were published recently [7 – 13].

The present work deals with impact strength at low temperatures, flexural modulus at room temperature, and dynamic mechanical properties in PP/EPDM systems filled with kaolin, $BaSO_4$, $BaSO_4 \cdot ZnS$ and ZnS respectively.

Optimization of impact strength and modulus finds special respect in this context.

Materials

PP	:	Vestolen@ P7000, Fa. Hüls AG melt flow index (DIN 53735, MFI 230/5) =10g/10min
EPDM	:	BUNA AP 447 G@, Fa. Hüls AG, semi crystalline
Kaolin	:	Icecap K@, Fa. Burgess
BaSO$_4$	:	Blanc fixe–micro@, Fa. Sachtleben
Lithopone	:	BaSO$_4$·ZnS, Lithopone DS@, 30% ZnS, Fa. Sachtleben
ZnS	:	Sachtolith HDS@, Fa. Sachtleben

@ are registered trademarks

Experimental

All three components were together premixed by tumble–blending and subsequently melt–blended in a Leistriz 2SE twin screw extruder with granulator. Standard samples were prepared by injection molding.

The Charpy impact strength of the unnotched samples was tested on a Frank impact testing machine according to DIN 53453 at –23^{0}C and –58^{0}C, respectively. The flexural modulus was measured according to DIN 53452 at room temperature on a Schenck–Trebel universal testing machine. Dynamic mechanical analysis was performed using a Myrenne torsion pendulum with free oscillation at 1 Hz.

The mixtures contained EPDM from 0 Vol% up to 20 Vol%. The same concentration range was used for the fillers.

Results

The dynamic mechanical behavior is shown in fig.1 and fig.2.

The storage modulus G'(T) is increased within the whole temperature range if the kaolin fraction is increased at a constant EPDM ratio. Increasing the EPDM content on the contrary decreases the storage modulus.

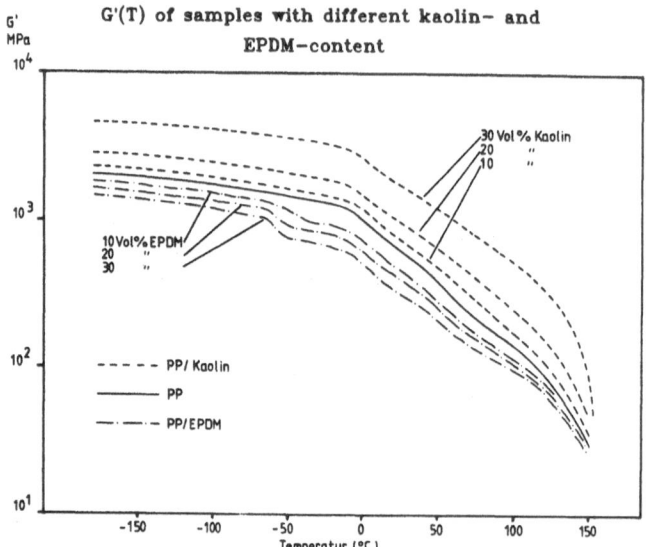

Figure 1: Storage modulus G'(T) of samples with different content of kaolin respectively EPDM

The glassy transition of PP is found at 5°C causing a strong decrease of G'(T) with further increasing temperature.

At −45°C the glassy transition of EPDM is located, and an additional low temperature transition of EPDM at −140°C is observed as a broad maximum in G"(T) and tanδ. The glassy transition of PP occurs almost unaltered if EPDM or filler is present.

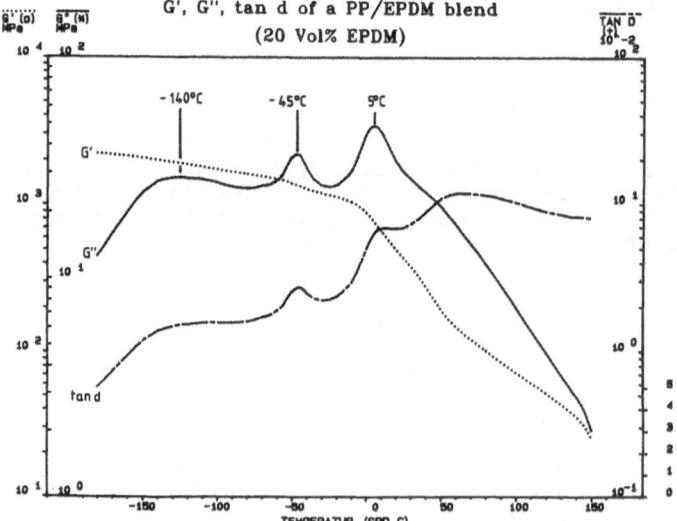

Figure 2: Storage modulus G'(T), loss modulus G"(T) and dumping factor tanδ of PP modified with 20 Vol% EPDM

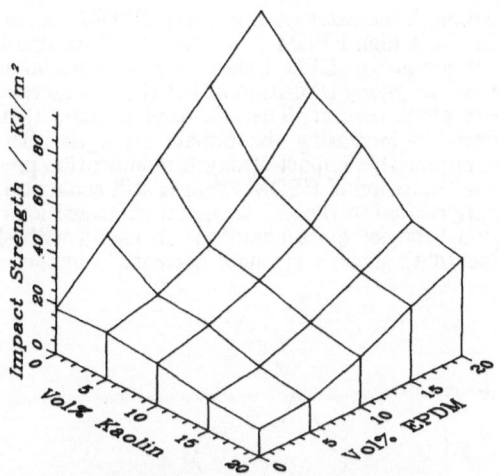

Impact strength at −23 C, filler Kaolin

Figure 3a: Impact strength as a function of EPDM and kaolin content, T =−23⁰C,

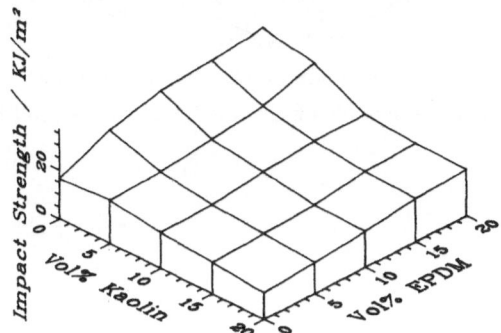

Impact strength at −58 C, filler Kaolin

Figure 3b: Impact strength as a function of EPDM and kaolin content, T =−58⁰C

The impact strength increases with growing EPDM content and decreases with the amount of kaolin. So, a high EPDM / low kaolin PP–composite shows best values of impact strength. In absence of EPDM there is almost no influence of the filler observable, so that below the glassy transition of PP there is no temperature dependence if there is no rubbery phase present. There is need of more than 5 Vol% EPDM to show significant effects in increasing the impact strength with rising temperature. Although there is an appreciable impact strength modification present at temperatures higher than the glassy transition of EPDM there is still some positive effect remaining below this temperature related to the low temperature transition of EPDM. Lithopone and Blanc fixe show a behavior quite analogous to kaolin with slightly lower values. On the contrary , Sachtolith shows a stronger decrease in impact strength as shown in figure 4.

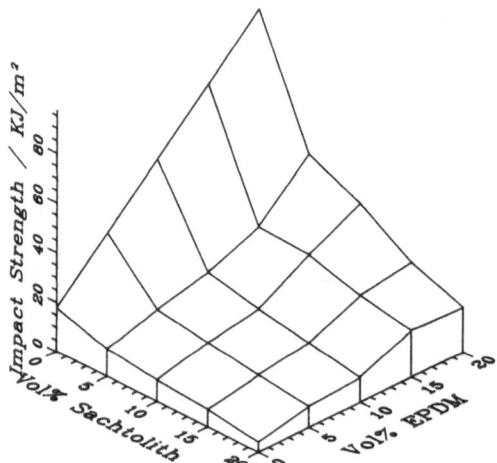

Impact strength at −23 C, filler Sachtolith

Figure 4:   Impact strength of blends from PP/EPDM/Sachtolith

The flexural modulus shows a behavior quite contrary to the composition dependence of the impact strength.

Figure 5 shows the influence of kaolin respectively EPDM content on the flexural modulus. With increasing kaolin content the flexural modulus increases while with increasing amount of EPDM a decrease is observed. High values of the flexural modulus thus are obtained in blends with a high amount of kaolin but a low content of the rubbery phase.

The other fillers studied do behave analogously: the flexural modulus is decreased from Blanc fixe over Lithopone to Sachtolith.

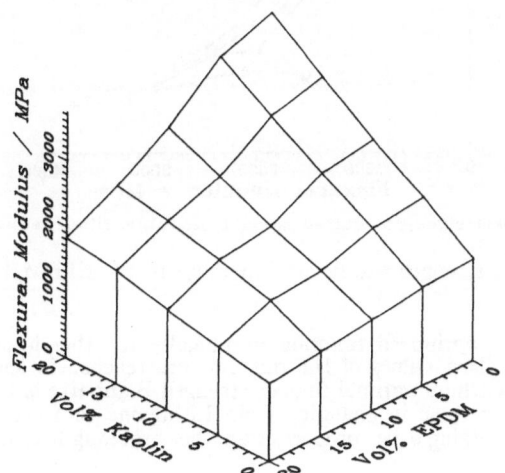

Flexural modulus at 20 C, filler Kaolin

Figure 5:   Flexural modulus PP/EPDM/kaolin

A combined figure containing flexural modulus and impact strength in an EPDM and kaolin modified PP shows the line with optimal combination of both properties as a hyperbolic shaped curve:

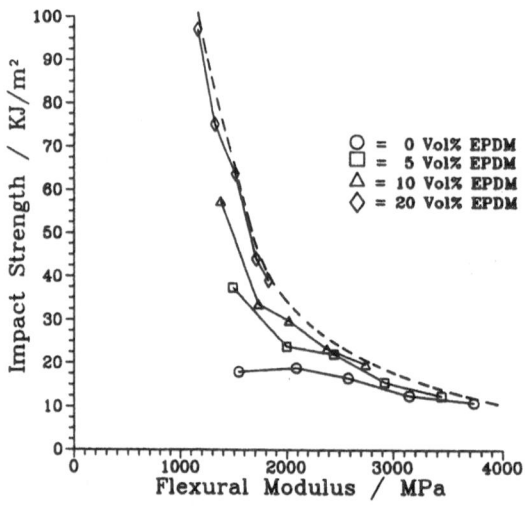

Impact strength plotted against modulus, filler Kaolin

Figure 6: Impact strength and flexural modulus in PP/EPDM/kaolin blends

Figure 7 shows this optimized function graphically for the different filler systems investigated. The highest values of the modulus are reached in kaolin containing systems. In order to reach an optimal impact strength Blanc fixe is the best choice. The optimal properties decrease in changing from Lithopone to Sachtolith. It has to be emphasized that combining other properties another "ranking list" might be obtained.

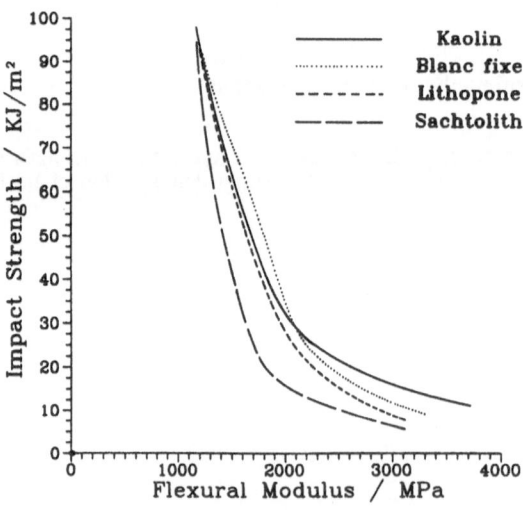

Curves with optimised properties

Figure 7: Traces of optimal properties of PP/EPDM/filler systems

Discussion

Impact strength as well as flexural modulus of PP/EPDM/filler blends can be improved simultaneously by the described compounding method.

From the impact strength / flexural modulus plots the combinations of optimized properties can be determined. They are described by a hyperbolic shaped curve.

The optimization effect of different fillers is strongly dependent on the particle size and geometry, furthermore factors like polymer adhesion, chemical and physical interaction as well as dispersion of the filler have to be taken into account.

The evaluation of SEM photographs of samples fractured at -23^0C shows good dispersion of the filler, agglomerates do not appear. The polymer–filler adhesion is not developed strongly. There remains no polymer on the filler surface visible by this technique.

The different behavior of the fillers is discussed primarily under the aspect of particle size and geometry.

Kaolin effects with its plate structure and an average particle size of 1μm the highest flexural modulus. Together with EPDM it shows in the region of high modulus (fig. 7) the optimized properties.

Blanc fixe has a grain structure and an average particle size of 0.7μm. At high impact strengths it has the best optimization effect.

Lithopone shows a particle size distribution with two maxima at 1μm and 0.35μm. The first corresponds to the particle size of $BaSO_4$ the second to that of ZnS. The optimization properties are reduced.

With a relative small particle size of 0.3μm Sachtolith shows the lowest values of the combined properties. A plate structure leads to better properties especially at high flexural modulus. Fernando obtained similar results for the plate structured filler mica. Small particle size as well as spherical structures seem to effect minor properties.

Obviously this point of view is a simplificating one, but the discussion is limited to these results until further investigations have been executed.

Conclusion

The impact to stiffness balance of PP is improved by simultaneous compounding with EPDM and filler. The impact strength plotted against flexural modulus allows to compare and evaluate the optimal properties of the different fillers.

The properties of the materials described in this paper can be adapted to the demands of different applications with the aid of optimization curves.

Literature

[1] Stamhuis, J.E.   Polymer Composites
                          5 (1984) 202–207

[2] Stamhuis, J.E.   Polymer Composites
                          9 (1988) 72–77

[3] Stamhuis, J.E.   Polymer Composites
                          9 (1988) 280–284

[4] Pukanszky, B.,   Polymer Composites et al.
                          7 (1986) 106–115

[5] Fernando, P.L.   Polymer Engineering and Science
                          28 (1988) 806–814

[6] Faulkner, D.L.   Journal of Applied Polymer Science
                          36 (1988) 467–480

[7] Kitamani, H.   Proceed. 4th Intern. Conf. Compos. Mat. ICCM–IV
                          Oct. 25–27, Tokyo 1982
[8] Comitov, P.G.   Europ. Polym. J.
    et al.   20 (1984) 405 ff
[9] Pukanszky, B.   in B. Sedlacek (ed), Polymer Composites
    et al.   De Gruyter, Berlin 1986
[10] Scott, C.   in H. Ishida, J.L. Koenig (eds), Composite Interphases
     et al.   Elsevier, London 1986, S.177 ff
[11] Scott, C.   2nd ann. conf., Comp. Inst., The Soc. of Plast. Ind., Inc
     et al.   Febr. 2–6, 1987, Session 19–E
[12] Scott, C.   J. Mat. Sci.
     et al.   22 (1987) 3963 ff
[13] Scott, C.   Rheol. Acta
     et al.   27 (1988) 273 ff

Acknowledgement

This study was financially supported by the "Arbeitsgemeinschaft Industrieller For-
schungsvereinigungen e.V." (AIF).
The Hüls AG is acknowledged for making available the polymer material, the Fa.
Sachtleben for the gift of their filler. Furthermore the authors want to thank Dr. Th.
Uhlenbroich, Fa. Sachtleben, for his engagement and helpful discussions.

INVESTIGATION OF THE TRANSCRYSTALLISED INTERPHASE IN
FIBRE-REINFORCED THERMOPLASTIC COMPOSITES

J.L. THOMASON and A.A. VAN ROOYEN
Koninklijke/Shell-Laboratorium, Amsterdam
(Shell Research B.V.)
Badhuisweg 3, 1031 CM Amsterdam, The Netherlands

ABSTRACT

The occurrence of transcrystallisation in fibre-reinforced polypropylene is
shown to depend on the type of fibre used. The list of fibres which transcrystallise polypropylene is not identical to that for other semicrystalline thermoplastics. Furthermore, the ability of aramid fibres and high-
modulus carbon fibres to induce transcrystallisation in polypropylene
depends on the crystallisation temperature and the polymer molecular
weight. With one particular polypropylene, at crystallisation temperatures
below 138 $^{\circ}$C, the growth rate of the transcrystallised region around three
different types of fibre in polypropylene was found to be the same as the
growth rate for the bulk spherulites. Above 138 $^{\circ}$C these fibres did not
nucleate transcrystallinity, although spherulites were still nucleated in
the bulk. Further more it was possible, through the application of shear,
to induce a crystalline region around fibres embedded in polypropylene at
temperatures where transcrystallisation was not obtained in quiescent
crystallisation. When studied using polarised light microscopy this region
could not be distinguished from a transcrystallised region. We therefore
propose that transcrystallisation and stress-induced crystallisation may be
related.

INTRODUCTION

A critical issue in the processing of semicrystalline thermoplastic compo-

sites is the microstructure or morphology of the matrix material. Morpho-

logical features such as degree of crystallinity, spherulite size, lamellae

thickness, and crystallite orientation have a profound effect on the ulti-

mate properties of the polymer matrix, and thus the composite. These fea-

tures are, in turn, affected by variation in the processing conditions. In

composites this situation is further complicated by the effect of the rein-

forcing fibres on the morphology of the matrix. In particular, when heterogeneous nucleation occurs with sufficiently high density along a fibres'
surface the resulting crystal growth is restricted to the lateral direction, so that a columnar layer develops around the fibre, known as transcrystallisation. This nucleation of a transcrystallised interphase around
the reinforcing fibre is thought to be central to the improvement of some
composite properties [1,2]. Although the presence of a transcrystallised
interphase has been reported to improve mechanical properties of some
fibre—reinforced polymer systems, the mechanisms by which transcrystallisation occurs is not fully understood. In particular there does not appear to
be a method by which its appearance in a particular fibre/matrix combination can be predicted. The fibre material toplogy, and surface caoting, and
the matrix type and thermal history have all been reported to affect transcrystallisation in these composites to some extent [3]. In an attempt to
clarify this situation, we have begun an investigation into the ability of
reinforcing fibres to induce transcrystallisation when incorporated in a
thermoplastic matrix. Some initial results from these experiments are
reported here.

EXPERIMENTAL

Isothermal crystallisation of the samples was carried out in nitrogen using
a Mettler FP52 hot—stage and observed under an Olympus BHS polarizing
microscope. Samples for microscopy were prepared using Shell SY6100 (MI=11)
and HY6100 (MI=2) grades polypropylene, and the fibres shown in Table 1.

TABLE 1
Transcrystallisation in SY6100 polypropylene at 135 $^{\circ}$C

Fibre nucleates transcrystalline interphase	
Yes	No
Grafil HM—U & HM—S	Grafil XA—U
Enka HM35	Enka ST & Enka IM
Apollo HM—U & HM—S	Apollo IM—U & IM—S
Thornel T50 (HM)	Thornel T40
Thornel P120 (pitch based)	
	Silenka P73
Twaron D1056	Silenka P62
Kevlar 49	Silenka 8045—1
	PPG 1062—TNT

Samples were held in the hot—stage for 5 minutes at 200 °C before being cooled at 10 °C/min to the isothermal crystallisation temperature. The crystallisation process was recorded with the aid of a video system. From these recordings growth rates at the fibre surface and in the bulk were determined.

For the shearing experiments we constructed a pulling motor which could be used in conjunction with the hot—stage microscope. Single fibres could be pulled (at 5—5000 μm/min) through a crystallising polymer melt while the effect on the morphology of the specimen was observed. The single fibres were pulled at a fixed velocity for a short time directly following the cooling step, and the effect on the crystallisation behaviour was examined.

RESULTS AND DISCUSSION

Table 1 indicates which of the fibres studied induce transcrystallisation in polypropylene. We note that, in common with other studies of trans—crystallisation in polymers such as polypropylene and nylon, high—modulus (HM) carbon fibres and aramid fibres induce transcrystallisation, whereas high—strength (HS) carbon fibres do not. Our results also show that four different types of glass fibre failed to transcrystallise polypropylene even though silane—coated glass fibres have been shown to produce trans—crystallisation to some extent in polypropylene. Figures 1a and 1b show the two typical morphologies observed in our samples. Figure 1a is split into a polarised micrograph, and a phase micrograph which shows more clearly the presence of the single glass fibre. It can be seen that the glass fibre has had no effect on the morphology of the polypropylene, which exhibits only spherulitic growth. Figure 1b shows a micrograph from a sample containing two HM carbon fibres close to each other. The transcrystalline region is seen as a white band of densely packed, radially oriented crystalline lamellae on both sides of the fibre. The transcrystallised region extends to approximately 180 μm from the fibres; the matrix spherulites have a similar radius, indicating that the rate of growth of the crystals in the spherulites and in the transcrystallised region is the same. However, it is quite clear that the nucleation density is much greater along the fibre surface than in the bulk. It should be also be noted that it is now somewhat unrealistic to refer to a "transcrystallised interphase" and a " bulk matrix" as the interphases now impinge on each other to form the major

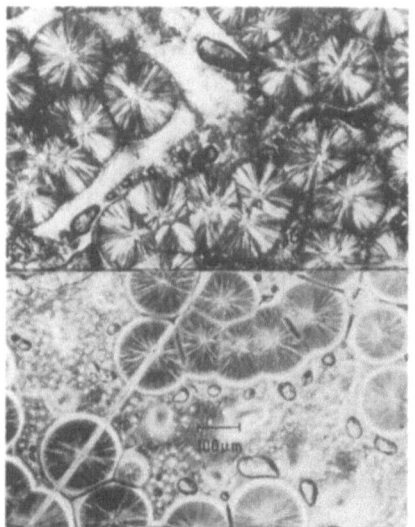 

FIG. 1a:
GLASS FIBRE

FIG. 1b:
HM CARBON FIBRES

fraction of the matrix. This may well be of great importance to the properties of 'practical' composites, even those containing a relatively low volume fraction of reinforcing fibres. Clearly, if the properties of transcrystallised thermoplastic are different from those of the other possible thermoplastic morphologies it becomes important to know whether a particular fibre will produce transcrystallisation in a composite.

The growth rates of the transcrystallised interphase around HM carbon fibres, and Twaron and Kevlar aramid fibres, are compared with bulk spherulite growth rates in Figure 2. It can be seen that at temperatures below 138 ^{o}C the crystallisation rates in the matrix and in the interphase regions around the three different fibres are identical with experimental error. This is in agreement with previously published findings [4,5] and is not unexpected because in both cases the growth of the crystallisation regions is due to secondary nucleation on the developed crystal faces and it seems unlikely that a polymer molecule can distinguish between a crystal face in a spherulite and a transcrystallised region. We can therefore assume that the crystallisation rates in the interphase and in the matrix should be the same at all temperatures, i.e. also at temperatures above 138 ^{o}C. However, no transcrystallised interphase was observed around the fibres in samples crystallised at temperatures above 138 ^{o}C, although the

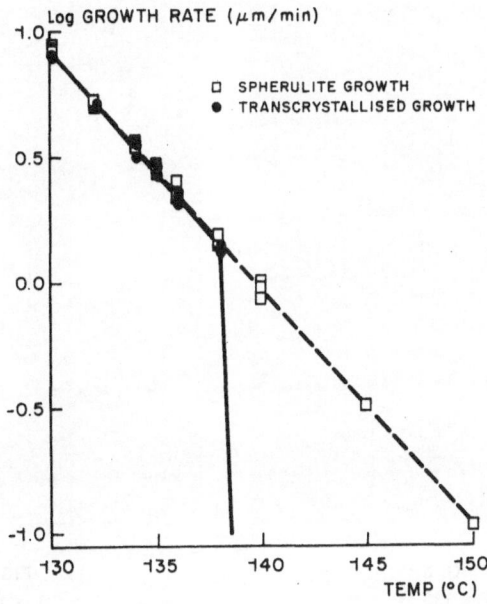

FIG. 2: GROWTH RATES vs CRYSTALLISATION TEMPERATURE

growth of spherulites continued. Figures 3a and 3b are polarised micrographs which show the difference in the polypropylene morphology around an
aramid fibre depending on the crystallisation temperature. That no transcrystallised region is observed above 138 °C is probably because no primary
nucleation occurs along the fibre surface above this temperature. The
explanation of this finding may will be one of the keys to understanding
the phenomenon of transcrystallisation.

There exist isolated examples in the literature showing that applied
stress can also influence or induce transcrystallisation [6,7]. Unlike
crystallisation under quiescent conditions, strain—induced crystallisation
of polymers is not very well understood; however, we feel that the effect
of applied stress in these systems merits close attention. Figure 4 shows
another aramid—polypropylene sample where, directly after cooling, at the
beginning of the isothermal period at 140 °C, the fibre was pulled at
5 mm/min for 12 seconds. The morphology of the polypropylene around the
fibre appears to be identical with the transcrystallised morphology
obtained at lower temperatures. Furthermore, it can be seen that this
morphology is also present in the 1 mm long region which the fibre was

FIG. 3a:	FIG. 3b:
T = 130 °C	T = 140 °C

FIG. 3: DEPENDENCE OF PP MORPHOLOGY ON
CRYSTALLISATION TEMPERATURE

pulled out of, showing that the presence of fibre surface is not necessary
throughout the crystallisation period to obtain this morphology. Through
the application of shear we obtained a transcrystallised interphase in
samples crystallised at temperatures up to 150 °C, which is well above the
temperature boundary for quiescent transcrystallisation. We have also
performed similar measurements using glass fibres and HS carbon fibres in
polypropylene. Once again the brief application of shear along the fibre-
matrix interface nucleated an apparently transcrystalline region around the
fibre. It should further be noted that using these fibres we have not
obtained transcrystallisation in polypropylene with quiescent crystallisation at any temperature.

These results minimally suggest that there could be a link between
transcrystallisation and shear-induced crystallisation. Certainly the two
phenomena cannot be distinguished using polarised light microscopy. This
idea is supported by the fact that we have not found a lower boundary in
pulling speed below which pulling the (aramid) fibre at 140 °C does not
induces crystallisation. Thus, pulling at 5 μ/min, the lowest speed as yet

FIG. 4
PULLED ARAMID FIBRE (x60)

FIG. 5
HM CARBON WITH HY6100 PP (x50)

available to us, still induces crystallisation around the fibre. It is
therefore conceivable that stresses induced at the fibre—matrix interface
during cooling (e.g. from mismatch in thermal expansion coefficient) may
be, at least partially, responsible for the phenomena of transcrystallisation. If this were the case, then transcrystallisation should be sensitive
to the polymer molecular weight and the temperature boundary for quiescent
transcrystallisation should be higher for higher molecular weights.
Figure 5 shows an aramid fibre embedded in HY6100 polypropylene (MI=2)
crystallised at 145 $^{\circ}$C. We see that transcrystallisation is obtained at
higher temperatures with a higher molecular weight polymer. Future work
will be directed to investigating this link between transcrystallisation
and stress—induced crystallisation.

REFERENCES

1. Burton, R.H. and Folkes, M.J., Plast Rubber Process. Appl., 1983, 3,
   129.

2. Peacock, J.A., in Composite Interfaces (Eds. H. Ishida and J.L. Koenig),
   Elsevier, New York 1986.

3. Bessel, T. and Shortall, J.B., J. Mater Sci., 1975, 10, 2035.

4. Chatterjee, A.M., Price, F.P. and Newman, S., J. Polym. Sci., Polym. Phys. Ed., 1975, 13, 2391.

5. Campbell, D. and Qayym, M.M., J. Polym. Sci., Polym. Phys. Ed., 1980, 18, 83.

6. Burton, R.H., Day, T.M, and Folkes, M.J., Polym Communi., 1984, 25, 361.

7. Misra, A., Angew. Makromol. Chem., 1983, 113, 113.

THE MOLECULAR MECHANISM OF ADHESION IMPROVEMENT OF 3-AMINOPROPYLTRI-ETHOXYSILANE IN GLASS REINFORCED NYLON MODELCOMPOSITES

L.W. JENNESKENS[*],H. ANGAD GAUR,A. VENEMA,H.E.C. SCHUURS and W.G.B. HUYSMANS
Akzo Research Laboratories Arnhem,Corporate Research,P.O. Box 9300,
6800 SB Arnhem,The Netherlands.

T.P. HUIJGEN,T.L. WEEDING and W.S. VEEMAN
Department of Molecular Spectroscopy,University of Nijmegen,Toernooiveld,
6525 ED Nijmegen,The Netherlands.

ABSTRACT

Selectively ^{13}C-enriched 3-(3-^{13}C)aminopropyltriethoxysilane ((3-^{13}C)-APS) is used to investigate silane-polymer reactions and/or interactions in glassbead reinforced Nylon (PA-6) modelcomposites. With ^{13}C MAS NMR and Pyrolysis GC/MS evidence is obtained for amide formation between -COOH PA-6 endgroups and -NH$_2$ groups of APS and interpenetration of PA-6 chains into the poly-APS network.

INTRODUCTION

Silane coupling agents of the structure $(RO)_3-Si-(CH_2)_n-Y$ are frequently applied as additives in inorganic filler reinforced polymers for the improvement of mechanical and physical properties of the final composite (1,2). Although the silane-filler adhesion mechanism is well established (1-4), the nature of silane-polymer reactions and/or interactions is still the subject of speculation. Various theories have been proposed to rationalize the silane-polymer adhesion mechanism. The Chemical Bonding (CB) theory states that covalent bonds are formed between the Y functionality and the polymer (5). In contrast, the Interpenetrating Polymer Network (IPN) theory states that silane-polymer adhesion is primarily due to the formation of an entangled network by diffusion of polymer chains into the poly-APS network near the filler surface (6). However, as a consequence of the small interfacial volume and the low level of silane (ca. 0.1-1.0% m/m, based on the filler)

usually applied, unambiguous spectroscopic (NMR,IR) detection of silane-polymer interactions is obscured by the filler and/or bulk polymer signals.
Notwithstanding, recent ^{13}C MAS NMR investigations of glassbeads, isolated
from APS pretreated glassbead reinforced Nylon (PA-6) modelcomposites, revealed that they contain a strongly adhered, insoluble PA-6 layer with physical properties different from the bulk PA-6 (4). The results are indicative for chemical and/or physical silane-polymer interactions. To enhance
the sensitivity and selectivity for the detection of these interactions
with ^{13}C MAS NMR, composites were prepared with selectively ^{13}C-enriched
3-(3-^{13}C)aminopropyltriethoxysilane ((3-^{13}C)-APS, Figure 1 (7)). In this
paper we report preliminary results of ^{13}C MAS NMR and Pyrolysis GC/MS investigations of isolated glassbeads from these composites with the aim to
elucidate the molecular silane-polymer adhesion mechanism.

$$-O)_3-Si-(CH_2)_2-^{13}CH_2-NH_2 + HOOC-PA-6 \longrightarrow -O)_3-Si-(CH_2)_2-^{13}CH_2-NH-(CO)-PA-6$$

Figure 1. Amide formation between -COOH PA-6 endgroups and -NH$_2$ groups of
APS.

MATERIALS AND METHODS

The PA-6 and glassbeads were the same as those used in a previous study (4).
Detailed procedures for the APS pretreatment of the glassbeads, modelcompo-
site preparation and isolation of the glassbeads from the composite, respectively, are reported in reference 4. 3-(3-^{13}C)Aminopropyltriethoxysilane
((3-^{13}C)-APS, ^{13}C-enrichment > 99%) was synthesized as described elsewhere
(7). For the ^{13}C MAS NMR experiments (Bruker CXP 300) pretreated (0.5% m/m,
based on glass) glassbeads isolated from the composite were used. Isolated
glassbeads, pretreated with 10% m/m natural abundance APS or (3-^{13}C)-APS before composite preparation were investigated by Pyrolysis (curie point; 10
sec., 600°C) cap. GC (OV-1, ρ 20 m., d_f 0.2x10^{-6} m., carrier gas He) MS
(Finnigan MAT 212).

RESULTS AND DISCUSSION

^{13}C MAS NMR. The high level of ^{13}C-enrichment at C3 in (3-^{13}C)-APS will enable its observation even after dilution by the PA-6. Amide formation between -COOH PA-6 endgroups and -NH$_2$ groups of APS can be most simply demonstrated by a change in chemical shift of C3 (Figure 1). Previous ^{13}C MAS

NMR investigations of poly-APS and the coupling product of poly-APS and he-
xanoic acid showed that amide formation leads to an upfield shift of C3 of
1.5 ppm (4). In order to observe a change in chemical shift of C3 it is desirable to have a low level of APS loading on the glassbeads; the number of
-NH$_2$ groups of APS that may react with -COOH PA-6 endgroups will then be a
significant fraction of the total. In Figure 2 the ^{13}C MAS NMR Bloch decay
spectrum of isolated glassbeads, pretreated with 0.5% (3-^{13}C)-APS before
composite preparation, is shown. At the C3 position two peaks, at 42.7 and
41.3 ppm, respectively, are distinguishable when resolution enhancement is
added before Fourier transformation. Although the resolution is only moderately better than the noise level, both chemical shift positions are in
agreement with those expected for C3 before and after amide formation (4).
Probably, some chemical bonding occurs (Figure 1).

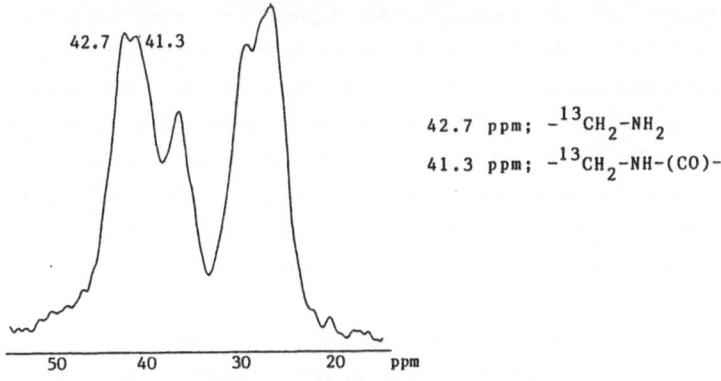

42.7 ppm; $-^{13}$CH$_2$-NH$_2$

41.3 ppm; $-^{13}$CH$_2$-NH-(CO)-

Figure 2. ^{13}C MAS NMR Bloch decay spectrum of isolated glassbeads containing
0.5% m/m (3-^{13}C)-APS.

Evidence for other silane-polymer interactions can be deduced from changes
in relaxation parameters. For APS on the glassbeads a two component structure is found, i.e. a rigid and a mobile fraction (Table, C3 43 ppm). Presumably, the former is poly-APS near the glass surface and the latter poly-APS
in the periphery. A similar analysis was done for the isolated,(3-^{13}C)-APS
pretreated glassbeads. Note that in this case at 43 ppm, in principle, three
components are present, i.e. PA-6, rigid- and mobile poly-APS, respectively.
Intriguingly, a two term fit gave the best result. Besides the PA-6 only the
rigid poly-APS could be observed (Table). These results can be rationalized
by invoking interpenetration (entanglement) of PA-6 into the mobile poly-APS

fraction; thus increasing its rigidity. A full account of these investigations will be reported elsewhere (8).

TABLE
Relaxation parameters determined at 43 ppm with ^{13}C MAS NMR.

Relaxation time	APS on glassbead		Isolated glassbead	
	rigid	mobile	APS	PA-6
T_2 ms	3.4	11	2.7	6.7
$T_{1\rho}(^{13}C)$ms	6.9	35	3.1	27
T_{IS} μs	28	307	24.1	336

Pyrolysis GC/MS. To obtain further support for amide formation two samples
of isolated glassbeads were subjected to Pyrolysis GC/MS. Again the ^{13}C-en-
riched carbon atom C3 of $(3-^{13}C)$-APS can be used to discriminate between
fragmentation products derived from PA-6, poly-APS and the coupling product,
respectively. Identical pyrograms were found for both samples. MS revealed
that two minor fragmentation products with $M^{+\cdot}$ 138 (HRMS $C_8H_{14}N_2$) and $M^{+\cdot}$
152 (HRMS $C_9H_{16}N_2$) contained ^{13}C-enrichment for 50 and 60%, respectively. On
the basis of their mass data and independent syntheses they were identified
as the azomethines 1 and 2 (Figure 3) (9). The position of the ^{13}C-label
could be deduced from a comparison of the fragmentation patterns of natural
abundance and ^{13}C-enriched 1 and 2, respectively. Since both compounds could
not be detected in the pyrogram of APS pretreated glassbeads, we conclude
that 1 and 2 have to be derived from the formal coupling product of -COOH
PA-6 endgroups and -NH$_2$ groups of APS. Despite the fact that the ^{13}C-enrich-
ment of $(3-^{13}C)$-APS was almost quantitative (7), only partial ^{13}C-enriched
1 and 2 is found. Careful analysis of the pyrogram of PA-6 itself showed
that also some 1 and 2 is formed. Currently we are investigating the mechanism of their formation. A full account will be reported in due course (10).

$$NC-(CH_2)_4-CH=N-^{13}CH_2-CH_2-R$$

1; R=H and 2; R=CH$_3$

Figure 3. Fragmentation products containing ^{13}C-enrichment.

CONCLUSIONS

With ^{13}C MAS NMR spectroscopy, Pyrolysis GC/MS and selectively ^{13}C-enriched (3-^{13}C)-APS as silane coupling agent direct evidence is obtained that the molecular mechanism of adhesion improvement of APS in glassbead reinforced Nylon (PA-6) modelcomposites is a consequence of a combination of covalent chemical bonding between -COOH PA-6 endgroups and $-NH_2$ groups of APS, and interpenetration (entanglement) of PA-6 chains into the poly-APS network near the inorganic filler surface.

REFERENCES

1. Plueddemann, E.P., Silane Coupling Agents, Plenum Press, New York, 1982.

2. Plueddemann, E.P., Interfaces in Polymer, Ceramic and Metal Matrix Composites, ed. Ishida, H., Elsevier Science Publishing Co., Inc., 1988, pp. 17-33.

3. Ishida, H. and Suzuki, Y., Composite Interfaces, eds. Ishida, H. and Koenig, J.L., Elsevier Science Publishing Co., Inc., 1986, pp. 317-27.

4. Weeding, T.L., Veeman, W.S., Jenneskens, L.W., Angad Gaur, H., Schuurs, H.E.C. and Huysmans, W.G.B., Macromolecules, 1989, 22, 706-14.

5. Rosen, M.J., J. Coatings Technol., 1978, 50, 70-82.

6. Plueddemann, E.P. and Stark, G.C., SPI 35th Ann. Tech. Conf. Reinf. Plast., 1980, 20-B, 264-69.

7. Jenneskens, L.W., Van den Berg, E.M.M., Heemskerk, B. and Lugtenburg, J., Recl. Trav. Chim. Pays Bas, 1988, 107, 627-30.

8. Huijgen, T.P., Angad Gaur, H., Weeding, T.L., Jenneskens, L.W., Schuurs, H.E.C., Veeman, W.S. and Huysmans, W.G.B., manuscript in preparation.

9. Fischer, M. and Djerassi, C., Chem. Ber., 1966, 99, 1541-57.

10. Jenneskens, L.W. and Venema, A., to be published.

Part 7

MISCELLANEOUS

EXTENSIONAL FLOW PREDICTIONS OF SOME DIFFERENTIAL MODELS

PAULA MOLDENAERS AND JAN MEWIS
Department of Chemical Engineering, K.U.Leuven
de Croylaan 2, 3030 Heverlee (Leuven), Belgium

ABSTRACT

A particular class of differential models is discussed with respect to their application in extensional flow. In the models under consideration nonlinearity is introduced by making the relaxation spectrum dependent on the flow history. This dependency is expressed by means of a kinetic equation for "structure" parameters which are associated with the different relaxation times. The Marrucci model is a possible starting point for such models. Using the available data for the IUPAC A LDPE, several models are evaluated in transient uniaxial extensional flow. It is shown that the discretization of the spectra is extremely critical. All the models associate the stress overshoot with the largest relaxation times which are not accurately known. A model with a strain rate dependent mobility is shown to describe the data well.

INTRODUCTION

Many rheological models have been suggested for polymer melts. Basically they can be divided in integral and differential types. In principle there is a strong analogy between the two groups. In practice, the various techniques to introduce nonlinearity in the two classes result in divergent predictions. Although considerable effort has been put in model evaluation, the possibilities and limitations of the various classes have seldomly been studied systematically [1-3].

It could be argued that nonlinear behaviour is caused by a change in structure on the molecular or intermolecular level. Each specific structure should then be characterized by a particular linear relaxation spectrum: the perturbation or superposition spectrum. This leads to a nonlinear model based on a variable linear spectrum [3-5]. It is not important here what molecular theory, e.g. the network or reptation model, is considered appropriate to explain the structure, but only that a linear spectrum can be attributed to each level of structure. This approach leads to differential models, which eventually can also be integrated. Contrary to other differential models they contain a kinetic equation to keep track of the time-dependent changes of the structure in transient experiments.

BASIC MODEL EQUATIONS

The rheological behaviour is represented by a discrete relaxation spectrum, i.e. a generalized Maxwell model, the parameters of which depend on the instantaneous level of structure:

$$\underline{\sigma} = \Sigma\underline{\sigma}_i \tag{1}$$

$$\frac{\underline{\sigma}_i}{G_i} + \lambda_i \frac{\delta}{\delta t} \frac{\underline{\sigma}_i}{G_i} = 2\lambda_i \underline{D} \tag{2}$$

with $\underline{\sigma}$ = stress tensor
$\underline{D}$ = rate of strain tensor
G_i = modulus at relaxatime time λ_i
δt = upper convective time derivative

There would be certain advantages in using continuous spectra but this would entail further complications without changing the essence of the approach. Following Marrucci and coworkers [5] the degree of structure is described by a series of parameters x_i. They indicate the fraction of the linear spectrum, at relaxation time λ_i, that is still active in a particular nonlinear condition. The same authors used a transient network picture to derive a structural dependence for G_i and λ_i:

$$G_i = G_{io}x_i \tag{3}$$

$$\lambda_i = \lambda_{io}x_i^{1.4} \tag{4}$$

where the subscript o denotes equilibrium values at zero shear rate, i.e. the linear region where all $x_i=1$.

Equations (1)-(4) have to be supplemented with a kinetic expression for the structural parameters. As in the models for thixotropy [6], the instantaneous rate of change for x_i is usually expressed as a function of the instantaneous flow conditions and the instantaneous structure. Lacking detailed information, first order kinetic equations for equilibrium reactions are normally used:

$$\frac{dx_i}{dt} = f_{i1}(1 - x_i) - f_{i2}x_i \tag{5}$$

In the original Marrucci model the rate constants are given by:

$$f_{i1} = 1/\lambda_i$$
$$f_{i2} = a(\mathrm{tr}\underline{\sigma}/2G_i)^{1/2}/\lambda_i \tag{6}$$

where a is an adjustable model parameter.

This describes the equilibrium situations well but seems to fail to simulate the structural changes during stress relaxation [7]. It can be shown that the Marrucci model, as well as a number of modifications of it, can be represented by the general expression:

$$\frac{dx_i}{dt} = (f_{i1}+f_{i2})(x_{i,eq} - x_i) \tag{7}$$

where $x_{i,eq}$ is the equilibrium value of the structural parameter pertaining to the instantaneous rate constants.

TRANSIENT EXTENSIONAL FLOW

If uniaxial extensional flow at a constant stretching rate $\dot{\varepsilon}$ is considered, assuming the sample to be unstrained and unstressed at time zero, eqns. (1)-(4) reduce to the following expressions for the transients of the stress components:

$$\frac{d\sigma_i^{11}}{dt} = 2G_{io}x_i\dot{\varepsilon} + \frac{\sigma_i^{11}}{x_i}\frac{dx_i}{dt} + 2\dot{\varepsilon}\sigma_i^{11} - \frac{\sigma_i^{11}}{\lambda_{io}x_i^{1.4}} \qquad (6)$$

$$\frac{d\sigma_i^{22}}{dt} = -G_{io}x_i\dot{\varepsilon} + \frac{\sigma_i^{22}}{x_i}\frac{dx_i}{dt} - \dot{\varepsilon}\sigma_i^{22} - \frac{\sigma_i^{22}}{\lambda_{io}x_i^{1.4}} \qquad (7)$$

Once a kinetic equation has been chosen, the model predictions can be compared with the experimental results published by Muenstedt and Laun [8] for IUPAC A LPDE (T=423K). The linear spectrum can be derived from available linear viscoelastic data. The required discretization is always somewhat arbitrary. Two possibilities, which have been used before [9,10], have been selected.

All the structural kinetics models which will be discussed here reduce to the Marrucci model for steady state flow. The value of the parameter a (0.4) has been obtained from nonlinear steady state shear flow data. The Marrucci model is now completely determined. The resulting predictions for IUPAC A are shown in figure 1. Unless otherwise specified the calculations are based on the discrete spectrum of ref. [9].

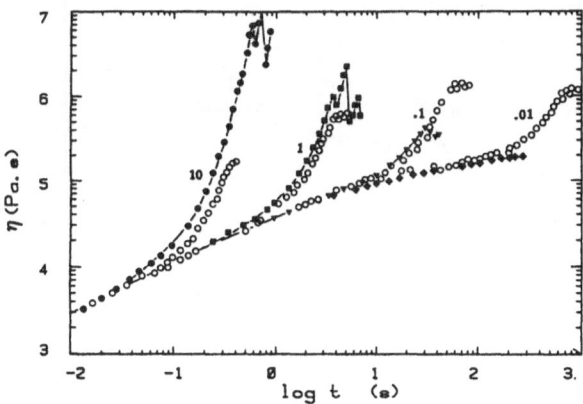

Figure 1. Simulation of IUPAC A extensional data with the Marrucci model (experiments: open symbols; calculated: filled symbols), parameter = $\dot{\varepsilon}$.

The model predicts high overshoot stresses which decrease again to an equilibrium value at higher strains. The predicted stress maxima are essentially rather broad but contain multiple peaks. This is a result of the discretization of the spectrum and is not a particular characteristic of the Marrucci model. The upturn in the stress curve is predicted well but this is associated with the linear behaviour and not with the quality of the nonlinear model. Neither the absolute value of the peak stresses nor their change with stretching rate, are predicted well except perhaps at the lowest stretching rates.

Deficiencies of the Marrucci model in transient extensional flow were reported earlier for the spinning of polymer solutions as well as in some transient shear flows [7,11]. De Cleyn and Mewis suggested the use of a purely kinetic parameter to correct the transients without affecting the equilibrium curves. It would appear as a pre-factor in eqns. (5) and (7). This adjustable parameter could reflect intrinsic differences in mobility between materials as compared at the same relaxation time. Obviously, the additional parameter makes it possible to improve the predictive power of a model. However, the erroneous increase of the peak stress with stretching rate still remains.

The Mewis-De Cleyn model can predict the data for low stretching rates quite well. In addition, spinning data for polymer solutions at various stretching rates also have been simulated adequately [11]. It could be argued that in both cases the intrinsic mobility of the chains is a dominant factor whereas at high stretching rates in melts the flow will affect the intrinsic mobility effect. Therefore the pre-factor in the kinetic equation has been modified by adding a second term which contains the second invariant of the strain rate tensor. The simulation of the IUPAC A data is shown in fig. 2.

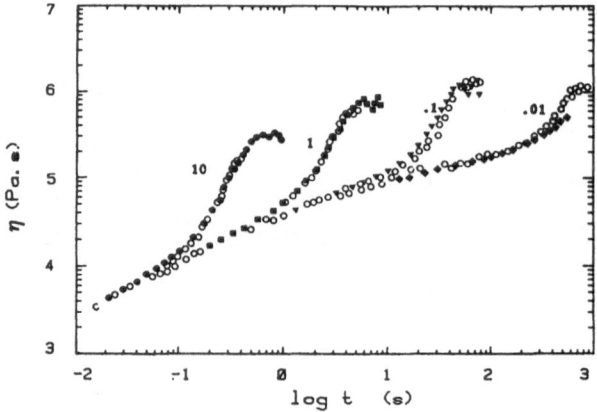

Figure 2. Simulation of IUPAC A with modified model, symbols as in fig. 1

The stress transients are described quite well over a wide range of stretching rates, including an initial increase and subsequent decrease of the peak stresses with stretching rate. Here, as well as in the Marrucci model and the original Mewis-De Cleyn model, the peak stresses at all stretching rates are caused by the slowest relaxational mechanisms. The latter are hardly measurable in the linear region where they occur at times of 100 to 1000 seconds. The two spectra which have been used to describe the IUPAC A differ somewhat in this region. The resulting difference in prediction for transient extensional flow is very pronounced. This kind of coupling between shear and extensional flow would be hardly noticeable in normal experiments, in agreement with the general experience.

It is concluded that structural kinetics models can predict transient extensional flow quite well. If a Marrucci type of model is used for the equilibrium behaviour a kinetic factor is required which depends on the flow. The models associate the peak stresses with the longest relaxation times which are very difficult to detect in shear flow.

REFERENCES

1. Giesekus, H., A simple constitutive equation for polymer fluids based on the concept of deformation-dependent tensorial mobility. J. Non-Newtonian Fluid Mech., 1982, 11, 69-109.

2. Khan, S.A. and Larson, R.G., Comparison of simple constitutive equations for polymer melts in shear and biaxial and uniaxial extensions. J. Rheol., 1987, 31, 207-234.

3. Mewis, J. and Denn, M.M., Constitutive equations based on the transient network concept. J. Non-Newtonian Fluid Mech., 1983, 12, 69-83.

4. Jongschaap, R.J.J., Derivation of the Marrucci model from transient-network theory. J. Non-Newtonian Fluid Mech., 1981, 8, 183-190.

5. Acierno, D., La Mantia, F.P., Marrucci, G. and Titomanlio G., A non-linear viscoelastic model with structure-dependent relaxation times. J. Non-Newtonian Fluid Mech., 1976, 1, 125-146.

6. Mewis, J., Thixotropy - A general review. J. Non-Newtonian Fluid Mech., 1979, 6, 1-20.

7. De Cleyn, G. and Mewis, J., A constitutive equation for polymer liquids: Application to shear flow. J. Non-Newtonian Fluid Mech., 1981, 9, 91-105.

8. Muenstedt, H. and Laun, H.M., Elongational behaviour of a low density polyethylene melt. Rheol. Acta, 1979, 18, 492-504.

9. Laun, H.M., Prediction of elastic strains of polymer melts in shear and elongation. J. Rheol., 1986, 30, 459-501.

10. Zuelle, B., Linster, J.J., Meissner, J. and Huerlimann, H.P., Deformation hardening and thinning in both elongation and shear of a low density polyethylene melt. J. Rheol., 1987, 31, 583-598.

11. Mewis, J. and De Cleyn, G., Shear history effects in spinning of polymers. A.I.Ch.E.Jl., 1982, 28, 900-907.

PLATEAU MODULUS AND ENTANGLEMENTS IN STYRENICS COPOLYMERS

P.LOMELLINI, A.G.ROSSI
Montedipe Research Center, Via Taliercio 13
46100 MANTOVA (Italy)

SYNOPSYS

The plateau modulus Go and the entanglement molecular weight Me of Styrene-co-Acrylonitrile (SAN) and Styrene-co-Methylmethacrylate (SMMA) random copolymers were measured as a function of composition. It comes out that a statistical copolymer can be treated as a "forced miscible blend".

EXPERIMENTAL

A series of SAN and one of SMMA copolymers with narrow chemical composition were prepared and investigated by means of a Rheometrics RMS-800 with oscillatory strain at various temperatures and frequencies in the plateau region. The Go values were determined as the G' at the minimum of tgδ as reported by Wu [1-3]. The Me values were estimated from Go using the equation Me = (ρ R T) / Go . Both the Me and the Go plots vs composition clearly show a non linear behaviour (see experimental points in figs.1,2,3).

DISCUSSION

In order to make an interpretation of the results we tried to adapt the models recently proposed by Wu [1-3] and Tsenoglou [4] describing Go vs composition in miscible blends. In our case blends of the homopolymers (PS-PMMA, PS-PAN) would give strong immiscibility due to the large differences in polarity. The physical idea is that a copolymer is like a blend "forced" to be compatible by means of the chemical linking between the two types of monomers. As interactions between different chemical species were described [1-4] to be responsible of Go behaviours in miscible blends we tried to apply this concept to copolymers. In what follows we use Gc as the Go of the copolymer, V1 and V2 as the molar fraction (instead of the volume fractions for blend) and G1, G2 as the Go of the homopolymers.

Tsenoglou's model

Within this scheme [4] Gc is given by

$$(1) \quad \sqrt{Gc} = V_1\sqrt{G_1}\left[1 + \varepsilon\left(V_2\sqrt{G_2} \Big/ V_1\sqrt{G_1}\right)\right]^{\pm 1/2} +$$

$$+ V_2\sqrt{G_2}\left[1 + \varepsilon\left(V_1\sqrt{G_1} \Big/ V_2\sqrt{G_2}\right)\right]^{\pm 1/2}$$

In eq.(1) the positive or negative exponents represent respectively increased or decreased entanglement density due to interactions between dissimilar species. The parameter ε accounts for the relative effect of these interactions on entanglements. $\varepsilon = 0$ is called the athermal case and describes the entanglements unaffected by interactions.

The figures 1,2,3 report the fits of the experimental data together with the athermal case. It appears a repulsive effect on entanglements (Me of the fit larger than Me of the athermal case; sign minus of the exponent of eq.1) for both the SAN and SMMA series. Least squares give ε = 0.427 for SAN and ε = 0.374 for SMMA. These values are higher than the ones estimated [4] for various miscible blends, indicating strong repulsive interactions. This is consistent with the fact that the corresponding homopolymers (PS-PAN, PS-PMMA) are immiscible.

Wu's model

In this model [1-3] Gc is given by

$$(2) \qquad Gc = V_1^2 G_1 + V_2^2 G_2 + 2V_1V_2 \, RT \, \frac{(\rho_1\rho_2)^{1/2}}{Me_{12}}$$

where Me_{12} is the molecular weight between entanglements of 1-2 type (i.e. between dissimilar species). The Go data were fitted with the equation (2) considering Me_{12} as a parameter. We found Me_{12} = 20000 for SAN and Me_{12} = 26000 for SMMA. The ratio of the 1-2 contact probability in the real blend (copolymer) to the contact probability of the unperturbed case is given by

$$(3) \qquad \lambda e = \frac{\dfrac{(\rho_1\rho_2)^{1/2}}{Me_{12}}}{1/2 \left[\dfrac{\rho_1}{Me_1} + \dfrac{\rho_2}{Me_2} \right]}$$

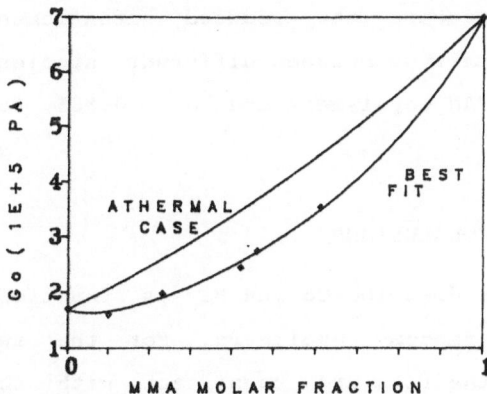

Fig. 1 — Go vs. composition
for SMMA copolymers
symbols = experimental data

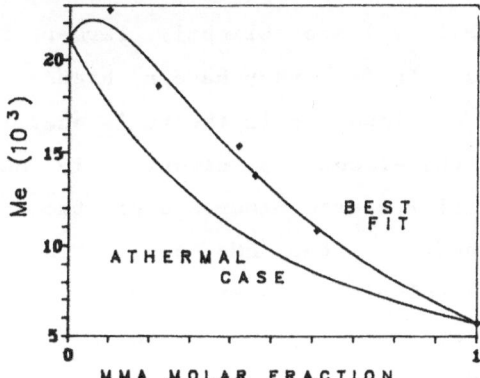

Fig. 2 — Me vs. composition
for SMMA copolymers
symbols = experimental data

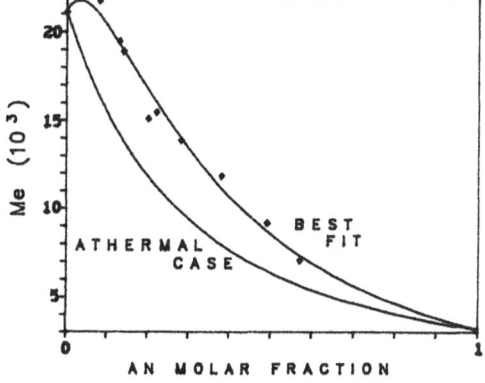

Fig. 3 — Me vs. composition
for SAN copolymers
symbols = experimental data

In other words λe represents the reduced entanglement probability due to interactions between different species. We found $\lambda e = 0.27$ for SAN copolymers and $\lambda e = 0.335$ for SMMA ones.

CONCLUSIONS

These models can suitably describe Go and Me as functions of the composition for random copolymers. For the two series we investigated, the fits are consistent with the repulsive interactions between the comonomers. Both the models suggest that the repulsive interactions (which affect entanglement probability) are slightly larger in the case of SAN copolymers . In fact they have a higher ε in the Tsenoglou 's model and lower λe in the Wu 's one. It should be underlined that the effect is stronger in SAN copolymers despite the smaller steric hindrance of the AN monomeric unit with respect to the MMA one. This is probably due to the higher polarity of Acrylonitrile which gives stronger repulsive effects and so stronger local chain stiffening which reduces the chain convolution.

REFERENCES

1) Wu S., J.Polym.Sci.,Polym.Phys.Ed. 1987 , 25, 557

2) Wu S., J.Polym.Sci.,Polym.Phys.Ed. 1987 , 25, 2511

3) Wu S., Polymer, 1987, 28, 1144

4) Tsenoglou C., J.Polym.Sci.,Polym.Phys.Ed. 1988, 26, 2329

PROCESSING OF THERMOTROPIC COPOLYESTERS

J. A. KINGMA, H. GEIJSELAERS
DSM Research
P.O. BOX 18, 6160 MD GELEEN, The Netherlands

ABSTRACT

For the injection moulding process the influence of the processing parameters on the mechanical properties is studied for two commercial available LCP's (Vectra A950 and Xydar SRT500).
It is shown that a qualitative prediction of the above relation is possible when the heat transfer process in the mould and a crystallization process in the LCP-melt is taken into account.

INTRODUCTION

This paper will address the injection moulding process of LCP's. Some papers have already appeared in this area [1,2,3,4]. In general they are directed towards a clarification of the LCP structure rather than towards an understanding of the relationship between processing conditions and material properties.
Our aim is to create an understanding of this relationship through the use of a simple model.

MATERIALS

Two commercial available copolyesters are studied:
Vectra A950 (Hoechst-Celanese) and Xydar SRT 500 (Amoco/Dartco).

The chemical composition and the melting point of these materials is listed in table I.

Table I: Chemical composition and melting point of the studied LCP's.

	HBA (mole%)	HNA (mole%)	TA (mole%)	BP (mole%)	T_m (°C)
Vectra A950	75	25	–	–	280
Xydar SRT500	50	–	25	25	400

HBA: p-hydroxybenzoic acid   HNA: 2,6-hydroxynaphtoic acid
TA: terephthalic acid   BP: p,p'-biphenol

EXPERIMENTAL

Both materials are dried prior to processing (4 hours, 150°C, vacuum, N_2).

For injection moulding an Arburg Allrounder (220-90-350) is used to prepare plaques with dimensions of 65x65x1.6 mm^3.

From the plaque bars are cut (65x12.75x1.6 mm^3) for further examination. Mechanical properties are obtained through a flexural test (ASTM D790), while the structure is studied through optical examination of the bars.

MODELLING

For a description of the injection phase an analytical solution of the heat transfer problem, as derived bij Janeschitz Kriegl, is chosen [5,6]. This model is based on penetration theory, assuming a linear temperature profile. Two extra terms to account for frictional heat generation and local heat transfer are added. This yields the following equation:

$$\lambda \frac{(Tl-Tm)}{\delta(t,x)} = \frac{\rho c(Ti-Tm)^2}{2(Tl-Tm)} \frac{\partial \delta(t,x)}{\partial t} + dP/dxHV + h_{//}(x)(Ti-Tl) \qquad (1)$$

$$\quad (1) \qquad\qquad (2) \qquad\qquad\qquad (3) \qquad\qquad (4)$$

For further details, see [6].

This model enables us to predict the thickness of the solidified layer formed during the injection phase (δC). It should be emphasized that, although this theory is rather naief and contains many approximations, the results obtained give a good indication of this thickness and are comparable with the result of more accurate numerical methods [7].

RESULTS

First the accuracy of the model will be established by comparing it's predictions with the experimentally observed material structure.

Figure 1 schematically depicts the results of an optical examination of a bar.

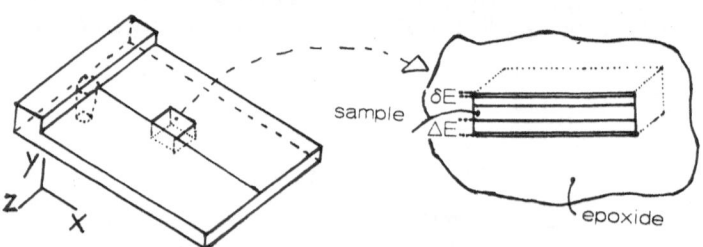

Figure 1: Schematical representation of the observed structure in LCP bars.

In most samples the four dark bands can be observed. These bands are related to orientation differences in the LCP [3]. Therefore these bands are thought to be related to the end of the injection phase and the end of the holding phase when a change in pressure occurs, which will influence the orientation of the molecules.

Figure 2 gives the result of a calculation for the reference condition of Vectra A950. The calculated thickness of the solidified layer at the end of the injection phase is compared with the experimentally observed position of the outer (first) dark band (δE). The two agree fairly well.

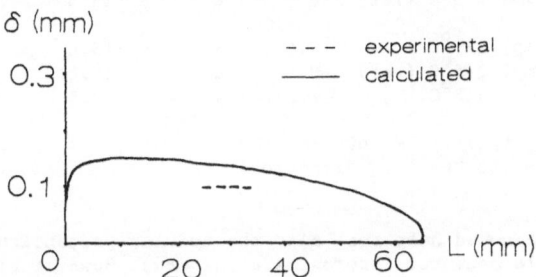

Figure 2: Vectra A950, calculated layer thickness and position of outer dark band.

The same agreement between experimental and theoretical results is obtained for the Xydar SRT500.

Results for the holding phase show a similar agreement between the position of the inner dark band and the calculated layer thickness for both materials as is obtained for the injection phase.

From these results it can be concluded that through the use of simple heat transfer models a reasonable prediction of the solidified layer thickness formed during the different stages in the injection moulding process can be obtained.

Therefore this approach can be used to predict the changes in layer thickness when the processing conditions are varied. Furthermore should it be possible to relate a change in layer thickness to changes in mechanical properties, since a change in layer thickness will affect the ratio between oriented and non-oriented material and therefore the mechanical properties.

This relationship is investigated for the injection speed, the mould temperature and the injection temperature.

Table III gives the results for Vectra A950.

The influence of the mould temperature is as expected: with a lower mould temperature a thicker solidified layer is predicted. This is confirmed by a shift in the position of the outer dark band. Furthermore is the change of the flexural modulus (parallel to the flow direction) in agreement with the predictions. The opposite trend is observed when the mould temperature is raised.

For the increased injection time the discrepancy between model and prediction is rather large. This is to be expected since one of the assumptions in the model (solidified layer << thickness of the plaque) obviously breaks down here.

The modulus and the predicted layer thickness showed no influence of the cylinder temperature therefore the position of the outer dark bands is not measured.

Table III: Vectra A950: variation of processing parameters

Parameter	δ_C (%)	δ_E (%)	Flex. Mod. // (GPa)
Mould temp. 200°C	3	*	10.7
Mould temp. 100°C (ref)	8	6	13.6
Mould temp. 22°C	10	7	15.3
Injection time 3 s	30	22	21.3
Injection time 0.3 s (ref)	8	6	13.6
Cylinder temp. 275°C	8.5	+	13.6
Cylinder temp. 295°C (ref)	8	6	13.6
Cylinder temp. 315°C	7.5	+	13.5

* = not detectable, + = not measured
// = parallel to the flow direction

It should be noted here that also the modulus perpendicular to the flow direction is measured (approximately 2 GPa). However since there is no influence of the processing conditions it will not be mentioned any further.

For the Xydar SRT500 the same influence of the processing conditions on the mechanical properties and the position of the outer dark band is observed.

These results show that the influence of processing conditions on the solidified layer thickness can be qualitatively predicted with a simple heat transfer model and the quality of the predictions is reasonable when compared to the experimentally observed layer structure. Furthermore it can be concluded that there exists a direct correlation between the solidified layer thickness and the mechanical properties of the LCP.

Some results however indicated that there is another process which influences the mechanical properties. This process will be discussed for the Xydar SRT500, although the same observation can be made for the Vectra. Table V gives the flexural modulus for the SRT500 whereby the cycle time is varied for different cylinder temperatures (the cycle time is varied by changing the mould opening time, which results in a change of the residence time of the LCP at high temperature).

Table V: Xydar SRT500: variation of cycle time

Parameter	Flex. Mod. // (GPa)
Cylinder temp. 370°C, mould open time 5 s	13.2
Cylinder temp. 370°C, mould open time 300 s	7.3
Cylinder temp. 390°C, mould open time 5 s (ref)	18.3
Cylinder temp. 390°C, mould open time 300 s	12.5
Cylinder temp. 410°C, mould open time 5 s	17.9
Cylinder temp. 410°C, mould open time 300 s	14.0

It is known that upon annealing the crystallinity of these materials increases [8]. Usually annealing is a seperate processing step but crystallization can also occur during injection moulding if the right conditions are present, which means processing close to or below the melting point of the LCP. This crystallization process results in a larger fraction of non-melted, crystalline material and therefore in a lower fraction of melted, orientable material and hence a drop in mechanical properties is expected.

To verify this hypothesis the following experiment is performed: A DSC-trace of the SRT500 is measured and the melting point is determined. Instead of cooling the material is kept at it's melting point for 300 seconds, then it is cooled and the melting point is measured again. An increase of the melting point of 10°C is observed, which confirms an increase in the crystallinity of the LCP.

It can therefore be concluded that the mechanical properties of injection moulded LCP articles are influenced by a heat transfer process and a crystallization process.

REFERENCES

1. T. Weng, A. Hiltner, and E. Baer, J. Mat. Sci., 1986, 21, 744-750.
2. G. Menges, T. Schacht, H. Becker, and S. Ott, Int. Pol. Proc., 1987, 2, 77-82.
3. D.J. Blundell et al., Polymer, 1988, 29, 1459-1467.
4. P.G. Hedmark et al., Pol. Eng. & Sci., 1988, 28(19), 1248-1259.
5. H. Janeschitz-Kriegl, Rheol. Acta, 1977, 16, 327-339.
6. H. Janeschtiz-Kriegl, Rheol. Acta, 1979, 18, 693-701.
7. H. van Wijngaarden, J.F. Dijksman, and P.Wesseling, J. Non-Newt. Fl. Mech., 1982, 11, 175-199.
8. Y.G. Lin, and H.H. Winter, Liq. Cryst., 1988, 3, 593-601.

List of Symbols

a	heat diffusivity of polymer [$m^2 s^{-1}$]
c_p	heat capacity of polymer [$Jg^{-1}K^{-1}$]
dp/dx	pressure drop in mould [Nm^{-3}]
$h_{//}$	local heat-transfer coefficient [$WK^{-1}m^{-2}$]
H	half height of mould [m]
L	length of mould [m]
t_i	injection time [s]
T	temperature [$^\circ$C]
T_c	cylinder temperature [$^\circ$C]
T_i	injection temperature [$^\circ$C]
T_1	no-flow temperature of polymer [$^\circ$C]
T_m	mould temperature [$^\circ$C]
v_i	flow front velocity [ms^{-1}]
α	heat transfer coefficient [$WK^{-1}m^{-2}$]
δ	thickness of solidified layer [m]
δC	calc. layer thickness/half height mould, injection phase [%]
δE	measured layer thickness/half height mould, injection phase [%]
ΔC	calc. layer thickness/half height mould, holding phase [%]
ΔE	measured layer thickness/half height mould, holding phase [%]
λ	heat conductivity of polymer [$JK^{-1}s^{-1}m^{-1}$]
ρ	density of polymer [gm^{-3}]

CURE MONITORING BY SIMULTANEOUS DIELECTRIC
AND DYNAMIC MECHANICAL MEASUREMENTS

R. E. WETTON & G. M. FOSTER
Polymer Laboratories Ltd.
The Technology Centre, Epinal Way, Loughborough, U.K.

M. DE BLOK
Polymer Laboratories Ltd.
Postbus 445, 3700 AK Zeist, The Netherlands

ABSTRACT

The new torsion head for the PL-Dynamic Mechanical Thermal Analyser allows
parallel plate rheometry measurements on thermosetting resins. Dielectric
measurements are made simultaneously on the same sample by linking the
PL-DETA measurement system to the rheometer plates as the dielectric
electrodes. This procedure allows the precise correlation of dielectric
parameters, ϵ', ϵ'' and tan δ, with dynamic mechanical viscosities if
required. Results show that cure effects occur later in time in G'' than
in the corresponding dielectric parameter ϵ''.

INTRODUCTION

There is considerable interest (1), (2) in using small dielectric sensors
to monitor the path of cure of thermosets in the production environment.
In previous publications (2), (3) we have discussed the results of model
experiments in which dynamic moduli and dielectric properties have been
separately studied. With absolute cure times there is always the
uncertainty in such parallel studies whether the induction time and then
the rates of cure themselves are identical in the two cases. In the
present paper this uncertainty has been removed by simultaneously
measuring the mechanical and dielectric property changes with the same
sample of epoxy during its cure in the new torsion head for the PL-Dynamic
Mechanical Thermal Analyser (DMTA).

EXPERIMENTAL

The new torsion head for the PL-DMTA is shown schematically in Figure 1. The torsion mode in particular facilitates the handling of liquid samples, albeit in these studies the samples will cure to solids. Using the parallel plate geometry with insulated electrodes, we have been able, rather simply, to use the PL-Dielectric Thermal Analyser (PL-DETA) to simultaneously study the dielectric changes while the mechanical measurements are being performed.

In order to keep reaction times long in these model studies, low temperatures of 30°C to 60°C were used, together with varying stoichiometric ratios of the diamine hardener mixed with the resin (Araldite 'Rapid', Ciba Geigy). The epoxies were mixed, then added to the parallel plate torsion cell with pre-heated plates. The gap between the plates was decreased until a 10mm diameter sample was generated with thickness approximately 3mm. This was ideal for both mechanical and dielectric studies. Simultaneious data was collected with multiplexed frequencies in both the DMTA and DETA experiments. The frequencies used were 0.3, 1, 3 and 10Hz in the DMTA and 1, 3, 10, 30 and 100kHz in the DETA.

RESULTS AND DISCUSSION

Loss peaks were always obtained in the shear loss modulus (G") versus cure time, but not always in mechanical tan δ. The data in Figure 2 (a-c) for example, is for 60/40 reactant ratio epoxy at 30°C. In 2 (a) G" produces systematic loss peaks with the high frequencies occurring earliest in time. The storage component increases smoothly in 2 (b), but tan δ values show a continuous fall from the high values of the liquid state to those for the cross-linked sample. Thus in the early stages of cure, frequency plane data has the characteristics of a Maxwell model. The dielectric loss parameters, ε" (dielectric loss) and tan δ, both exhibited peaks. Again the high frequency peaks occurred at shorter times and as expected the ε" peaks were displaced to shorter times than the corresponding tan δ peaks. The dielectric constant (ε') values in Figure 3 (c) show a continuous, but not smooth decrease.

Loci of the G" and ε" peak positions as a function of cure time are shown in Figure 4. The loci are reasonably parallel, but the dielectric peaks are relatively later in cure time. This shows the importance of performing a simultaneous experiment. We had previously reported from separate dielectric and mechanical studies that dielectric loss peaks correlated well or occurred slightly earlier in time relative to the mechanical loss peak locations. The present data are obtained at sufficiently low temperatures that genuine dipole vitrification effects are being observed and not the overwhelming conductivity effects which dominate at high temperatures/low frequencies. The 1Hz mechanical loss (G") peak occurs at the same cure time as the 1kHz dielectric loss (tan δ) peak.

368

CONCLUSIONS

Simultaneous Dynamic Mechanical and Dielectric measurements have been made very easily by combining DETA and DMTA measurements on the DMTA torsion head using a parallel plate assembly. Such a system could be valuable in performing incoming quality control measurements on resin batches to define the dielectric equivalence of the mechanical properties.

REFERENCES

1 Kranbuehl, D., Hoff, M., Haverty, P. and Hamilton, T.
   Proceedings of 16th NATAS Conference Washington 1987 p.70.

2 Wetton, R.E., Morton, M.R., Rowe, A.M. and Easter, G.M.
   6th International Conference on Composite Materials,
   Imperial College, London, July 1987.

3 Wetton, R.E., Foster, G.M., Gearing, J.W.E., Van-de-Velde, G. and
   Richmond, J.C. 9th International SAMPE Conference, Milan, June 1988.

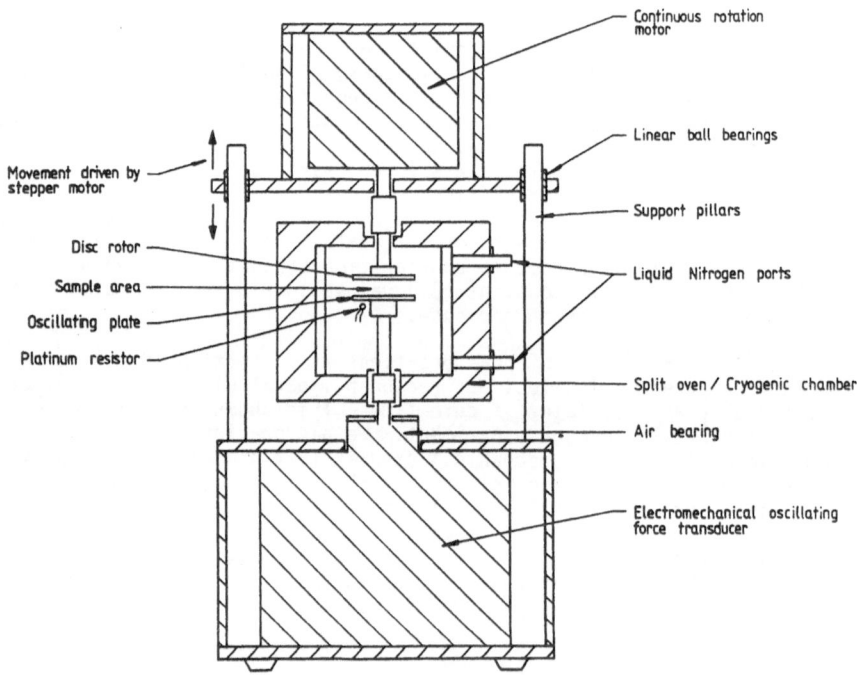

Figure 1. PL-DMTA Torsion Head

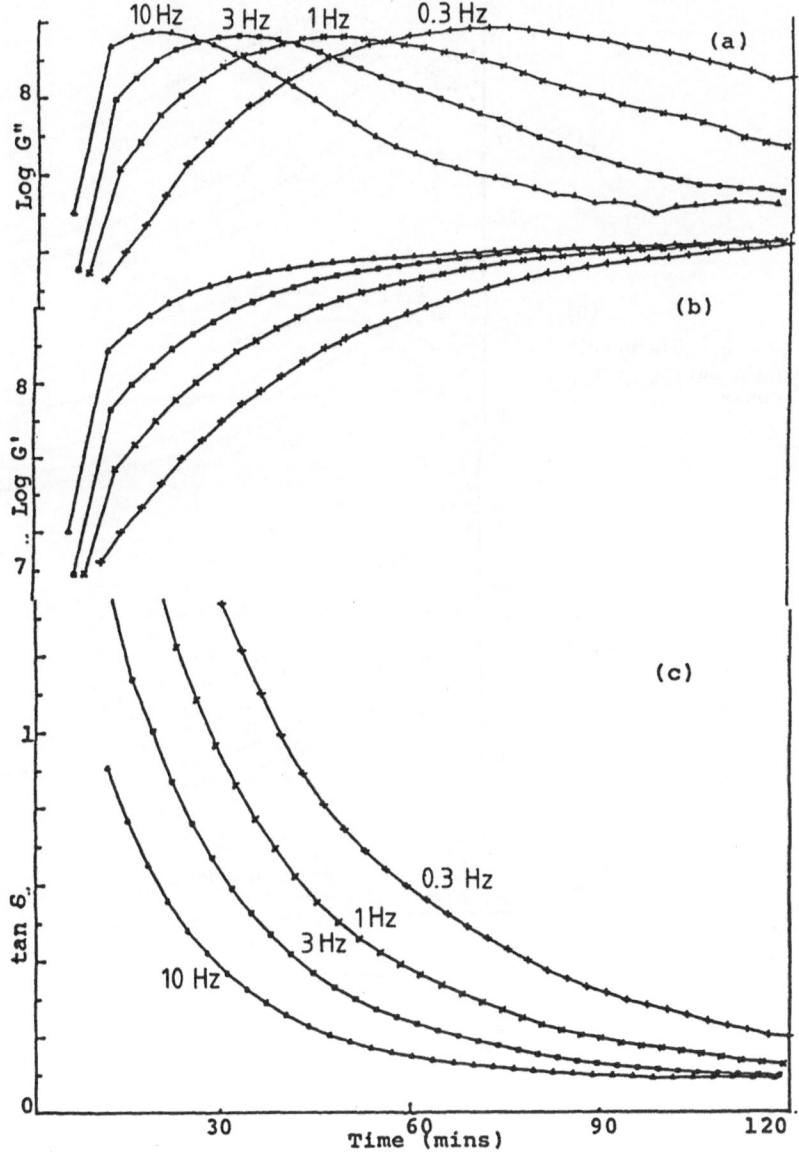

Figure 2. Dynamic modulus and tan δ during epoxy cure

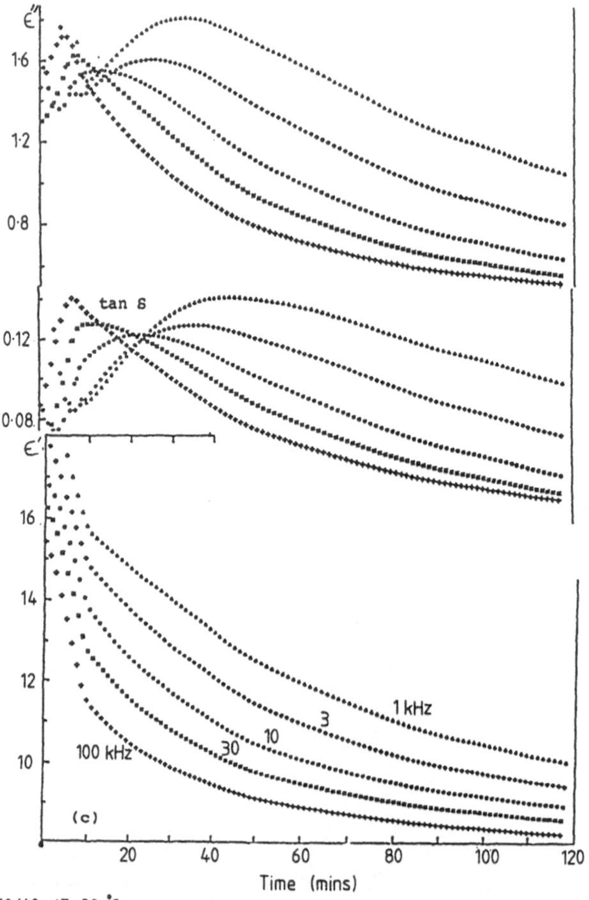

ϵ''

1·6

1·2

0·8

(a)

(b)

tan δ

0·12

0·08

ϵ'

16

14

12

(c)

10

1 kHz

3

10

30

100 kHz

(c)

20   40   60   80   100   120

Time (mins)

Figure 3. Dielectric changes during cure of epoxy

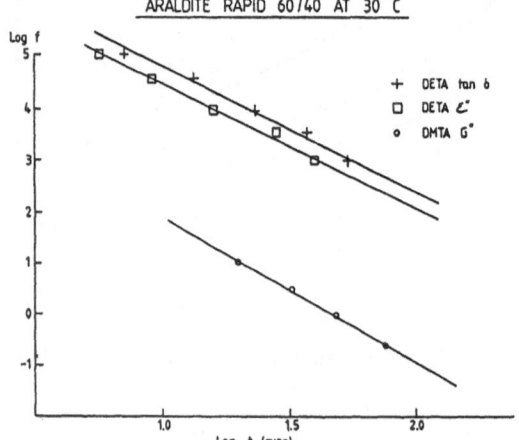

ARALDITE RAPID 60/40 AT 30 °C

Log f

5

4

3

2

1

0

-1

+ DETA tan δ
□ DETA ϵ''
○ DMTA G″

1.0   1.5   2.0

Log t (mins)

Figure 4. Comparison of Dielectric and Dynamic Mechanical loss peak locations

MOISTURE SORPTION MECHANISM OF KEVLAR FIBERS

HIROMICHI KAWAI, MITSUHIRO FUKUDA, MARIKO MIYAGAWA, AND MIYUKI OCHI,
DEPARTMENT OF PRACTICAL LIFE STUDIES, FACULTY OF TEACHER EDUCATION,
HYOGO UNIVERSITY OF TEACHER EDUCATION, HYOGO 673-14, JAPAN

ABSTRACT

The moisture sorption characteristics of an aromatic polyamide fiber, Regular Kevlar, were investigated in terms of the moisture sorption isotherm
and the heat of moisture sorption, both at 30 $^\circ$C, and compared with those
of an aliphatic polyamide fiber, nylon 6 fiber, having nearly the same
molar concentration of moisture adsoptive sites of peptide group. Despite
of much higher degree of crystallinity in Regular Kevlar than in nylon 6
fiber, both fibers sorbed moisture in almost the same extent as high as
around 7 to 8% regain at saturation. The sorption characteristics of Regular
Kevlar were intended to explain in terms of particular emphases of either
paracrystalline nature of the material associated with an unusually large
value of crystal imperfection factor attaing as large as 8 or existence of
micro-voids or -cleavages whose internal surfaces acting as adsorptive
surface but not being detected as noncrystalline region by X-ray diffraction.

INTRODUCTION

So far as the X-ray diffraction is concerned, the Kevlar fibers are revealed
to have very high degree of crystallinity, X_x, almost monophase structure of
crystalline or paracrystalline material associated with far less degree of
noncrystallinity, $(1 - X_x)$. Nevertheless, they have occasionally rather high
degrees of moisture sorptivity, as high as nylon 6 and 6-6 fibers, attaining
to 7 to 8% regain at saturation. Actually, the maximum value of adsorbed
water in monolayer fashion (Langmuir's) per unit mass of dry material nor-
malized by the degree of noncrystallinity, $v_m/(1 - X_x)$, is found to be 0.31
mole/100 gram of dry material for Regular Kevlar, which is much larger than
0.17 mole/100 gram of dry material for nylon 6 fiber. It is difficult to
understand the moisture sorption behavior of Kevlar fibers in terms of usual
concept of semicrystalline hydrophilic polymers in which the moisture is
mostly adsorbed on water accessible sites in noncrystalline region.

TEST SPECIMENS AND EXPERIMENTAL PROCEDURES

A wet-spun poly(p-phenylene terephthalamide) fiber, Du Pont Regular Kevlar, and a melt-spun polycaproamide fiber, Toray nylon 6 fiber, furnished from factories, were washed by a Soxhlet extractor using CCl_4 several times, leached in boiling water, air-dried, stored in P_2O_5 desiccator for at least two weeks, and further dried in vacuum oven at 90 °C for two days to prepare the bone-dried specimen for the measurement of moisture sorption isotherm and the heat of moisture sorption at 30 °C. The isotherm was determined by two types of gravimetric method; mainly by a weighing bottle method and partly by a sorption balance method with quatz spring in vacuum, and the heat of moisture sorption (differential heat of moisture sorption) was evaluated from the heat of wetting determined by a heat-transfer type calorimeter.[1]

EXPERIMENTAL RESULTS AND DISCUSSION

Fig. 1 shows the moisture sorption isotherm of Regular Kevlar at 30 °C and its analysis in terms of the B.E.T.'s multilayer adsorption model.[2] Fig. 2 shows the isotherms of nylon 6 fiber at three different temperatures of 15, 30 and 40 °C. As can be seen in the figures, the isotherm is of sigmoidal shape and typical as adsorption behavior of Langmuir's monolayer or its expanded B.E.T.'s multilayer adsorption. It is also seen in Fig. 2 for the nylon 6 fiber that the higher the temperature, the smaller is the moisture

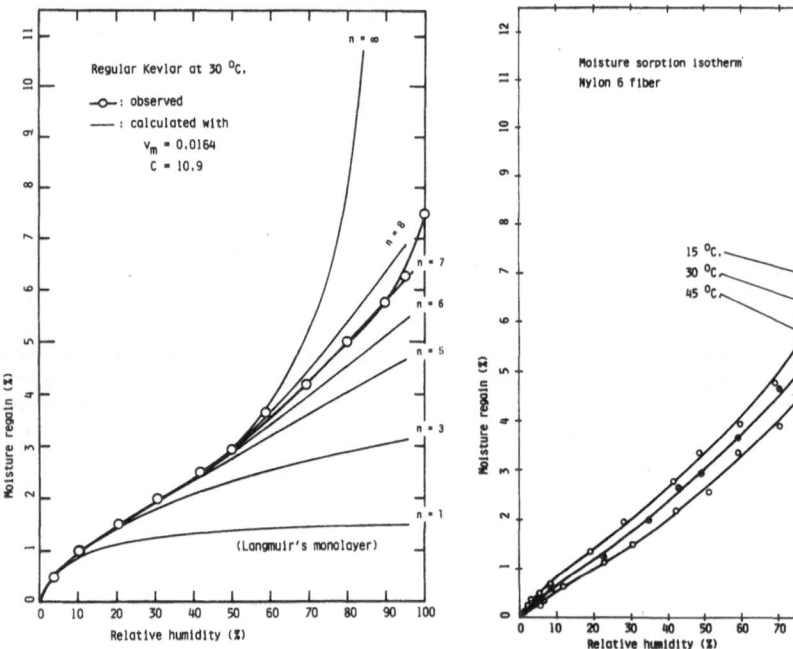

Fig. 1. Sorption isotherm of Regular Kevlar and its analysis by B.E.T.'s model.

Fig. 2. Sorption isotherms of nylon 6 fiber at three different temperatures.

up-take,1) suggesting the moisture sorption to be of exothermic process for both of the specimens, as discussed later in terms of the heat of moisture sorption.

Table I. Moisture Sorption and Some Physical Properties of Regular Kevlar and Nylon 6 Fiber at 30 $^{\circ}$C.

Specimen	Bulk density ρ (gr/cm^3)	X-ray crystallinity X_x (%)	B.E.T. parameters			Moisture regains at 95% relative humidity		
			v_m	C	n_{max}	n=1 (%)	$n_{max} \geq n>1$ (%)	$n>n_{max}$ (%)
Regular Kevlar	1.445	72.2	0.0164	10.9	6~7	1.5	4.3	0.7
Nylon 6	1.146	39.0	0.0200	3.70	6	1.5	4.9	0.8

Table I lists some physical properties of the specimens, such as the bulk density ρ and the degree of crystallinity in weight fraction X_x determined from X-ray diffraction by Ruland method,3) as well as numerical results of analysis of the isotherms in terms of the B.E.T.'s multilayer adsorption model, where v_m is a maximum volume of adsorbed water in monolayer fashion per unit mass of dry material, C is adsorptive energy factor, and n_{max} is a maximum number of multilayers n with) which the calculated isotherm is the closest but never exceeds the observed isotherm.

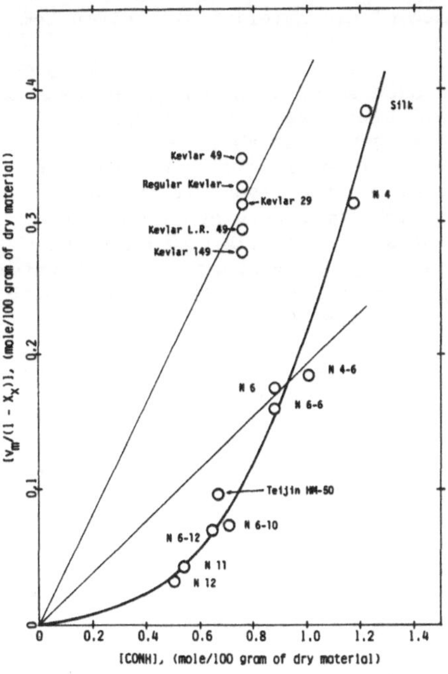

In Fig. 3 are shown the plots of $v_m/(1 - X_x)$ against concentration of peptide groups both in molar concentration per 100 gram of dry material for a natural polypeptide fiber, degummed silk, and a series of synthetic polyamide fibers including Kevlar fibers and Teijin HM-50 fiber; (50/50)(paraphenylenediamine/3-4' diaminodiphenylether)copolymer. The plots can be represented by a single curve, as demonstrated in the figure, for a series of aliphatic nylons including silk fibroin and Teijin HM-50. The slope of the curve must manifest the accessibility of water molecules per peptide group, concluding that the denser the distribution of the peptide groups along the backbone chain the higher is the water accessibility of each peptide group. In contrast, the plots for the Kevlar fibers including the Regular Kevlar deviate very much toward up-side from the single curve. The deviation can not simply interpreted in terms of the much larger water accessibility of peptide group in aromatic polyamide

Fig. 3. Plots of $v_m/(1 - X_x)$ against (CONH) concentration for a series of polyamide fibers.

than in aliphatic polyamide, because of the fact that the plot for the Teijin HM-50 well follows the single curve.4)

Table II. Heat of Wetting W and Differential Heat of Moisture Sorption Q_L for Regular Kevlar, Nylon 6 Fiber and A Normal Viscose Rayon at Dryness ($\alpha=0$) and 30 $^\circ$C.

Specien	$W(\alpha=0)$ (cal/g of dry material)	$W(\alpha=0)/(1 - X_x)$ (cal/g of dry amorphous material)	$Q_L(\alpha=0)$ (cal/g of liquid water)
Regular Kevlar	6.22 ± 0.08	22.4	362 ± 28
Nylon 6 fiber	5.82 ± 0.08	9.44	258 ± 36
Normal viscose rayon	23.8 ± 0.20	34.5	341 ± 22

Table II shows calorimetric results of the heat of moisture sorption for Regular Kevlar together with those for nylon 6 fiber and a normal viscose rayon, all at 30 $^\circ$C.1) The heat of wetting $W(\alpha=0)$ is an integral heat evolved per unit mass of dry material, when bone-dried specimen with moisture regain $\alpha = 0$ is dipped into liquid water, and is definitely exothermic for these particular specimens. $Q_L(\alpha=0)$ is a differential heat defined as $Q_L(\alpha=0) = - dW/d\alpha|_{\alpha=0}$,5) and is the heat evolved when one gram of liquid water is added to an infinite mass of dry material at a given temperature of T.

The value of $Q_L(\alpha=0)$, which can be understood as being independent on the degree of crystallinity, is about 1.5 times larger for Regular Kevlar than for nylon 6 fiber. The value attains as high as about 350 cal/gram of liquid water (16.9 kcal/mole of water vapour) and is almost the same as that for a regenerated cellulose of normal viscose rayon. It is noted that the value of 16.9 kcal/mole is much too high for single hydrogen bond but is reasonable for double hydrogen bond of water molecule to adsorptive sites, and be suggested that in the stage of primary adsorption each water molecule must be directly linked to a pair of sutably placed adsorptive sites on adjacent chains in the amorphous region.

Furthermore, if we simply define the water accessibility of the peptide group by the slope of thin straight line, as demonstrated in Fig. 3, the accessibility could be found as 0.41 and 0.19 moles of water per mole of the peptide group for Kevlar fibers and the nylon 6, 6-6, and 4-6 fibers, respectively. That is, the accessibility to be about 2.2 times larger for the Kevlar fibers than for the nylon fibers.

The above stoichiometry of aromatic and aliphatic polyamide interaction with water molecule may deduce a conclusion that the water adsorptivity, manifested by the differential heat of moisture sorption at dryness as well as by the number of water molecules adsorbed in monolayer fashion by the peptide group, is considerably higher for aromatic polyamide than for aliphatic polyamide. However, it may be criticized that the number of water molecules must be overestimated due to the normalization of the value of v_m by the X-ray noncrystallinity, $(1 - X_x)$. Because the wide-angle X-ray diffraction is not necessarily sensitive enough for detecting any micro-voids or -cleavages whose internal surfaces must be also active for the water adsorption.

In addtion, the degree of crystallinity, X_x, determined by Ruland's method for Kevlar fibers was found to be extremely high, but to be associated with unusually large value of the crystal lattice imperfection factor, k, attaining to 8 in contrast to 4 to 5 for the nylon fibers.[3] In Table III are listed the degree of crystallinity X_x as functions of the lattice imperfection factor k and of the integration interval of scattering vector s. The factor k was determined so as to give X_x as identically as possible with each other irrespctive of the inegration interval of s; i.e., k to be 8 and X_x to be 72.2% for a particular specimen of Regular Kevlar.

Table III. Degree of Crystallinity, X_x, determined by Ruland Method as Functions of Crystal Lattice Imperfection Factor, k, and of Integration Interval of Scattering Vector, s, for Regular Kevlar

s^*	k = 0	k = 3	k = 5	k = 6	k = 7	k = 8
0.05 - 0.35	57.7	63.2	66.6	68.2	69.9	71.5
0.05 - 0.50	50.7	59.8	65.1	67.4	69.6	71.6
0.05 - 0.72	46.7	60.8	67.4	70.1	72.4	73.4

* $s = (2/\lambda)\sin\theta$ Mean: 72.2%

These facts suggest the so-called crystalline region in Kevlar fibers to be much more paracrystalline nature than that in the nylon fibers, and may invalidate the two-phase concept that the moisture is mainly sorbed in the noncrystalline region but not in the crystalline region. In other words the moisture sorption of Kevlar fibers must be interpreted in terms of the degree of paracrystallinity, rather than the degree of noncrystallinity basing on a simple two-phase hypothesis. A more detailed investigation on the moisture dependences of the crystal lattice constants is now being carried out.

Recent preliminary investigation of the small-angle X-ray scattering by Hashimoto et al.[6] has revealed that the equatorial scattering from the Kevlar fibers, especially from Regular Kevlar and Kevlar 49, clearly show the existence of a shoulder in scattering intensity distribution at around q = 0.13 and 0.08 (A^{-1}), respectively, in contrast to monotonous decrease in the meridian scattering with increase in scattering vector q. These must arise from periodic electron density fluctuation in the direction perpendicular to the fiber axis, possibly related to inter-fibrilar micro-voids.

REFERENCES

1) M. Fukuda et al., J. Soc. Fiber Sci. & Tech., Japan, 43, T567 (1987).
2) S. Brunauer et al., J. Amer. Chem. Soc., 60, 309 (1938).
3) W. Ruland, Acta Cryst., A27, 73 (1971).
4) M. Miyagawa et al., J. Soc. Fiber Sci. & Tech., Japan, 43, T57 (1987).
5) W.H. Rees, J. Text. Inst., 39, T351 (1948).
6) T. Hashimoto, H. Hasegawa, and K. Saijo, private communication.

MOBILE CHARGE CARRIERS IN PULSE IRRADIATED POLYETHYLENE

MATTHIJS P. DE HAAS AND ANDRIES HUMMEL
Radiation Chemistry Department,
Interfaculty Reactor Institute, Technical University Delft,
Mekelweg 15, 2629 JB Delft, The Netherlands

ABSTRACT

The electrical conductivity of polyethylene induced by pulses of high energy radiation has been studied on a nanosecond time scale using the time resolved microwave technique (TRMC).

In ultra high molecular weight polyethylene (UHMWPE) samples, with a high degree of crystallinity, rapid charge migration is observed to persist for approximately 10 ns after pulse irradiation at room temperature.

The temperature dependence of the radiation induced conductivity is found to be small in the region between the glass-transition temperature Tg and the melting point Tm. Close to Tm the lifetime and magnitude of the radiation induced conductivity is reduced substantially, indicating that the rapid charge migration occurs in the crystalline regions of polyethylene.

An anisotropy ratio of 15 for the electrical conductivity parallel and perpendicular to the polymer chain direction has been observed.

The mobility of the charge carriers is estimated to be approximately 2×10^{-4} $m^2V^{-1}s^{-1}$ on the basis of an assumed value of the yield of escaped charge carriers of 0.1 $(100\ eV)^{-1}$.

INTRODUCTION

The study of the behaviour of the charge carriers formed in polyethylene by high-energy radiation is of importance for gaining an understanding of the electrical properties of polymer materials as well as of the chemical effects of radiation.

A considerable body of literature exists on the study of charge transport and trapping in irradiated material by means of measurement of the DC conductivity [1,2]. Application of conventional DC conductivity techniques for the study of charge migration in solid insulators is often complicated because of field induced space charge problems and the need of good electrical contact with the sample.

In this work we have studied on a nanosecond time scale the temperature dependence of the electrical conductivity resulting from radiation induced conductivity in several polyethylene samples using the time resolved microwave conductivity (TRMC) technique. With this method polarization problems and the requirement of good contacts are absent.

EXPERIMENTAL

The change in conductivity on irradiation was monitored by measuring the change in reflected microwave power on a nanosecond time scale. For small perturbations the relative change in absorbed microwave power is proportional to the change in conductivity of the sample. This time resolved microwave conductivity (TRMC) method has been described fully elsewhere [2,4-6]. In the present experiment microwaves in the Ka band (26.5-38 GHz) were used. The time response of the detection was 1 ns.

Samples were irradiated with 3 MeV electrons from a Van de Graaff accelerator using pulses of 2 and 5 ns duration and beam currents up to 4 A. The samples were contained in a microwave conductivity cell consisting of a short-circuited, 14 mm length of 3.55x7.1 mm i.d. waveguide. The cell was positioned in a cryostat, where the temperature could be regulated between 80 K and 450 K. The samples were irradiated over a length of 10 mm from the short-circuit. The dose of the irradiation pulse was calibrated using a microwave cell filled with CO_2 gas [3]. In this way the dose at the position of the cell was determined to be 300 J/m^3 (0.3 Gy) per nC beam charge. The total dose, accumulated in a sample during a study of the temperature dependence of the radiation induced conductivity, was less than 3 kJ/m^3.

The samples were positioned against the short-circuit end of the microwave cell. In order to reduce the lifetime of electrons that are produced in the irradiated air inside the waveguide, a small amount of the electron attaching agent sulfur hexafluoride was added into the waveguide. In this way, spurious signals during the irradiation pulse were eliminated.

Several polyethylene samples have been investigated:
- A block of low density polyethylene (LDPE) (Mn=22000 g/mol, Mw=3x10⁵

g/mol, 27 CH₃/1000 C, Tm-112.5 °C) with dimensions 3.5x7.1x20 mm³ was obtained from the Zentralinstitut für Isotopen- und Strahlenforschung, Leipzig, DDR.

- A block of ultra high molecular weight polyethylene (UHMWPE) (Hifax 1900 from Hercules, Mn=5x10⁵ g/mol, Mw=4x10⁶ g/mol, Tm=138 °C) was prepared by compression moulding at 200 °C and 64 MPa for 10 hours in a nitrogen atmosphere and was cooled down slowly to room temperature. This sample was obtained from the Department of Polymer Chemistry, University of Groningen.

- The gel-spun hot-drawn UHMWPE fibers used in this study were Spectra 1000 (Allied Fibers, Tm≈150°C) with a thickness of ≈25 μm. One sample consisted of 150 mg of these fibers, randomly oriented in the microwave cell. The other sample consisted of oriented fibers. The oriented sample was obtained as follows: An oriented bundle of UHMWPE fibers was immersed into the basic resin GMA (glycol methacrylate monomer, polyethylene glycol and hydroquinone) of the embedding kit 2218-500 Historesin from LKB and evacuated during 6 hours to release air bubbles. The GMA was polymerized by γ radiation using a ⁶⁰Co source (total dose 4.5 MJ/m³). From this UHMWPE-GMA material blocks were machined with dimensions of 3.55x3.55x7.1 mm³ and with the fiber orientation along one of the short axes. Due to shrinkage of the sample while polymerizing, the accuracy of the orientation could be reduced to approximately 5 degrees. The cell was filled with 4 of these blocks, with the fibers either in the parallel or perpendicular direction of the electric field.

- Samples of UHMWPE powder used in this study were Hostalen GUR-412 (Hoechst Ruhrchemie Mn=2x10⁵ g/mol, Mw=1.5x10⁶ g/mol). The UHMWPE fiber and powder samples were obtained from the Department of Polymer Technology, Eindhoven University of Technology.

The gel-spun hot-drawn fibers had a degree of crystallinity of at least 95%, for the other samples this was much lower. The density of the polyethylene in the powder and fiber samples was approximately 0.5 g/cm³.

RESULTS AND DISCUSSION

In all polyethylene samples radiation induced conductivity transients have been observed. The average lifetime ranged from 8 to 15 ns for the different UHMWPE samples to less than 0.1 ns in LDPE. In Figure 1 is shown the temperature dependence of the end of pulse conductivity in a block of UHMWPE. The height at the end of the pulse and the lifetime did not vary much in the range of -120 °C to approximately 10 °C below the melting point. For temperatures close to or above the melting point the end of pulse height was found to be approximately 6 x smaller than on room temperature and the lifetime was reduced to less than 0.2 ns.

When the melted sample was cooled down to room temperature within 15 minutes, the small signal remained unchanged; 48 hours after the "melting" no increase in signal height or lifetime could be detected. However, when the melted sample was cooled down to room temperature slowly (1 °C per 5 minutes) to allow recrystallisation, almost the same conductivity transient has been observed as before the heating of the sample.

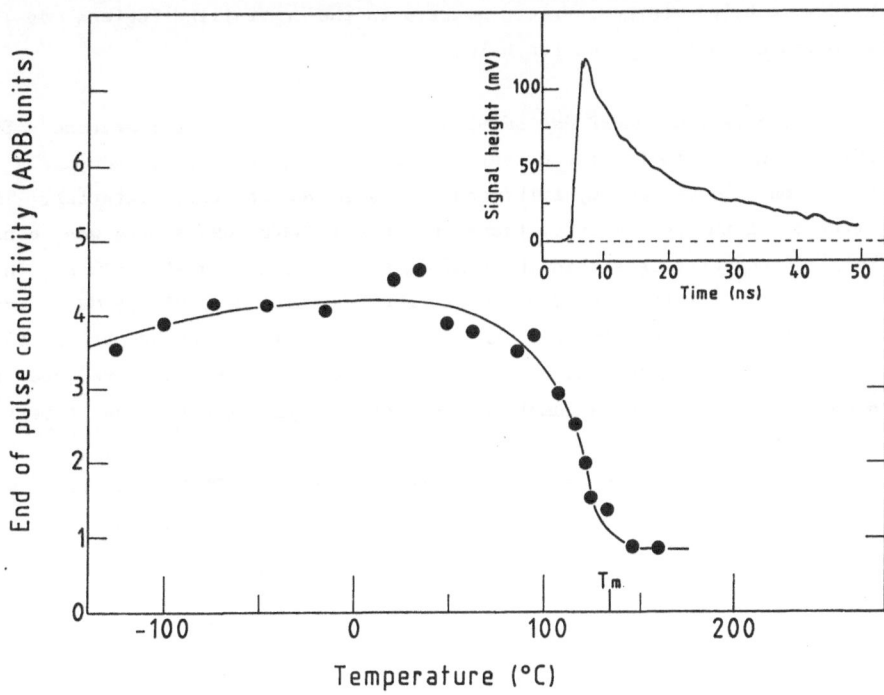

Figure 1. The temperature dependence of the end of pulse conductivity as observed in a block of UHMWPE. A 2 ns pulse of 3 MeV electrons was used. The beam charge was 6 nC, the dose was 2x10³ J/m³. The inset shows the transient microwave conductivity signal at room temperature.

Also in the other UHMWPE samples little temperature dependence of the height at the end of the pulse and the lifetime was observed. At room temperature the radiation induced transients in crystalline UHMWPE fibers were found to be approximately 3 times larger than in the block or powder.

In the LDPE block only a small signal with the shape of the irradiation pulse has been observed in the temperature range from -100 °C to 150 °C, the size of this signal was at all temperatures approximately the same as the signal in the molten UHMWPE samples.

It is known that the crystalline regions in LDPE are considerable smaller than in UHMWPE. The melting and cooling will perturb the crystalline regions. The results strongly support the conclusion that the highly mobile charge carriers migrate in the crystalline regions as is also indicated by the results below.

The oriented UHMWPE-GMA sample was studied at room temperature. The GMA was chosen to fix the orientation of the UHMWPE fibers, because no radiation induced conductivity could be observed in this material. The lifetime of the conductivity transient in the UHMWPE-GMA sample was 6 ns. This is somewhat less than the 9 ns in the original UHMWPE fiber. This reduction in lifetime could be due to the radiation dose of 4.5 MJ/m^3 that was applied to polymerize the GMA. It was found in the UHMWPE block that the lifetime and height of the conductivity transient was reduced by accumulated dose. An accumulated dose of 10 MJ/m^3 reduced the lifetime from 10 to 5 ns.

The conductivity transient in the oriented UHMWPE/PMMA sample was measured for the fibers parallel, $\sigma_{//}$, and perpendicular, σ_+, to the electric field direction of the microwaves. With the fibers oriented parallel to the field the transient is found to be a factor of 15 larger than with the fibers perpendicular to the field. Probably the anisotropy in the conductivity is even larger. A deviation in the orientation direction of the fibers in the sample of 5 degrees will yield a ratio of $\sin(85°)/\sin(5°)=11.4$. Therefore we believe that the factor 15 shows that the orientation of the fibers in the present sample is probably 3.8° and therefore that the anisotropy of $\sigma_{//}/\sigma_+$ could be larger than 15. It should be mentioned here that an anisotropy of a factor >10^3 has been observed in the thermal conductivity in stretch-oriented UHMWPE [7].

The radiation induced conductivity observed is related to μ, the sum of the mobilities of the charge carriers (either the excess electron, the electron hole or possibly both) by

$$\sigma = e\mu gD \qquad (1)$$

where e is the charge of the electron, g is the number of charge carriers formed that contribute to the conductivity per unit of energy absorbed and D is the energy deposited per unit volume.

The value found for $\sigma/D = e\mu g$ at the end of the 2 ns pulse is 2×10^{-7} $\Omega^{-1}m^{-1}J^{-1}m^3$. In order to determine the mobility of the charge carriers, g for UHMWPE is assumed [8] to be 0.1 $(100 \text{ eV})^{-1}$. In this way we find for the mobility a value of 2×10^{-4} $m^2V^{-1}s^{-1}$. The value of 0.1 $(100 \text{ eV})^{-1}$ for the yield of escaped charges has been estimated on the basis of experiments with low molecular weight hydrocarbon solids, without anisotropy. Since we have observed a large anisotropy in the electrical conduction the value assumed for g could be erroneous. The effect of anisotropy of the mobility on the geminate recombination will have to be considered.

The results demonstrate that migrating charge carriers with an appreciable mobility can be observed in polyethylene. We do not know at present whether the migrating charges are excess electrons or electron holes or both. Further work on the nature of the charged species, the trapping process and the role of these species in the chemical effects of high energy radiation are in progress.

ACKNOWLEDGEMENTS

We would like to express our thanks to Prof. Dr. A.J. Pennings and Dr.Ir. D.J. Dijkstra for providing the UHMWPE blocks, to Dr. O. Brede for providing the LDPE blocks, to Dr. N.A.J.M. van Aerle for providing the UHMWPE fiber and powder samples and carrying out the DSC experiments and to Prof. Dr. P.J. Lemstra and Dr. N.M.M. Overbergh for the helpful discussions about the interpretation of the results. We acknowledge the preparation of the sample with the oriented fibers by H.C. de Leng and J. van Leeuwen.

REFERENCES

1. Wintle, H.J., Campbell, F.J., Engineering Dielectrics, volume IIA, (ASTM Special Technical Publication 783, Philadelphia, 1983), R. Bartnikas and R.M. Eichhorn, eds., pp. 239-54 and pp. 619-62.
2. Warman, J.M., and Haas, M.P. de, Time-resolved conductivity techniques, dc to microwave, IRI report Nr. 134-88-01, 1988 (to appear in Pulse Radiolysis of Irradiated Systems, CRC Press, ed. Y. Tabata).
3. Warman, J.M., and Haas, M.P. de, The use of CO_2 gas as an in-situ dosimeter for dc and microwave pulse-radiolysis experiments, Radiat. Phys. Chem. **32** (1988), 31-53.
4. Haas, M.P. de, The measurement of electrical conductance in irradiated dielectric liquids with nanosecond time resolution, PhD Thesis, University of Leyden, (Delft University Press, Delft, 1977).
5. Infelta, P.P.,Haas, M.P. de and Warman, J.M., The study of the transient conductivity of pulse irradiated dielectric liquids on a nanosecond timescale using microwaves, Radiat. Phys. Chem., **10** (1977) 353-65.
6. Warman, J.M., The Study of Fast Processes and Transient Species by Electron Pulse Radiolysis, (Reidel, Dordrecht, 1982) J.H. Baxendale and F. Busi eds., pp. 129-61.
7. Choy, C.L., Leung, W.P., Thermal conductivity of ultradrawn polyethylene, J. Polym. Sci: Polym. Phys. Ed., **21**, (1983), 1243-46.
8. Haas, M.P. de and Hummel, A., Charge migration in irradiated polyethylene, IEEE-DEIS, Transactions on Electrical Insulation, **24**, (1989), to be published.

THERMOPLASTIC AROMATIC POLYAMIDES:
ADVANCED MATRIX MATERIALS FOR
REINFORCED PLASTICS AND COMPOSITES

D J Sikkema
Akzo Research Laboratories Arnhem
P O Box 9300
6800 SB Arnhem, the Netherlands

ABSTRACT

Amorphous fully aromatic polyamides can be prepared by multicomponent copolymerization. Injection mouldable products were polymerized via an acidolytic process, i.e. in the melt. The products show T_g up to 290 ^{o}C, respectable mechanical properties and they do not absorb common organic solvents. Similar amorphous aromatic polyamides were polymerized in solution, wet-spun, solution-impregnated and otherwise combined with reinforcing yarns in order to develop unidirectionally and bidirectionally reinforced composite intermediates.

INTRODUCTION

The search for thermoplastic materials that may be used at elevated temperatures has received much attention in recent years. Materials that are relatively difficult to prepare like polyetherimide and polyetheretherketone have been introduced as High Temperature Plastics, and fabricated into advanced composite objects. Our aim has been the development of materials that would be fairly inexpensive by virtue of simple raw materials and a relatively simple polymerization process.

The fact that the glass transition temperature defines a significant drop in the mechanical properties in all polymers except the exceedingly crystalline, suggests a high T_g as an all-important goal in the search for new HTPs. If one confines oneself to amorphous polymers, T_g may be targeted as high as about 100 ^{o}C below the highest temperatures that one could contemplate for thermoplastic forming work with the polymer.

The traditional drawback of amorphous polymers is their ready dissolution or at least swelling in various solvents.

RESULTS AND DISCUSSION

Our approach has been to look for amorphous fully aromatic polyamides, to optimize both T_g (by virtue of their rigid chains) and solvent resistance (by virtue of their high polarity and chain-chain interaction). It turned out that random copolymerization employing at least four monomers (in the four-comonomer case: in close to equimolar amounts) was necessary and sufficient to render the aramid polyemer amorphous – the monomers had to be selected to an extent of at least half the total mass, from the group that would lead to some form of kinking in the chain.

Most of the work has been performed with the system 3-aminobenzoic acid, 4-aminobenzoic acid, isophthalic acid, bis(4-aminophenyl)methane or sulfone. These monomers, in about equimolar quantities (and of course the diamine(s) and diacid(s) in stoichiometric quantities with respect to each other) were polymerized by an acidolytic scheme. Normally they were treated with acetic anhydride in slight excess relative to the amines present and - after the exothermic amine acetylation subsided - heated to remove acetic acid, ultimately in a high vacuum. Other anhydrides can of course be used. This means a melt polycondensation approach; although this is not without its pitfalls (high temperatures, high viscosities, high vacuum) we envision economic benefits, compared with polymerizations requiring work in solution. Whereas it is difficult to satisfactorily copolymerize aminoacids in solution (this requires the cumbersome preparation of the hydrochlorides of amino acid chlorides, or using the expensive phosphorylation polymerization procedure [2] rather than acid chlorides), the choice of monomers expands to include aminoacids by the acidolysis process. On the other hand, the high reaction temperature at relatively high acidity limits the monomer choice to exclude monomers that would give degradative or cross-linking reactionsunder such conditions, such as m-phenylene diamine.

The properties of the aromatic products, which so far have been prepared at a fairly small scale only, i.e. laboratory and limited semitechnical scale, are promising. T_g values up to 290 $^{\circ}$C, excellent solvent resistance (in terms of swelling, or rather non-swelling) to e.g. ASTM oils, toluene, alcohols, chlorinated hydrocarbons, ethers and esters and attractive (bending) modulus and strength were recorded. It came as no surprise, for rigid materials like the present polymers, that tensile evaluations were difficult (clamp-induced failure). On failure, the unreinforced glassy materials break without crazing or fibrillating. (Table 1)

Naturally, we are investigating the possibilities to combine these HTPs with reinforcing fibres. Recipe variation is undertaken in response to findings in this area; it is clear that the present approach allows much room for such variation.

Advanced composites, requiring continuous filament reinforcement, are a second objective. Many of the problems in continuous yarn impregnation that one faces with the high viscosity high temperature melt can be circumvented when working with polymer solutions.

Consequently, multicomponent solution polymerization was investigated. Most work centered on copolymers from 4,4´-methylene dianiline, p-phenylene diamine, isophthalic and terephthalic acids. It proved easy to wetspin the solutions prepared in the polymerization experiments to produce yarns or films that we are now combining with reinforcing yarns into UD tapes.[3] Of course, other methods of fabricating UD tapes are under investigation too: solvent impregnation and techniques employing powdered polymer either from the melt or the solution polymerization route. Unreinforced samples of the solution made polymer showed mechanical and chemical properties somewhat better than the product from the melt route; especially elongation proved much better (Table 1). Although much work remains to be done to optimize the composite workpieces that we have prepared until now, the outlook is very promising, judging by the adhesion already achieved between this HTP and reinforcing yarns (Table 2).

LITERATURE

1. D J Sikkema, US 4,758,651
2. N Yamazaki, M Matsumoto and F Higashi, J Pol Sci, Chem 13, 1373 (1975)
   W R Krigbaum, R Kotek and Y Minara, J Pol Sci, Chem 22, 4045 (1984)
3. D J Sikkema, NL 8702221

Table 1

	Start Tg, °C, DSC	Peak tan δ, °C, dynamic	E modulus, GPa	bending strength MPa	elongation, % (30 % GF)	solvent resistance	impact resistance	T inj. moulding, °C
PC	125	140	2.3	62	2.3	--	+	300
PSO	150	165	2.6	70	1.5	--	+	370
PEEK	150	165	3.5	101	1.6	+	+	400
PEI	203	230	3.3	94	1.8	−	+	400
PPS	80?	100	4.1	74	0.9	+	−	320
PES	185	230	2.6	85	2.0	--	+	370
HTP − 1 A	235	265	4.0	105	3.2 *	+	−	350
HTP − 1 B	260	288	4.1	110	3.5 *	+	−	375
HTP − 2	270	302	4.5	130	25 *	+		380 **

* unfilled
** compression moulding rather than injection moulding

Table 2.

Single fibre pull-out tension τ_i (MPa)
$\tau_i = (r/2\ell).\sigma_{fail}$

	Regular Twaron	Adhesion activated Twaron
Epoxy LY556/ HT 972	27.2	51.9
Ultem 1000	17.4	15.1
Akzo HTP-2	24.3	43.0

Like epoxy resins, our polyamide HTP is easier to adhere
to fibres by reactive bonding than a polyimide.

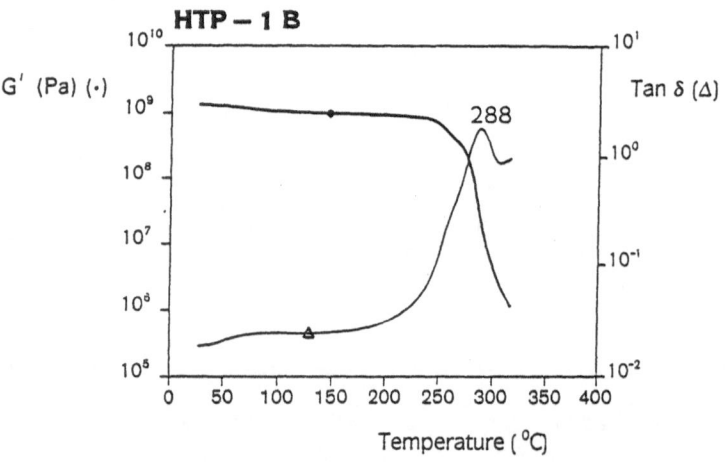

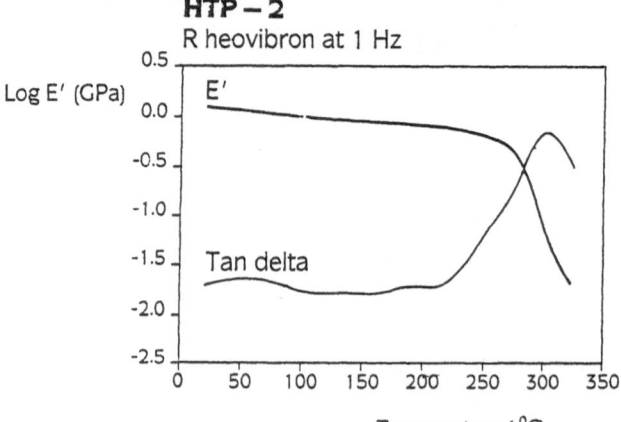

A MODEL STUDY ON THE ADSORPTION OF SMALL MOLECULES ON ORGANIC MOLECULES AND POLYMERS

SHANG ZHANG LIU, YANG GAO, LIN YOU WU,
BAO QIANG YU, and SHU MIN JIANG
Department of Chemistry, Liaoning University,
No. 4, 3Block, Chuongshan West Road,
Huanggu District, 110036 Shenyang,
PR China

ABSTRACT

The CNDO/2 method has been employed to study the chemisorption of H_2, O_2 and CO on pyrrole and model macromolecules built up from pyrrole units. Different orientations of the adsorbed molecule relative to the heterocyclic hydrocarbon have been considered. The intermolecular distance has been optimized with respect to the energy for each composite system. The depence of the interaction energy on the intermolecular distance and on the number of pyrrolyl units has been investigated. The adsorbed energy and the optimized steric configuration of small molecules on polypyrrole are obtained. These are of interest to understand the interactions between adsorbates and polymer surfaces and the binding situation of the molecular cluster complexes, as well as the catalysis of a new type of molecular clusters.

INTRODUCTION

Structures and properties on conducting organic polymers have received considerable attention both theoretically and experimentally in recent years. The present authors and relative authors have successely reported the conducting polymers can be used as chemical and electrochemical catalysts (1,2), and the adsorption of small gaseous molecules on polymers has been investigated using different quantum chemical methods (3). The partial results very much coincide with some experimental values. In this work results are reported on the chemsorption of H_2, O_2 and CO on pyrrole and models for the corresponding periodic macromolecule by performing CNDO/2 calculations. A series of interesting phenomena are discussed, and a new molecular cluster catalyst is found. Of special interest was the question

as to whether the properties of a molecular complex, e.g. binding energy, charge transfer and equilibrium geometry, can be
transferred when the same molecule is adsorbed on a polymer. A
further intention was to obtain information on the cooperative
effect in the adsorption process, which may be energetically
favored (autocooperative), imepeded (anticooperative) or not
influenced at all.

MATERIALS AND METHOD

The heterocyclic hydrocarbon pyrrole has been chosen to examine
complex formation with H_2, O_2 and CO. The periodic polymer
built up from the &-pyrrole units ($-C_4H_3N-$) would be poly(&-
pyrrole). As a model for this macromolecule &-dipyrrole has
been used. A series of molecular calculations for the system
pyrrolyl-(pyrrole)$_n$-pyrrolyl with n = 1,2,3 has shown that the
electronic charge distribution in the pyrrole units has effectively converged for n = 1. Furthermore the contribution to the
total energy per C_4H_3 subunit did not change very much as n increased. The adsorption of one and two CO molecules has been
studied for dipyrrole.

Standard bond lengths and bond angles (4) were used for
all molecules. It was assumed that the internal structure of
the components in the complex remains the same as in the isolated molecules. For each chosen configuration of the admole-
cule relative to the hydrocarbon (shown in fig.1) the binding
energy has been computed for different intermolecular distances
to obtain the equilibrium position for stable adducts. The

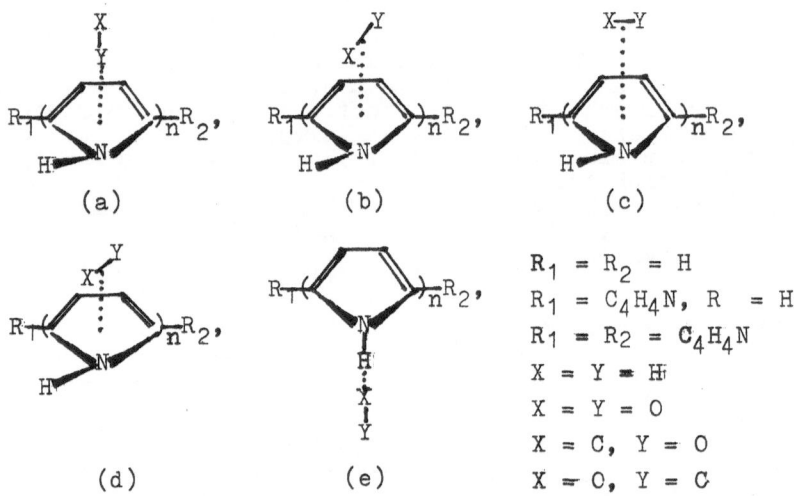

$$R_1 = R_2 = H$$
$$R_1 = C_4H_4N, \quad R = H$$
$$R_1 = R_2 = C_4H_4N$$
$$X = Y = H$$
$$X = Y = O$$
$$X = C, \quad Y = O$$
$$X = O, \quad Y = C$$

Fig.1. The structure of the investigated adsorption complexes.

CNDO/2 method has been employed for the molecular clusters involving H_2, O_2 and CO. Pople's CNDO/2 parameters were used (5).

RESULTS

1. Adsorption of H_2

Two configurations of the adsorbate H_2 relative to the hetero-cyclic hydrocarbon which are shown in fig.1. In the first structure bond between the two H atoms parallel with the pyr-role ring plane and the midpoint of the bond corresponds with the center of the ring plane. In the second the bond is perpen-dicular to the ring plane and the distance is measured from the ring plane to the nearest H atom. It has been found only under the configuration (a) case the complexes formed H_2 with $(-C_4H_3-N-)_n$ (n = 1-3) units are stable. The minimum binding energy is -0.00271, -0.00291, and -0.00295 au for the adsorbed bondlength of about 2.00 Å, respectively. From these can know these bind-ing energy were same order of magnitude with the physical heat.

2. Adsorpion of O_2

Owing to the particularity of O_2 molecule five configurations of the adsorbate O_2 relative the pyrrole ring are chosen. The calculatins were shown that each configuration there is a mini-mum point. The corresponding bonding energy is -0.35635, -0.44534, and -0.51784 au for $C_4H_5N + O_2$ ((a)-(c)), and -0.38512, -0.31232, -0.72286, -0.88056, and -1.03126 au for $C_4H_4N-C_4H_3N-C_4H_4N + O_2$ ((a)-(c)) and ((d)-(e)) at a adsorbed bond distance of about 1.5 Å.

3. Adsorption of CO

As for the H_2 and O_2 adsorbates equivalent relative orienta-tions of the CO molecule with respect to the pyrrole ring plane have been investigated. One configuration of CO is character-ized by the orientation of the CO bond parallels to the ring plane at the corresponding center position of the ring plane. As the diatomic adsorbate consists of two different atoms there are two possibilities of placing the CO molecule perpendicular to the ring plane, either with the C or the O atom pointing to the plane. The molecular cluster complexes were more complex. The minimum binding energy is listed in table 1. From table 1 can see that the perpendicular configuration of CO is more sta-ble than parallel that, but $(-C_4H_3N-)_nC\overset{O}{}$ is more stable than $(-C_4H_3N-)_nO\overset{C}{}$.

DISCUSSION

1. Hydrogen adducts

From the calculations can show that the adsorption of H_2 is a physical adsorption, and can use as well known Lernard-Jones's equation to describe. The bonding energy slightly increases with the number of the ring. The repelling energy of the H_2 molecule on polypyrrole is quite small. The charges from pyr-role transfer to the H_2 molecule were about 0.0083. A similar trend is found when the H_2 molecule is parallel to the ring plane.

2. Oxygen adducts

390

TABLE 1

The binding energies (BE, in au) of CO complexes with
$(-C_4H_3N-)_n$ (n = 1-3) as a function of the intermolecular
distance (d)(in Å)

System	d	BE
$C_4H_5N^O_C(1)$	1.25	−0.25896
$C_4H_5N^C_O(1)$	2.50	0.00529
$C_4H_5NOC(2)$	1.75	−0.12388
$C_4H_5NCO(2)$	1.75	−0.08234
$C_4H_5NCO(3)$	1.50	−0.14920
$C_4H_4N-C_4H_4N^O_C(1)$	1.25	−0.28358
$C_4H_4N-C_4H_4NOC(2)$	1.75	−0.14418
$C_4H_4N-C_4H_4NCO(2)$	1.75	−0.10118
$C_4H_4N-C_4H_4NCO(3)$	1.75	−0.13148
$C_4H_4N-C_4H_3N-C_4H_4N^O_C(1)$	1.25	−0.26876
$C_4H_4N-C_4H_3N-C_4H_4N^C_O(1)$	2.50	−0.00016
$C_4H_4N-C_4H_3N-C_4H_4NOC(2)$	1.75	−0.13156
$C_4H_4N-C_4H_3N-C_4H_4NCO(2)$	1.75	−0.08846
$C_4H_4N-C_4H_4N-C_4H_4NCO(3)$	1.70	−0.14466
$C_4H_4N-C_4H_4N^{OO}_{CC}(1)$	1.50	−0.14404
$C_4H_4N-C_4H_4N^{CO}_{CO}(2)$	1.75	−0.28574
$C_4H_4N-C_4H_4N^{CO}_{CO}(3)$	1.75	−0.32504
$C_4H_4N-C_4H_4N^{OO}_{CC}(1)$	1.25	−0.55004

The charges from the pyrrole ring transfer to the O_2 molecule
increase with the number of ring, but the C charges at the ⍺
position of the pyrrole ring are evidently increasing than before the adsorption of the O_2 molecule. At the same time the
positive charges on the pyrrole ring increase, too. It can see
that the adsorption of the O_2 molecule is the chemical adsorption. This is in agreement with some experiments which the oxidation can be taken place at the ⍺ position of the pyrrole
ring.

3. Carbon monoxide adducts

The adsorption of the CO molecule is a chemisorption, too. Comparing the calculated energies (see table 1) for the complex of two CO molecules with dipyrrolyl with the energies of the complexes formed between a single CO and a dipyrrolyl molecule, we can conclude that in both cases an anticooperative effect occurs. Namely, in these systems the repulsive interaction energy of two CO molecules with dipyrrolyl is larger than twice the energy of a single CO adsorbed to dipyrrolyl. It merits notice, the energy for the unoccupied π^* orbital in near to the lone pair electrons for $\sigma(nb)_2$. The CO molecule can be both the electron donor and acceptor, and pyrrole is a bond structure with many electrons also can accepte and backdonate CO electrons, therefore, the charges transfer from the CO to be smaller. Having experimental information is known that heterocyclic systems may form complex with non-polar molecules which are apparently stabilized through dispersion interactions or oxidations. For these can obtain satisfactory interpretation.

Although calculations were performed with the CNDO/2 method suggest that our results give the right trends.

CONCLUSIONS

1. The molecular complexes are obtained using the CNDO/2 method. The results on the adsorption of H_2 and O_2 on pyrrole momomer and polypyrrole very well cocincide with relative to experimental facts.

2. These are of interest to understand the interactions between adsorbate and polymer surfaces and the binding situations of molecular cluster complexes, as well as interpret the catalysis of this type of the molecular cluster catalysts.

REFERENCES

1. Eugene, G., Troussal, L.B.K. and Agnes, S., Procédé de préparation du methanol par hydrogénation du monoxyde de carbone en présence d'un catalyster constitué par un polymère organique conducteur tel que polypyrrole, EP 0130895, 1985, pp. 1-18.
2. Liu, S.Z., Wu, L.Y., Zhang, B.F., and Yu, B.Q., Alcohol synthesis by PPCP molecular cluster catalysts, Proceedings of the Third International Meeting on Polymer Science and Technology, ed. P.J. Lemstra and L.A. Kleintjens, Elsevier Applied Science Publisher, London, to be published, 1988.
3. Liu, S.Z., Otto, P., Ladik, J., and Gies, M., A model study on the adsorption of hydrogen, oxygen and carbon monoxide on organic molecules and polymers, Chem. Phys. Lett., 1987, 134, 133-8
4. Sutton, L.E., Pyrrole, in: Table of Interatomic Distances and Configurations of Molecules and Ions, Chemic Society, London, 1958, pp. M164.
5. Pople, J.A. and Beveridge, D.L., The CNDO/2 parameterization, Approximate Molecular Orbital Theory, McGraw-Hill Book Company, New York, 1970, pp. 75-84.

STUDY ON THE ADSORPTION OF SMALL MOLECULES ON HIGH CONDUCTING DOPED POLYPYRROLE BY CHROMATOGRAPHY

SHANG ZHANG LIU, LIN YOU WU, YANG GAO,
BAO QIANG YU, and SHU MIN JIANG
Department of Chemistry, Liaoning University,
No. 4, 3Block, Chuongshan West Road,
Huanggu District, 110036 Shenyang,
PR China

ABSTRACT

The behavior of the adsorption of H_2, O_2 and CO on highly conducting doped polypyrrole has been studied using the gas chromatographic technique. A series of pictures which adsorbed amount and potential energy change correspondingly with temperature and the distance between the interaction of the adsorbate and adsorbent are obtained. The results are shown that the behavior of the adsorption is related to doped constituents and properties, as well as topologic property and distance between the molecules. From the adsorption of H_2 and CO on the polypyrrole can see that due to the existence of H the amounts of CO adsorbed increases apparently. H_2 seems to be a "scavenger" or "cleaner" of the hole, but the exsistance of CO inhibits the adsorption of H_2. The relative adsorbed equations are obtained.

INTRODUCTION

At present, it is not many that the reports on the adsorption of small gaseous molecules on "pure" polypyrrole(PPPy) and doped polypyrrole(DPPy). We found that a series of catalysts of the molecular cluster complex containing conducting doped polypyrrole with transition metal organic phosphine can be applied to the syntheses of low and high carbon alcohol (1) and a very well catalytic activity is observed. It is well known that the original synthesis of alcohol was procured under high temperature and pressure. The conditions were very harsh. However, the synthesis of alcohol could be realized under normal conditions using the molecular cluster catalyst with the polypyrrole backbone (PPMC). The reactions contain gas-solid and gas-solid-liquid phases. First, small molecules are adsorbed and activated. It is clear that the studies on the behavior of the adsorption of H_2, O_2 and CO on PPPy and DPPy are very interesting to find the adsorbed lows and interpret this type of catalytic reaction mechanism. The present paper reported the corresponding experimental and theoretic results.

MATERIALS AND METHOD

Apparatuses and chemicals
Chromatograph: Model GC-1026, Beijing.
Ultraviolet-Visible Spectrophotometer: Model UV-240, Japan.
Fourier Transform Infrared Spectrograph: Model FT-IR 1730, USA.
Elementary Analyzer: Model EA-1106, Italy.
Chromatographic Column: 3-mm diameter, 2m length, 5-A molecular sieve (60-80 mesh).
The adsorbed temperature of small gaseous molecules on PPy-$FeCl_2$ and PPy-BF_4: 25-100°C.
Pretreating: all reagents monomer pyrrole, acetonitrile, electrolytes, etc. were purified before use. High-purity gas or crrier gas is used.
DPPy: Both chemical and electrochemical syntheses for the above both adsorbents can see analogous references (2,3). Their compositions and properties were determinated through element analysis, UV, IR, DTA, MS, conductivity, etc., and they were confirmed to be the desirable adsorbent. Then were grinded into the fine powders with 80-100 mesh for the use of the determinations and spares.

Adsorbed procedure
The prepared DPPy are put into a microreactor for the adsorbed experiment. When the corresponding gas was passed through the adsorpbent bed the amounts before and after the adsorption were same in the gas chromatograph, i.e. the adsorption was fully saturated. Successively the repeat can be obtained. A series of pictures which the adsorbed amounts changed with temperature from 25 to 100°C were obtained.

The CNDO/2 method for the experimental "trace" and confirmation has been reported in another work (4), here the details are ignored.

RESULTS

From experiments and quantum chemical calculations can be known that the geometric configuration of PPy-$FeCl_2$ (a) and PPy-BF_4 (b) (see later discussion in fig.1) is a kind of DPPy.

The results are listed in table 1 and table 2.

DISCUSSION

From the adsorbed pictures can be known that the DPPys obtained from the chemical synthesis and electrochemical one are quite different. The adsorbed ability for the latter is stronger than one for the former. Also, the adsorbed ability of PPy-BF_4 is stronger than one of PPy-$FeCl_2$ under the same conditions and one time larger than the latter. H_2 seems to be a "scavenger" or

TABLE 1

The comparison on ratio of C:H:N obtained by elementary analysis with theoretic one

| Species of DPPy | C:H:N | |
	Experimental value[a]	Theoretic value
PPy-FeCl$_2$	53.50:3.69:14.79	52.89:3.77:15.40
PPy-BF$_4$	49.25:3.43:20.21	50.38:4.58:19.58

a) Average value for 3 times.

TABLE 2

Adsorbed spectra for different gas on PPy-FeCl$_2$ and PPy-BF$_4$

| Adsorbed conditions | PPy-FeCl$_2$ | | | PPy-BF$_4$ | | |
	H$_2$	O$_2$	CO	H$_2$	O$_2$	CO
Temperature from 25-100°C						
Temp. °C for max. peak	55	56	64	36	40	47 (86)
Height for max. peak (in ml/g)	1.30	0.88	0.95	3.10	0.86	0.80 (0.90)
Coadsorption of H$_2$ and CO Height for max. peak (in ml/g)	no but 0.43→0		1.05 (40°C)	no but 0.48→0		1.25 (70°C)

"cleaner" of the hole, but the existence of CO inhibits the adsorption of H$_2$. The amounts of the adsorbed CO apparently increases with the adsorption of H$_2$. What happen these cases? We can see that the electronegativity of H$_2$ is stronger than that of CO; from the knowledges of the chemical structure and the calculating results by CNDO/2 (4), also, the conductivity of PPy-BF$_4$ is larger than that of PPy-FeCl$_2$, i.e. the greater the dipol moment of PPy-BF$_4$, the lower the adsorption temperature. Therefore the interactions between H$_2$, CO and PPy$^+$-BF$_4^-$ are easier than those of PPy$^+$-FeCl$_2^-$. The former adsorbed temperature is lower and the adsorbed amounts are higher.

The energy the dipole-dipole interaction can be calculated by

$$E_{d-d} \propto - \frac{M_a M_b}{R^6}$$

where M_a and M_b are the dipole moment of a and b, R is the

distance **between** two dipole center (5). The coadsorbed structures of H_2 and CO can be described as follows:

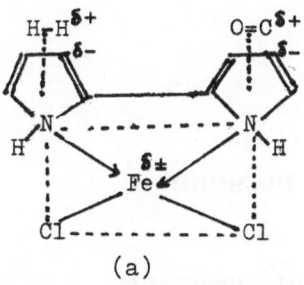 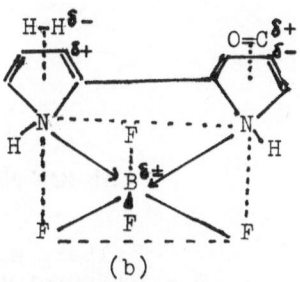

(a)   (b)

Then the mechanism of $CO^{\delta+}$ hydrogenation could be H_2 gas
reacts with $CO^{\delta+}$ through a transient state on $Fe^{\delta\pm}(B^{\delta\pm})$ or CO
gas reacts with adsorbed $H^{\delta-}$ through a transient state on $Fe^{\delta-}($
$B^{\delta-})$ or both CO and H_2 simultaneously pass through corresponding transient states react each other. So DPPy is applied to
the syntheses of methanol and other alcohol to be reasonable.

CONCLUSIONS

1. There is a remarkable adsorbed ability both DPPys for small
molecules. The adsorbed order: PPy-BF_4 PPy-$FeCl_2$; the electrochemical synthetic DPPy chemical one.

2. The greater the dipole moment or the conductivity of DPPy
the more the adsorbed amounts.

3. A series of the pictures are obtained and can use a corresponding adsorbed equation to describe.

4. The adsorbed behavior for DPPy belong to a new type.

REFERENCES

1. Liu, S.Z., Wu, L.Y., Zhang, B.F. and Yu, B.Q., Alcohol synthesis
   by PPCP molecular cluster catalysts, Proceedings of the Third
   International Meeting on Polymer Science and Technology, ed. P.J.
   Lemstra and L.A. Kleintjens, Elsevier Applied Science Publishers,
   London, to be published, 1988.
2. Armes, S.P., Optimum reaction conditions for the polymerization of
   pyrrole by iron(III) chloride in aqueous solution, Synth. Met.,
   1987, 20, 365-71.
3. Díaz, A.F. and Kanazwa, K.K., Electrochemical polymerization of
   polypyrrole, J.C.S. Chem. Comm., 1979, 14, 635-6.
4. Liu, S.Z., Gao, Y., Wu, L.Y., Yu, B.Q. and Jiang, S.M., Model
   study on the adsorption of small molecules on pure polypyrrole,
   Rolduc Polymer Meeting-4, 1989, April 23-27, The Netherlands,
   submitted
5. Gasser, R.P.H., An introduction to chemisorption and catalysis by
   Metals, Clarendon Press, Oxford, 1985, pp. 2-3.

CARBON NETS: "DUTCH DIAMOND"

Albert H. Alberts*,
Laboratory for Organic Chemistry,
University of Groningen,Nijenborgh 16,
Groningen, The Netherlands 9747AG.

ABSTRACT

An ideal (hypothetical) threedimensional diamond-like network derived from the (elusive) monomer tetraethynyl methane, C9H4, is entitled "Dutch Diamond". This network is an example of a class of polymeric all-carbon compounds,defined here as "carbon nets". The theoretical and empirical state of the art in this field is reviewed. The facile auto-polymerization of a silylated CH-monomer, 1,1-diethynyl ethene,is reported as a first example of a practical approach to a carbon net. Selective polymerisation of the central triple bond in silylated 1,3,5-hexatriyne is projected and a convenient synthetic route to this monomer is reported.

INTRODUCTION

The special properties of natural diamond have inspired scientists to attempt to develop preparative procedures for more then a century. The first successfull preparation of a diamond-like substance has to be attributed to J.B.Hannay of Scotland in 1880. His procedure involved lithium and graphite. At the time his results were deemed fraudulent mainly by the verdict of Sir Walter Rayleigh.The Hannay-crystals, exhibited in the British Historical Museum, were investigated by Bannister and Lonsdale in 1943 by X-ray techniques and established to have diamond-like character. Hannay, 38 years after his death in 1931 and about 15 years after the diclosure of the patents of General Electric and Siemens, was fully rehabilitated by Flint (1)(2).

The idea of preparing diamond-like solids from monomers with a high C:H ratio under less drastic conditions then those of the conversion of graphite to diamond, was born around 1960 at the time of the invention of the so-called Hay catalyst, a system consisting of Cu(I), a diamine, oxygen in acetone for the oxidative coupling of acetylenes (3).Linear polymers of aromatic ethynyl compounds were prepared and converted to carbon films and fibres. An economically viable process is elusive, but industrial workers (Hercules, IBM) are still active in the field.(4).

A vast amount of data has been accumulated on various processes of deposition of diamond-like coatings by plasma condensation, shock wave, laser techniques, etc. (5)

The first synthetic allotrope of carbon was reported by Sladkov in 1968, who prepared a linear oxidative polymer of acetylene (6), "karbin":

$$\text{---}C\equiv C\text{---}C\equiv C\text{---}C\equiv C\text{---}$$

Difficulties with the reproduction of the procedure and correct characterization of the material have caused that the product is often called: "chaoite" (7).

A polymeric cubane-type carbon, "C8", was reported by Strel' nitskii in 1978 (8).

Intensive theoretical work on "carbon nets",hypothetical two- and threedimensional sp² and sp3 networks, has been done by Russian workers and the group of Roald Hoffmann. Hoffmann predicted a dense metallic synthetic allotrope of carbon in 1983 (9). Recent theoretical work on the C8 polycubane carbon endowed this material with the highest density ever, 4.1 gram/cmð,a 15% increase relative to natural diamond.(10).

The approach involving aromatic polyethynyl compounds has been revived after publication of hexaethynyl benzene 1 by Vollhardt in 1986 (11). The groups of Whitesides at Harvard (12) and Stille at Colorado (13) have published syntheses of new monomers and polymerizations to glassy carbon-like materials.

1

Our own efforts (initiated april 1987)(16) are also directed towards the syntheses and polymerization of new monomers with extreme C:H ratio. On of our targets was (is) the hitherto unknown tetraethynyl methane, 2, C_9H_4:

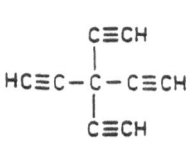

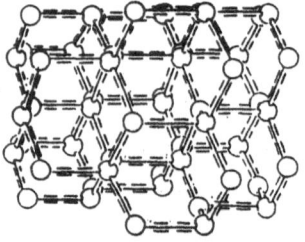

$2,\ C_9H_4$

Drawn naively in two dimensions, 2 resembles a Dutch "windmill". Ideally, oxidative coupling of all acetylenes of this monomer would lead to a synthetic allotrope of carbon, a diamondiyn network, that we call "Dutch Diamond". Of course in laboratory practice one will probably obtain an irregular network, but even if an average of two acetylene functions in 2 are coupled, a (white) polymer with a C_9H_2 stochiometry is far too great a temptation for an experimentalist. Not to mention possibilities of heat/pressure treatment of the product.

It is certainly encouraging to note that X-ray studies of other "giant tetraeders" (1,3,5,7 tetracarboxylic acid of adamantane and methane tetracarboxylic acid) reveal that these monomers form highly synmetric "superdiamondoid" lattices in the solid state(15).

RESULTS AND DISCUSSION

Monomer 2 remaining elusive, we concentrated on the synthesis of other highly unsaturated precursors. We developed viable synthetic routes to tetraethynyl ethene 3, and 1,1-diethynyl ethene 4a:

3

4a, R= H
4b, R = Me₃Si-

Preliminary experiments led to the conclusion that both 3 and 4a are highly reactive monomers. Compound 3 polymerized rapidly in solution at ambient temperature and pressure without a catalyst to form a carbon film. We noted that the silylated derivative 4b autopolymerized under very mild conditions (20-80°C, neat, 4 days- 1 hr) to a linear polymer 5 in which the geminal diethynyl substitution pattern is still intact (16) (M.W. 4 000- 14 000):

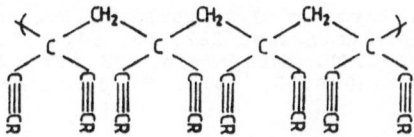

The autopolymerizations of monomers 3 and 4b are unique in the sense that they produce carbon films without contamination of residual catalyst.

The ability of silyl groups to protect triple bonds in polymerization reactions seems to be general (17)(18). This prompted us to explore the selective polymerization of the central triple bond in silylated hexatriyne 6 :

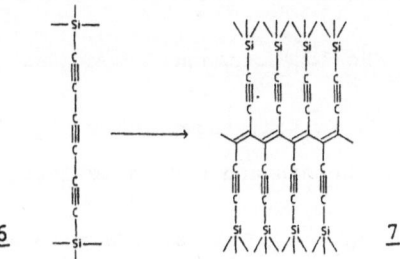

6   7

In the perspective of development of new conducting polymers and advanced polymerization techniques of acetylene (19) a structure like 7 is a valuable contribution (20) (21). It closely resembles a so-called "Little" polymer, proposed by W.A.Little in 1964 . Based on an exiton-electron coupling mechanism it was predicted to be an organic superconductor with a critical temperature above room temperature.

This research was supported by STW-grant "Carbon Nets".0771 in close cooperation with Philips Research BV, Eindhoven, The Netherlands (dr.E.W.Meyer, dr.E.E.Havinga). We thank prof. dr. Hans Wynberg for his incessant support.

References.

(1) The complete Hannay-mystery is revealed in an article by R. Ferreira,Cienca e Cultura Sao Paulo, 35, 1827 (1983).E.P.Flint's rehabilitation was published in Chem.&Ind 1968, 1618.

(2)The story of General Electric's "Diamond Maker", Scient.Amer. 1973, 675.

(3)L.Hay, General Electric patent, cited Chem. Abstr. 60,7953 (1963).

(4) D.J.Dawson,W.W. Fleming, J.R.Lyerla, J.Economy in "Reactive Oligomers", ACS Symposium Series 282, pg. 63. M.Flandera, C.Y.Lin US Pat. 4 258 079, 1981. M.Flandera, US Pat. 4273 906, 1981. L.C.Cessna, US Pat. 3882 073, 1975. H. Jabloner US Pat. 4 097 460, 1978.

(5) Y. Deryagin, Russ. Chem. Rev. 53, 435 (1984).

(6) A.M. Sladkov, et. al., Izv. Akad. Nauk. SSR Ser Khim. 1968, 2697.

(7) R. Hoffmann, T. Balaban, M. Kertesz, J.Am.Chem.Soc. 109, 6742, 1987.

(8) A.S.Bakai, V.E.Strel' nitskii, cited in ref. 10.

(9)R.Hoffmann, T. Hughbanks, M.Kertesz, P.Bird, J.Am.Chem.Soc. 106, 1135 (1984).

(10)R.J.Johnston and Roald Hoffmann, J.Am.Chem.Soc. 111, 810 (1989).

(11) C.P.K.Vollhardt, J. Fritch, Angew, Chem. 98, 270, (1986).

(12) M.R.Callstrom, T.X.Neenan and G.M.Whitesides, Macromolecules 21,3530 (1988).

(13) D.R.Rutherford and J.K.Stille, Macromolecules,21, 3532 (1988).

(14) A.H.Alberts and H. Wynberg, J.C.S.Chem.Comm. 1988, 875.

(15) O. Ermer, J.Am.Chem.Soc. 110, 3747 (1988).

(16) A.H.Alberts, J.Am.Chem.Soc. in press, 1989.

(17) E. Tsuchid, et. al., New Polym. Mat. vol. 1, 1, pg. 1 (1988).

(18) I.Kanedo N. Hagihara, Polym. Lett. 9, 275 (1971).

(19) W.A.Little, Phys.Rev. A134, 1416 (1964), Phys. Rev. B13, 4766 (1976).

(20)H. Naarmann, D. Theophilou, Synth. Met. 22,1 (1988).

(21) A convenient one-pot synthesis for monomer 6 was developed in our laboratory: A.H.Alberts, Rec.Trav.Chim. in press, 1989.

NOTES ADDED IN PROOF

* present adress of the author: Department of Polymer Chemistry, Technical University Eindhoven, P.O.Box 513, 5600MB Eindhoven, The Netherlands, prof. P.J.Lemstra, supervising.

Prof. Stille died in an airplane crash in the summer of 1988.

New (hypothetical) macrocyclic all-carbon compounds and eventual two-dimensional C-nets arising from these precursors were projected by F.Diederich et.al.,Science 245, 1088,1989. highlighted by J.K.Fraser-Stoddart, Nature 342, 483, 1989.

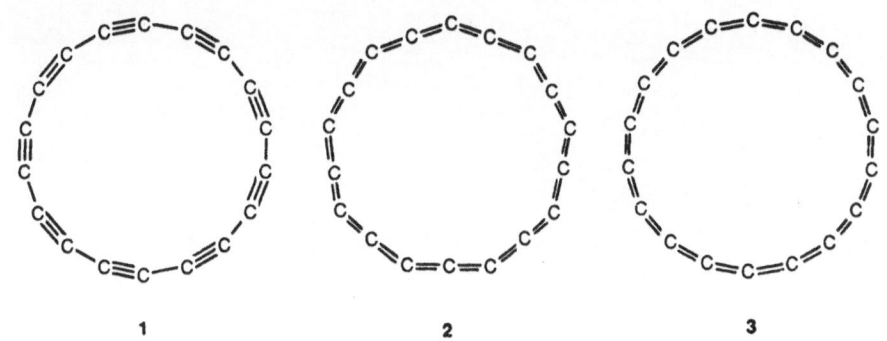

1   2   3

INDEX OF CONTRIBUTORS

SUBJECT INDEX